D1594550

CITY WATER, CITY LIFE

CITY WATER, CITY LIFE

Water and the Infrastructure of Ideas in
Urbanizing Philadelphia, Boston, and Chicago

CARL SMITH

The University of Chicago Press
Chicago and London

Carl Smith is the Franklyn Bliss Snyder Professor of English and American Studies and professor of history at Northwestern University. His books include three prize-winning volumes: *Chicago and the American Literary Imagination, 1880–1920; Urban Disorder and the Shape of Belief: The Great Chicago Fire, the Haymarket Bomb, and the Model Town of Pullman;* and *The Plan of Chicago: Daniel Burnham and the Remaking of the American City.*

The University of Chicago Press, Chicago 60637
The University of Chicago Press, Ltd., London
© 2013 by Carl Smith
All rights reserved. Published 2013.
Printed in the United States of America

22 21 20 19 18 17 16 15 14 13   1 2 3 4 5

ISBN-13: 978-0-226-02251-2 (cloth)
ISBN-13: 978-0-226-02265-9 (e-book)

Library of Congress Cataloging-in-Publication Data

Smith, Carl S., author.
  City water, city life: water and the infrastructure of ideas in urbanizing
Philadelphia, Boston, and Chicago / Carl Smith.
      pages cm
  Includes bibliographical references and index.
  ISBN 978-0-226-02251-2 (cloth: alkaline paper)—ISBN 978-0-226-02265-9
(e-book)   I. Municipal water supply—United States—History—19th century.
2. Waterworks—United States—History—19th century.   3. Urbanization—United
States—History—19th century.   I. Title.
  TD223.S64 2013
  363.6'1097309034—dc23   2012043193

♾ This paper meets the requirements of ANSI/NISO Z39.48-1992 (Permanence of Paper).

To Jane

CONTENTS

ACKNOWLEDGMENTS

This book has taken its time, and it has evolved a great deal in the writing. Its origins date back to an essay on the San Francisco earthquake and fire of 1906, which encouraged me to study disaster in the rapidly urbanizing United States. As my thinking progressed, the focus shifted from disaster in particular to how city people desired to exert imaginative as well as physical control over the general daily disorder of urban experience. With this in mind, I planned to write a book about the literal and figurative dimensions of fire and water in nineteenth-century American city life. I soon realized that anxieties about incendiaries of all sorts in highly volatile Chicago deserved a volume of its own, and that water would have to wait. Several other projects and duties, not to mention the decision to deal with three cities instead of one, further delayed this book, but here it is at last.

I had a lot of help along the way. At the beginning, middle, and end of the process, I received essential major assistance in the form of a Lloyd Lewis/National Endowment for the Humanities Fellowship from the

Newberry Library, an American Council of Learned Societies Fellowship, and the R. Stanton Avery Distinguished Fellowship from the Henry E. Huntington Library. For the opportunity to work in specific collections, I am grateful for the support provided by a Mellon Research Fellowship from the Center for the Study of New England History at the Massachusetts Historical Society and a Visiting Research Fellowship in Early American History and Culture from the Library Company of Philadelphia and the Historical Society of Pennsylvania. A Rockefeller Foundation Bellagio Study and Conference Center Fellowship enabled me to do a month of very intense writing in a splendid setting and amidst wonderful colleagues.

I have relied greatly on the skilled research assistance of a series of very capable Northwestern University students, notably Abigail Masory, Courtney Podraza, Kathryn Burns-Howard, Sherri Berger, Kathryn Schumaker, Lindsay Shadrick Dunbar, Nathan Enfield, Nora Gannon, and Sarah Collins. I owe thanks to Northwestern's Institute for Policy Research and the university's Nicholas D. Chabraja Center for Historical Studies for a substantial portion of the funding to compensate these assistants, and to the Judd A. and Marjorie Weinberg College of Arts and Science at Northwestern for the balance.

Like all scholars, I would be at a major loss without the expertise of the staffs at the many libraries, archives, and collections at which I have done research, among them the Massachusetts Historical Society, the Boston Athenæum, the Society for the Preservation of New England Antiquities, the Boston Public Library, the Library Company of Philadelphia, the Historical Society of Pennsylvania, the Philadelphia Water Company, the Philadelphia History Museum at the Atwater Kent, the Archives of Fairmount Park, the Fairmount Water Works Interpretive Center, the Pennsylvania Academy of the Fine Arts, the Philadelphia Museum of Art, the Franklin Institute, the Chicago Public Library, the Chicago History Museum, the Newberry Library, the Henry E. Huntington Library, and the libraries of Harvard University and especially those of the University of California at Berkeley and of Northwestern University. In this regard, I am very glad to have the opportunity to thank Peter Drummey, John

Aubrey, Lesley Martin, Laura Stalker, David Mihaly, Daniel Lewis, Cornelia King, Phillip Lapsansky, Sarah Weatherwax, and Thomas McMahon. Special gratitude is due to Frederick E. Hoxie, now at the University of Illinois, who as Vice President for Research and Education at the Newberry Library during my fellowship year there was such a gracious and thoughtful host to scholars in residence; Robert C. Ritchie, W. M. Keck Foundation Director of Research Emeritus at the Huntington Library, who created such an exceptional working environment and intellectual community; and Russell Lewis, Executive Vice President and Chief Historian at the Chicago History Museum, who so consistently and generously made that institution's splendid resources available to me. My year at the Huntington would not have been as richly rewarding and enjoyable without the advice, assistance, and friendship of Susi Levin.

I have bent the ear and benefited from the wisdom and learning of dozens of scholars. I would particularly like to acknowledge Tridib Banerjee, Paul Boyer, Lawrence Buell, Alexander Butkus, Louis Cain, Robert Coen, Michael Ebner, Robin Einhorn, Philip Ethington, Joseph Ferrie, Leon Fink, Jane Mork Gibson, James Grossman, Paul Groth, Neil Harris, David Henkin, William Howarth, Richard Hutson, Kevin Leonard, Adam Levine, Margaretta Lovell, Betsy Mendelsohn, Kathryn Olesko, Samuel Otter, Dominic Pacyga, Theodore Porter, Michael Rawson, Janice Reiff, Robert Righter, Nicholas Rogers, Mary Ryan, Nancy Seasholes, Sherry Smith, John Stilgoe, Eric Sundquist, Joel Tarr, Mary Terrall, John Van Horne, David Van Zanten, Sam Bass Warner Jr., Elliott West, and Caroline Winterer. I cannot thank enough those who read an earlier version of the entire manuscript and offered both general and detailed suggestions that have made the book much stronger. These include Henry Binford, James N. Green, Ann Durkin Keating, Martin Melosi, Harold Platt, Jeremy Smith, and Wendy Woloson.

Robert Devens, my editor at the University of Chicago Press, has provided the kind of encouragement, support, and patience that mean so much to any author, while Russell Damian has offered essential and timely assistance in the preparation of the manuscript. Erin DeWitt has once again proved a peerless partner in editing it.

It has long been my inestimable great fortune to have had Jane Smith's unblinking editorial eye watching over my attempts to say something that is at once interesting, important, and clear. To whatever extent I have succeeded, it is in significant part attributable to her unparalleled combination of insight, intelligence, generosity, and good sense, as well as her deep devotion to the reader as well as the writer of these pages.

1

INTRODUCTION

City Water, City Life

Water and meditation are wedded for ever.
—HERMAN MELVILLE, *Moby-Dick*

Water is all things to all people. It is a universal necessity, whether for drinking, cooking, sanitation, transportation, manufacturing, or fighting fire. It is the primary component of the human body and of the earth's surface, so that life is inconceivable without it. Water is a bearer of aesthetic, symbolic, and sacramental meaning in every culture, central to so many rites, from baptism of newborns to cleansing of the dead. Water imagery is omnipresent in all forms of expression, suggesting an intimacy between the flow of water and that of language, music, and drawing, not to mention consciousness, which was famously described by William James as a "stream."[1] Water is the locus and principle of experience itself, with all its uncertainties and contingencies, the "tide in the affairs of men" that abandons Shakespeare's Brutus on the shore of defeat, the infidelity Othello misperceives in Desdemona when he accuses her of being "false as water."[2] Water is, indeed, a bottomless well of metaphors. Individual existence has frequently been likened to a journey along a river, social and economic life to flows of people, goods, money, and information. In Melville's *Moby-Dick* (1851), water is the medium of Ishmael's multi-level journey, in the course of which he discovers endlessly reverberating correspondences between matter and mind. Water is always irreducibly figurative and inescapably literal, constructed and real, fabulous and mundane, and profoundly cultural.[3]

City Water, City Life is a meditation on how Americans living in leading cities meditated on water in the formative period of this country's great age of urbanization. While I focus on the establishment of water-supply systems in Philadelphia, Boston, and Chicago, which took place serially between the 1790s and the 1860s, this is neither a technological nor an environmental history as such. It is, rather, an intellectual and cultural study whose broadest purpose is to analyze how people thought and spoke about urbanization as they participated in it. It is based on two premises: that cities are built out of ideas as much as they are of timber, bricks, and stone, and that the discussion of city water is a kind of a universal solvent that reveals this in striking ways. "The mind now thinks; now acts; and each fit reproduces the other," Emerson declared in "The American Scholar" (1837).[4] In urbanizing America, certain ways of thinking about water, the words and actions through which these ways of thinking were expressed, and the physical environment of the city mutually conditioned and constituted one another.

The City as an Infrastructure of Ideas

In the pages that follow, I argue that a city is as much an *infrastructure of ideas* as it is a gathering of people, a layout of streets, an arrangement of buildings, or a collection of political, economic, and social institutions. The infrastructure of ideas neither precedes nor follows the building of a physical and social infrastructure, but is inseparable from them. An urban reservoir or pumping station is a work of hydraulic engineering, but in its design and the way it is managed it also expresses the beliefs, values, and aspirations of the city that created it. It negotiates not only practical matters of physics and thirst, finance and health, but also philosophical questions of secular and sacred, real and ideal, immanent and transcendent.

Between 1790 and 1870, the United States shifted irreversibly from an overwhelmingly rural to an increasingly urban society. In 1790 only 5 cities in the new nation had a population of 10,000 or more. This jumped to 23 by 1830 and then to 63 by 1850 and 168 by 1870. Between 1830 and 1870, when the momentum of urbanization became unmistakable, the number of places with at least 25,000 people went from 7 to 52, and the percentage of the population classified as urban climbed from 8.8 to 25.7 percent (it would tip past 50 percent by 1920). In 1830 the United States could boast of only one city, New York, with 100,000 residents or more. By 1870 there were 14. Philadelphia's population went from 28,522 in 1790 to 674,022 in 1870 (when it was the country's second largest city, after New York), Boston's from 18,320 to 250,526 (seventh), and Chicago's from a smattering of people to 298,977 (fifth—it would be second by 1890).[5]

The question of how these and other cities would obtain their water became more and more urgent as they grew. A history of Philadelphia's water system prepared in 1860 by the city's department of water observed, "The water supply to a great city is necessarily one of the most important and interesting features, upon which depends, to a greater extent, possibly, than any of its other advantages, either natural or artificial, its ultimate growth and prosperity."[6] As in other times and places, ambitious nineteenth-century cities in the United States legitimized themselves as vital and commanding by their ability to control water on a heroic scale.

Those who lived in these cities certainly conceptualized what they were doing in abstract terms as they confronted the more concrete dimensions of their water needs and of city life in general. As city dwellers worked out their multi-faceted physical relationship with water, they came to terms as well with timeless questions of what is the best conception of how to live, how to realize that conception, and how possibility and actuality determine one another. As the water histories of Philadelphia, Boston, and Chicago indicate, those who lived in these cities

frequently considered the deeper implications of what they were saying
and doing as they dealt with their need for water. The construction of
waterworks and of the infrastructure of ideas they embodied did not
develop smoothly, in a single direction, with sharp finality, or with anything approaching unanimity, even if virtually all city people accepted
and internalized, however uncertainly, the process of urbanization. A
crucial component of this process was the fashioning of a dynamic and
conflicted imaginative as well as physical place.

Four Questions, Three Cities

After presenting an overview of the histories of the building of the early
waterworks systems in Philadelphia, Boston, and Chicago, I explore four
broad and overlapping sets of questions posed by urbanizing America's
meditation on water.

*1. How did people living in these cities understand citizenship at a time of
unprecedented urban growth and change?*

City dwellers recognized themselves as participants in a complex society at the moment when they realized that the quantity of water from
the local well, stream, or pump no longer met their personal needs and
commercial demands; that the quality of water at hand was jeopardized
by privies, cesspools, and industry; and that a fire or fever that broke out
even many blocks away might quickly spread, threatening their own lives
and livelihoods. The need for a large, dependable, and accessible supply
of clean water raised multiple concerns about individual and collective
priorities. To build and manage a system that could provide this water
required the expansion of heretofore limited city government and an accompanying large increase in the municipal budget. The imperatives of
water brought to the fore conflicting ideas of the public good, including
disagreements over what resources should be provided, and by whom,
to that elusive entity, "the people," in a burgeoning capitalist democracy
whose members were fiercely devoted to freedom of individual action
and increasingly divided politically, even as they became more dependent upon one another.

2. How did urban Americans conceptualize the cityscape they were build-ing and inhabiting in relation to the natural world it was displacing?

City water blurred the line between nature and the built environment. Throughout the nineteenth century, cities expended enormous effort and resources to dredge, dam, drain, and redirect nearby and distant riv-ers and lakes, creating artificial replacements in the form of aqueducts and reservoirs. Water is, of course, a naturally occurring compound, but waterworks systems denaturalized it into an apparently manufactured commodity that was sold, delivered, used, and discarded. To be reliant on a network of water pipes defined more viscerally than did walking or riding through streets the condition of being "on the grid," that is, within a human-made world separate from nature.[7] Given how radical the shift was from fetching water from a natural source to taking it from a tap, it is remarkable how quickly city people came to view the new state of things as normal and, for lack of a better word, natural.

Meanwhile, the American nation's presumed identification with na-ture, as well as a timeless concern with the dehumanizing tendencies of urban life, drove efforts to protect or restore the presence of the natural world in the city, even if that presence was nearly as fabricated as the grid. Accompanying the general transformation of the landscape into the cityscape was the reconstruction of nature in aqueducts and reser-voirs, and, more programmatically, in gardens and parks that contained artificial ponds, streams, lakes, lagoons, and fountains. All of these were intended to revitalize the urban citizen and public life.

3. How did people living in these three cities perceive the relationship be-tween their physical well-being and that of the city?

To go on the urban water grid was to connect one's human body to that of the so-called body of the city. The close connection between the healthy individual and the healthy city was irrefutable at those harrow-ing moments when an epidemic or pollution of the drinking source con-vinced city people that they must build a new or improved waterworks. The water metaphors employed in engineering reports, public speeches, booster rhetoric, and cultural commentary imagined the city as a living human body that corresponded to the bodies of those that inhabited it.

At the same time, those with strong class and ethnic prejudices viewed some portions of the population—most often the poor and the foreign-born, as well as the diseased—as dirty bodies that required sanitizing. The bathing, temperance, and water-cure movements looked to new waterworks as part of the battle to cleanse, control, and heal the individual body, both for its own sake and for that of a well-regulated and righteous social body. While depending on city water meant tethering oneself to the urban collective, however, those who could afford indoor plumbing or take the water cure attempted to protect themselves from contamination by isolating themselves from the urban body.

4. *Where did city people locate contemporary city life in the flow of time and history?*

The transformations that accompanied sudden urbanization caused city people to contemplate their place in history. When officials calculated the optimum capacity of a waterworks, they believed they were determining their city's destiny. But building a large and expensive waterworks also raised the sensitive subject of urban debt. The need to construct capital-intensive waterworks that would supposedly foster a glorious future meant the assumption of indebtedness on a scale previously unimagined. While many citizens argued that the present had no right to encumber the future with such financial obligations, proponents of new systems spoke of a willingness to incur water debt as a civic duty. Imagining the urban prospect in terms of water was also in some respects a retrospective act. This was true especially when it entailed framing the present in terms of the storied civilizations (and waterworks) of an earlier age, often classical Rome, even when this framing derived from an exceptionalist outlook that wished to see American urban life as following an ever-upward trajectory that distinguished one's own city from other great metropolises of the past that had long since gone into decline.

While *City Water, City Life* examines Philadelphia, Boston, and Chicago throughout the period it covers, it focuses on each of them mainly in the years when it built its first successful water system: Philadelphia between the 1790s and the early 1820s, Boston from the mid-1820s to about 1850, Chicago starting in the 1840s and ending around 1870. A "successful" waterworks is defined as one that by broad agreement has met the technological challenge of delivering water of acceptable quantity and quality and that offers a reasonable expectation of financial sustainability. Even before each city had a successful system up and running, its leaders realized that more work and expense than they had anticipated lay ahead, and that meeting a large city's water needs was always a work-in-progress that would never be finally done.

The three cities discussed here were chosen because of both their differences and their similarities. The early water histories of Philadelphia, Boston, and Chicago have their distinctive particulars, so examining them together provides more range and variety than would an account of a single city. These histories involve not only different settings, populations, and cultural moments, but also three different kinds of water sources—a river (Philadelphia), an aqueduct (Boston), and a lake (Chicago). Still, there are enough resemblances among the three individual stories that they combine into one narrative, and by the end of the period covered here the three systems were in many important respects more like each other than they were like themselves at an earlier stage. This is because while Philadelphia, Boston, and Chicago were in the process of developing their waterworks, they studied and shared expertise with other growing urban industrial centers of the United States and western Europe. With this in mind, I occasionally refer to contemporary events in these other cities, especially New York's much-examined water history.

Sources and Voices

City Water, City Life concentrates on the public discussion of the two terms in its title. It examines mainly print sources, including surveys prepared by engineers and health officials, statements and reports by legislative committees and departments of water and public works, public addresses that were subsequently published, newspapers, periodicals, and a wide variety of writing that ranges from advice manuals to poetry. This study also analyzes paintings and prints, sculpture, and, of course, the built environment of the city, notably the components of waterworks systems, as well as the many ceremonies and celebrations mounted at all stages of the construction process.

These sources expressed the ideas mainly of a small elite. The justification for concentrating on such voices rests precisely on their status. The people heard here dominated discussions of city water, and they did so by dint of their wealth, expertise, and positions of power, responsibility, and influence. They were almost exclusively men, and, whether they were born in this country or abroad, they were overwhelmingly of northern and western European Protestant background. They were engaged with issues that mattered to everyone, however, and their major proposals often required the endorsement of voters (albeit the electorate was exclusively male). And they were hardly a monolithic group in their political and social outlooks. They included Federalists, Democrats of the period's several evolving types, Whigs, Republicans, and members of other parties and interest groups. They strongly differed in their opinions on the proper direction of urban society. The only thing on which they agreed was that, like it or not, city life was changing constantly and would continue to do so. They worked to control this change as they thought best by having their ideas integrated into the infrastructure of the city they were building, so that the tasks of managing water and containing the fluidity of urban experience were closely related.

The different ideas through which people framed their understanding of city water revealed the period's shifting mix of beliefs. Among these beliefs were an evangelical Protestantism divided in its view of human capability but united in its emphasis on the importance of the moral implications of human action; a post-Enlightenment rationalism that championed the accumulation, organization, and application of secular knowledge and expertise, frequently through the evolving disciplines of engineering, public health, and statistics; a political outlook whose enthusiasm for democracy was often qualified by a sense, whether based on class and ethnic prejudice or sympathetic observation (or both), that large portions of the population were unable to look after their own best interests, let alone those of the city as a whole; an aggressively entrepreneurial spirit and unapologetically materialistic drive; and a devotion to fine feeling, in modes ranging from sentimentality to metaphysics. A particular person might see city life through more than one intellectual, aesthetic, or philosophical framework, even if this led to contradictions.

The history of city water, like all of city life, entailed a ceaseless interplay between human beings and the urban worlds they were building. I generally neither applaud nor criticize the people studied here for what they did or did not say, do, and achieve. I usually give them the benefit of the doubt in assuming that they thought they were acting for the best of others as well as themselves. But it should come as no surprise that the ideas they expressed sometimes revealed bigotry and self-interest, as well as open-mindedness and altruism.

Without suspending hindsight entirely, my analysis thus largely avoids retrospective evaluations, either to praise innovative techniques and progressive social ideas or to condemn mistakes attributable to ignorance of discoveries that would come later or narrowness of thinking that was characteristic of the time. In the period this book covers, for example, though the problems presented by pollution were evident, few paid sufficient attention to sewerage or waste management, and the development and acceptance of the germ theory lay in the future. I try to see city life as it was happening through the eyes of the people whose views and actions I examine, as they attempted to figure out what to do and think about the world they were making as they were making it.

2

THE RIVER, THE AQUEDUCT, AND THE LAKE

Bringing Water to Philadelphia, Boston, and Chicago

Every settlement must have water. Even before William Penn visited the New World in 1682, he decided that Philadelphia, the urban seat of the commonwealth of Pennsylvania, would be located in a stretch of land blessed with numerous streams and springs and nestled between the Delaware River on the east and the Schuylkill River on the west. The Puritans who founded the Massachusetts Bay Colony a half century earlier briefly set up their community in Charlestown, but they soon crossed the Charles River to the hilly Shawmut Peninsula, where they established the town of Boston. They were drawn by word of the "excellent springs of good water, which there abounded."[1] Chicago is situated where the Chicago River meets Lake Michigan, at the southwestern edge of the Great Lakes, the largest body of surface fresh water in the world.

Even with water close at hand, getting it to where it would be used was still a challenging and sometimes costly proposition. Water is heavy—a single gallon weighs just over eight pounds—and transporting it even a short distance is an awkward and burdensome chore. Before water sys-

tems were built, a person with no other way to obtain water conveniently
might purchase it from an itinerant waterman who dispensed it from a
large barrel mounted on wheels. Boston constructed its first reservoir
in the middle of the seventeenth century. Called the Conduit, this was a
shallow basin about twelve feet on a side to which water was piped from
nearby wells and springs, to be used both for fighting fire and for domestic purposes.[2] By the end of the eighteenth century, residents of means
might take their water from the Boston Aqueduct Company, established
in 1795, which delivered it through wooden pipes from Jamaica Pond in
West Roxbury, about five miles southwest of the State House, to certain
portions of the city.

Urban expansion raised the demand for water as it diminished and
degraded the natural supply. Growing cities built over some sources, de-
pleted others, and compromised them all. With his characteristic civic-
minded foresight, Benjamin Franklin offered both his native Boston
and his adopted Philadelphia a solution to this problem. Franklin had
learned from his long periods of residence in Europe that "buildings and
pavements" prevented rainwater from "soaking into the Earth and re-
newing and purifying the Springs," so that "the water of wells must gradually grow worse, and in time be unfit for use." When he died in 1790,
he willed £1,000 sterling apiece to the two cities, specifying that these
funds be loaned to promising young craftsmen of the kind he had once
been. Franklin explained that the original endowment plus the interest
it would earn as the loans were repaid would amount over time to a large
sum, which Boston and Philadelphia should devote to building the water systems he knew they would inevitably need. He recommended that
Philadelphia construct a gravity-driven aqueduct from the Wissahickon
Creek, which flowed into the Schuylkill River a few miles northwest of
the city.[3]

Over the next few decades, many others would confirm Franklin's
observation on the unhappy effects of urbanization on the local water
supply. A long letter sent in 1802 to the Philadelphia *Aurora and General
Advertiser* by "A Citizen" noted, "It is found by experience in the growth
and establishment of many large cities, that the pure springs of nature

become corrupt in proportion as the population encreases [sic]; hence it has been found necessary, for the preservation of health in such cities, to introduce by art a supply of pure and wholesome water, from some neighboring sources." While "Philadelphia, in its infant state, no doubt enjoyed as pure water, as any city in the world," "A Citizen" advised, "the time is long past since the natural purity of the springs has been polluted from the necessity or causes which cannot be dispensed with."[4] Dr. James Mease's 1811 city guide, *The Picture of Philadelphia*, reported that an assay of 220 gallons of water drawn from a downtown source found it laden with chalk, saltpeter, magnesia, and salt. Mease concluded, "From the number of causes serving to contaminate the springs in all cities, the water may be reasonably supposed to be impure and of a disagreeable taste."[5]

In his 1838 inaugural address, Boston mayor Samuel A. Eliot deemed "the introduction of an abundant supply of water" as "the most interesting and important" issue before the City Council. "Springs fail, or the water from them becomes impure; and the supply of rain water is more and more affected by the increased consumption of bituminous coal, and other causes of impurity,"[6] Eliot explained. Among the most objectionable of these "other causes" was leakage from privy vaults that seeped into what groundwater remained available. Chicago's well water likewise soon became contaminated. As for the most convenient source for many people, the Chicago River, the *Chicago Tribune* accurately described it as "the common receptacle for all the filth of the city."[7] Action to improve the local water supply might be delayed, but it could not be avoided.[8]

The construction of the first successful water systems by Philadelphia, Boston, and Chicago involved an extraordinary group of accomplishments. The difficulties encountered were daunting, demanding prodigious amounts of ingenuity, persistence, labor, and money. The paths by which cities faced these difficulties did not trace a straightforward march of progress, a series of wise and confident strides to a widely acknowledged best possible solution. A comment made by the Chicago water commissioners in the mid-1850s applied all too well to every major waterworks project of the previous half century: "It is hardly to be expected that a work of this kind would be carried through in such a manner that there will not appear many things that might be improved upon a second trial."[9] In all three cities there were mistakes and oversights, short-lived "solutions," alternative roads not taken, and controversy and debates, as well as impressive and even daring achievements.

Philadelphia: The Centre Square and Fairmount Works

A series of health catastrophes, and the fear of more to follow, convinced Philadelphians to erect the first truly comprehensive waterworks system in a major American city. During the summer and early autumn of 1793, yellow fever killed an estimated five thousand people, or about 10 percent of the residents of Philadelphia and adjoining towns. Close to half the citizenry, including President George Washington (Philadelphia was then the national capital), literally ran for their lives, fleeing the death-haunted metropolis for the surrounding countryside.[10] Yellow fever revisited Philadelphia in four of the next five summers; the devastation and panic in 1798 were comparable to five years earlier.

In 1797 "an unprecedented number of the most respectable citizens of Philadelphia" petitioned the members of the city's Select and Common Councils (referred to collectively as the Councils), imploring them, "as Fathers of the City, as Guardians of the Poor, and the health and prosperity of their Fellow Citizens in general," to authorize the construction of a water supply that could be used to cleanse the city and, presumably, reduce its susceptibility to the fever. It would also fulfill the essential need of improving protection from fire, which threatened every city. The petitioners contended that there was "no object of use or ornament to which a liberal proportion of the city Funds can be more acceptably applied."[11] The Councils, to which the city charter granted more executive authority than to the mayor, responded by creating the Joint Committee on Bringing Water to the City. This came to be known as the Watering Committee, an exceptionally powerful body consisting of four members from each of the Councils' two chambers.[12] The committee's 1798 report agreed with the petitioners, stating that there was "no object to which the funds of the city could be more acceptably appropriated, inasmuch as such a supply of water is now thought essential to the health of the community, and one of the means most effectual to prevent or mitigate the return of the late contagious sickness."[13]

The Watering Committee initially considered a plan put forth by the Delaware and Schuylkill Canal Company, which had been incorporated in 1791 but still was far from realizing the project from which it took its name. The company, badly in need of funds, declared that the yet-unbuilt canal would do Philadelphia a double service by furnishing water as it facilitated local trade. The committee's reservations soon prompted its members to hire thirty-four-year-old British architect and engineer Benjamin Latrobe, still early in his illustrious American career, to advise them on how to proceed.[14] Latrobe was familiar with the poor quality of Philadelphia water. In the spring of 1798 he remarked in his journal that while houses on the edge of the city's settled area had "admirable Water," elsewhere "the Water is not to be drank [sic], and it is worst in the most crouded [sic] neighbourhoods," where it tasted "as if it contained putrid matter."[15] Latrobe very quickly reviewed the plausible choices (including Wissahickon Creek, as Franklin had suggested) and concluded that the Schuylkill River was the best available source.

As to how to deliver the Schuylkill's water to Philadelphians, Latrobe was enthusiastic about his answer: steam engines. Steam was still an immature technology, however, especially on this side of the Atlantic. Though the earliest steam engine dates to the close of the seventeenth century, the broad practical application of steam power did not begin until Scotsman James Watt's improvements of the 1760s, and Latrobe's plan was boldly innovative.[16] He prescribed two engines. The first, on the east bank of the Schuylkill at Chestnut Street, would raise water from the river. This supply would flow through a mile-long underground conduit to Centre (sometimes spelled Center) Square, which, as its name suggests, was located at the geographical center of the city, at the intersection of Broad and High (now Market) Streets.[17] At this time Philadelphia civic, commercial, and residential life was oriented toward the Delaware, and almost all homes, businesses, and public buildings were several blocks to the east of Centre Square.

The second engine, located in the square, would lift the water to storage

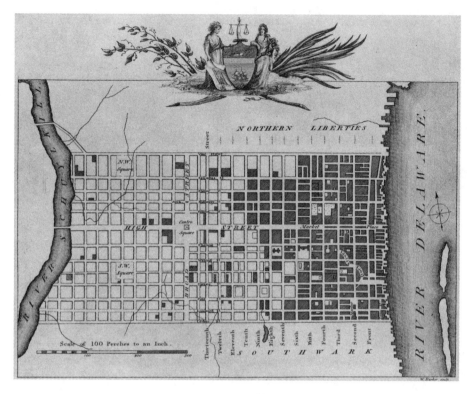

Figure 2.1. This map was included with a series of views of Philadelphia first published by artist and engraver William Birch in 1800. The settled portions of the city's grid are shaded in, revealing the relative isolation at the time of Centre Square, at the intersection of High (now Market) and Broad Streets. One can also see the projected Northwest (now Logan) and Southwest (now Rittenhouse) Squares. At the city's western edge is the Schuylkill River, from which Philadelphia would start drawing its water the following year. The intake was at Chestnut Street, a block south of High Street. On the east is the Delaware River. The Fairmount works would be located along the Schuylkill just northwest of the grid. Courtesy of the Library Company of Philadelphia.

tanks forty feet above the ground, and then gravity would drive it through bored-out pine logs to every part of the city. Homes and businesses that connected to the system would pay an annual fee for water, which would be available for free at public hydrants. The city would meanwhile also use the water for cleaning the streets and for putting out fires. Latrobe estimated the cost at $127,000: $75,000 for the two engines and the labor and materials required to bring Schuylkill water to Centre Square, another $52,000 to accomplish the first stage of piping it through the city. He predicted that the system would serve at least 4,000 private customers paying an average charge of $10 annually, so that the waterworks would soon recoup its costs and start turning a profit.[18]

The committee endorsed this proposal. As its members explained, the "style" in which Latrobe presented his plan "evinced his clear conception of the subject, and excited a degree of confidence in his ability."[19] They hired him to supervise the construction, replacing an engineer who had favored the canal. Latrobe in turn asked Nicholas Roosevelt to manufacture the engines at the latter's factory in New Jersey. Roosevelt (a member of the same family that subsequently produced two presidents) would later join with Robert Fulton in the development of the steamboat. He would also become Latrobe's son-in-law.

The construction process was fraught. Even before it began, Latrobe faced sharp criticism from investors in the Delaware and Schuylkill Canal, who were outraged by the city's rejection of their plan. They published a pamphlet declaring their canal the best choice for watering the city and describing Latrobe's proposal as misleading on how much it would cost and how long it would take to build.[20] After Latrobe responded with a defense of his design in general and of steam engines in particular, a supporter of the canal issued an anonymous *ad hominem* attack that excoriated him for his "officious interference, and ostentation of *professional abilities* [which it called "*unknown*" and "*untried*"], with his *doubts* and *fears*" about the canal. This critic ridiculed Latrobe's "confused and enormously expensive project of '*aerial Castles*, and elevated *Reservoirs*, of *different stories*, Fountains, Baths, &c.'"[21] Deeply insulted and eager to "protect [his] opinions and assertions against misrepresentation," Latrobe replied with another pamphlet, this a full eighteen pages of detailed argument, dismissing the remarks about "aerial castles" and "elevated reservoirs" as "an amusing proof of the gaiety of the writer's disposition."[22]

The recurrence of yellow fever in 1799, which again scattered the local population, slowed the pace of the building, as did Roosevelt's failure to stay on schedule. Latrobe's budget estimates, as the canal company had charged, proved to be much too optimistic. To supplement the funds the city had borrowed to pay for the works, it levied a very unpopular tax. This drew sharp complaints, some from proponents of the canal, that the Watering Committee had acted very foolishly in hiring Latrobe and choosing steam power. Latrobe published yet another defense of his plan and its use of steam engines. While earlier models were "justly considered as dangerous, . . . and now and then they did a little mischief," he wrote, "[a] steam engine is, at present, as tame and innocent as a clock."[23]

In its 1799 report, the committee admitted to numerous problems but argued for pushing on. The members emphasized the positive as much as they could, restating the compelling reasons why the works was being built and why they had hired Latrobe. The report reminded critics that the committee was doing what the public had asked and that, in respect to expense, "a loss to the city in a single visitation" of yellow fever, not to mention a major fire, would possibly be higher than what Philadelphia would have to pay to construct Latrobe's entire system. Aided by personal loans from members of the committee, which were needed to meet "embarrassments," construction proceeded.[24] The new works were at last in service by the end of January 1801. While Roosevelt's engines were its most significant mechanical feature, the system's distinctive emblem was the sixty-foot-square stone building, three stories high and clad in white marble, that Latrobe designed to house (and hide) the Centre Square engine and the two storage tanks positioned above it. It featured Doric columns and a domed roof with an opening at the top, recalling the oculus of the Pantheon in Rome, whose purpose was to allow smoke from the engine to escape.

Latrobe scholars and historians of technology agree that the Philadelphia waterworks was his "greatest and most successful engineering work," a milestone in the evolution of American civil engineering, but it was a very mixed success in practical terms.[25] At first the works and the engine house were widely popular. "It is amusing to observe the crowd round the hydrant opposite to my door," Quaker merchant and Watering Committee member Thomas Pym Cope wrote in his diary the day after Schuylkill water began to course beneath Philadelphia's streets. "Everyone must have a pull at the lever & even when the water flows, many seem as if they could not believe it."[26] In 1802 the committee stated that the system was delivering about 400,000 gallons per day.[27] Very soon, however, significant shortcomings became evident. The engines were hard to maintain and maddeningly unreliable, requiring the repeated replacement and refinement of major and minor parts. Since the engines were arranged in series, if either one stopped, whether for planned maintenance or by accident, Philadelphians faced the inconvenience of having no running water, not to mention the peril of being without protection against fire. Despite Latrobe's assurances about the engines' safety, in April 1801 two men who entered the large boiler in Centre Square in order to repair it were suffocated to death, and it had to be torn apart to retrieve their bodies.[28]

Other faults beset the system. The capacity of the storage tanks was too small—demand could exhaust them in half an hour. Meanwhile, the wooden pipes began to rot and leak. And, as the canal company's owners predicted, the works cost far more than expected to build and operate. In order to allow for some leeway beyond Latrobe's estimate of $127,000, the committee budgeted $150,000, but by 1801 expenses to date exceeded $220,000.[29] At the close of 1803, total costs were close to $300,000—and climbing. Meanwhile, revenues were meager. Far fewer Philadelphians had connected to the system than Latrobe had thought would do so. The income for 1802 and 1803 was under $1,500.[30]

In spite of these problems and all the second-guessing they provoked, the need for the works was so great and the investment already made so considerable that there was no acceptable alternative. In his letter to the *Aurora* early in 1802, "A Citizen" made this point, reminding those who wished to suspend funding the unfinished Centre Square system that Philadelphia's other water sources were so miserably inadequate that there was no option but to persevere. "What kind of policy is that which would now put a stop to the work[?]" he asked. "Do not good policy and justice require that it should be complete[?]"[31]

The mechanically and financially disappointing system continued to evoke impatience, anxiety, and anger. The *United States Gazette* of August 10, 1805, criticized the managers of the works for failing to provide either warnings before or explanations after the frequent shutdowns. "If it be necessary, from any cause whatever," the *Gazette* inquired, "to stop the water two or three times a week, can it not be done at stated periods, so that those who wish may furnish themselves with a quantity before hand?" The cost and unreliability of the steam engines had already made the city's relationship with Latrobe and Roosevelt sharply adversarial. Watering Committee member Cope, who had been so pleased and amused by Philadelphians' initial delight in their new water supply, developed a profound distrust of the two men.[32]

The bad feeling became public. The committee's 1801 report, without mentioning Latrobe by name (it referred to him as "the Engineer"), sniped that after he had received virtually his entire fee of $6,350, he had asked for almost $2,000 more, "on grounds which the committee deemed inadmissible." A purposeful footnote pointed out how much higher actual costs were than Latrobe's estimate.[33] Mutually recriminating letters appeared in the *Aurora*. Latrobe complained of the "very evident pains [that] have been taken . . . to cast a shade over my character as a man." In response to the implication that he had either intentionally or incompetently misled the city with an unrealistically low estimate of costs, Latrobe invited "any impartial judge [to] look at the work . . . and say whether human judgement be adequate to the exact calculation of the expence [*sic*] of a work so various in its nature, so much dependent on weather, and on the industry and talents of so many individuals—a work which has struggled through [an] embarrassment [of] funds, and has been interrupted by yellow fever.—And this may safely be asserted, that if a large sum has been expended, it has been expended faithfully, and has accomplished its purpose."[34] In 1803 he told a correspondent that the problems with the pipes stemmed from the fact that Cope and fellow committee member Henry Drinker had purchased them from "improper and inadequate agents."[35] The dissatisfaction on both sides had led to an acrimonious parting of the ways the year before.

Annual proceeds did not pass $10,000 until 1810, and they rose only slowly for the next several years.[36] By the end of September 1811, with both Latrobe and Roosevelt long gone, Philadelphia had invested more than $500,000 in the Centre Square works, with very mixed results. The annual income of the works was $12,163, while the expenditures were $29,702. All of the 28-plus miles of wooden pipe that had been laid would have to be replaced, and there were only 2,127 paying customers in this city of well over 50,000 people. The number of subscribers, about half Latrobe's projection, was disappointing even if the number of actual users of the water was much higher.[37]

The best thing one could say about the system was that it did seem to improve the health and fire safety of Philadelphians. Although yellow fever returned several times in the early years of the nineteenth century, the city appeared to be less prone than it had been to severe attacks by this and other diseases. A few weeks after the waterworks opened, a blaze on Second Street between Market and Chestnut burned on, the *Aurora* reported, because the nearest hydrant was out of order, but an account of another fire on Market between Fourth and Fifth Streets several months later declared that "every good citizen present, who values the convenience and safety of the city, must have observed with pleasure the astonishing effects produced by the Schuylkill water." In 1806 a story on another fire, which broke out between the Bank of the United States and the Bank of Pennsylvania, ended by reflecting, "We cannot but congratulate our fellow citizens on this occasion, upon the acquisition of the Schuylkill water," without which the fire would have "laid one fourth part of the city in ashes." It noted that "in this one night the city has saved more by the use of this water than the whole expense of introducing it."[38]

Nonetheless, since there was little prospect that the works would ever break even, let alone pay back construction costs, there was a possibility that the system might have to shut down. The Watering Committee requested an assessment of the situation from the former and current waterworks superintendents, John Davis and Frederick Graff. Davis, who succeeded Latrobe, served from 1802 to 1805, when he left to run Baltimore's water company, which had been founded the year before. Graff, who had received his training as Latrobe's assistant, would direct the Philadelphia system almost continuously from 1805 until his death in 1847. He was said to have sometimes worked all night overseeing repairs on one or another of the engines of the original system because he wanted to avoid interruptions during the daytime, when demand was

high. Graff was a significant figure in the early history of American urban waterworks, not only for his long, loyal, and often trying service to
Philadelphia, but also because, like several directors of major systems, he
consulted for dozens of other cities.[39]

Davis and Graff submitted what the Watering Committee termed "an
ample corroboration of the opinion entertained by many of our fellow
citizens, that the present plan of watering the City is miserably deficient
in the essential requisites of security and economy." The Centre Square
works was rife with "radical defects," and, "even if put into complete repair," would never be satisfactory. The two men advised that "no time
ought to be lost in preparing for the adoption of a better system."[40] Graff
proposed moving the water intake about a mile farther up the Schuylkill,
to the base of a large hill known as Fairmount. From that spot, new and
better engines would pump the water to a capacious reservoir atop the
hill, which was taller than the highest point in the city, and gravity would
take things from there.[41]

Constructing an intake, pumps, and reservoir at this site was a tough
job. For starters, the city had to blast away rock to clear a place for the
waterworks buildings and engines to stand. About two weeks after construction on this new system began, a local resident predicted that it
would prove another costly failure. A letter to the *Aurora* from "A Citizen" (it is not clear whether this was the same person who used this
frequently employed pseudonym a decade earlier) began by recalling the
city's recent waterworks history. While the Centre Square system "promised permanency and an ample supply of water," he stated, the people
of Philadelphia had "nevertheless been disappointed at the hazard of
our safety, and been frequently put to great inconveniences." "A Citizen"
predicted that a new steam-driven works, like the one it would replace,
would function inefficiently and prove "a whirlpool to the wealth of the
city." He suggested that Philadelphia tap a more distant and pristine
supply—echoing Franklin, he recommended Wissahickon Creek—that
was of sufficient elevation so that gravity alone, with no assistance from
costly and trouble-prone steam engines, would be all that was required
to drive the water.[42]

In November 1812, another letter from "A Citizen" posed an entirely new alternative, one that he claimed would furnish Philadelphia "any quantity of the purest quality, and in the easiest manner; and at a comparatively moderate expense." He expressed astonishment that Philadelphia's stubborn fixation on steam had caused it to overlook the fact that the current of the Schuylkill River could be harnessed to propel its own water up Fairmount. "A Citizen" claimed not to understand why "the citizens of Philadelphia should be so regardless of their own interest, as to suffer a power [i.e., the motive force of the Schuylkill], worth thousands annually, to roll idly away merely for the want of improvement." After all, many American cities and towns were using local waterpower productively, "and Philadelphia, by only attending to what immediately belongs to herself, may in this respect, be placed on a footing with, if not surpass, the most favored among them." The city simply needed to dam the river at Fairmount in order to divert it over drive wheels that would pump water to the reservoir far more cheaply and dependably than steam engines.[43]

In 1812 the city was still not ready to abandon its financial and imaginative investment in steam. It stuck with Graff's scheme, on which construction began in the fall of 1815. The cost by 1818 was over $320,000. A key new element was an innovative high-pressure engine designed by the brilliant Philadelphia inventor Oliver Evans.[44] The project also included the building of the first few of what would be a group of elegant structures along the river, to enclose the machinery, afford workspace for the system's managers and employees, and provide shaded viewpoints for visitors. Like Latrobe's engine house in Centre Square, these structures were in the neoclassical style. Their aesthetic appeal could not, however, compensate for the fact that the expensive technological upgrade was another disappointment. Although the new engines worked better than the ones in Centre Square, they were still costly to run and unreliable to operate, and they could not keep up with the growing demand for water.

Graff and the committee decided that their best option after all was what "A Citizen" had proposed. By mid-April 1819, contractors began erecting a dam across the Schuylkill to elevate the level of the river and divert its current to pump water to the Fairmount reservoir. In the same year, the Watering Committee approved Graff's request that the wooden pipes be gradually replaced with iron ones.[45] The earliest of these were imported from England, but soon Philadelphia would begin using pipes made in the city, which its manufacturers would also sell to other American systems.

The switch to waterpower from steam was another complex undertaking. Since the Schuylkill is an estuary, Graff observed the shifting level of the river carefully to determine what adjustments needed to be made to accommodate this. In addition, while the actual building of the dam would prove to be difficult in itself, so was obtaining the requisite legal rights. This needed the consent of two mill proprietors who would lose their power source once the dam was constructed, and of the Schuylkill Navigation Company, to which the legislature had recently granted the authority to construct a canal in the river. To gain this consent, the city had to pledge to pay the mill owners $150,000 and to build a lock for the canal next to the dam.[46] The agreement worked for about a decade, but then the company would precipitate a legal confrontation when, over the city's strong objections, it abruptly began construction on a second lock by the first one.[47]

By 1822 the new waterworks was ready. After all the problems of the previous two decades, the Watering Committee was eager to announce a triumph. "The great work at Fair Mount," it stated, was "now brought to a conclusion so far as it is contemplated at present to extend it."[48] Its 1824 report announced that the sixteen-foot diameter waterwheels were even more efficient and economical than predicted. While previously it had cost $206 to raise 3,375,000 gallons of water a day, this could now be done for $4, a saving of over 98 percent. By adding more waterwheels, the capacity could be increased to 10 million gallons per day, and daily operating costs would be only $10, which was more than $500 less than it would cost to pump the equivalent volume with steam engines.[49] By 1831 the committee could boast, "Since the erection of the water-power in 1822, the City and Districts [i.e., the neighboring communities to which the city could now sell its surplus] have had a constant supply of wholesome water, without one day's intermission."[50]

Figure 2.2. In this peaceful scene, fishermen try their luck from the shore of the Schuylkill opposite the Fairmount waterworks. The dam is visible on the left. Behind the works is the hill known as Fairmount. Carefully landscaped rustic paths lead to the top, where the reservoir was located and which is now the site of the Philadelphia Museum of Art. Courtesy of the Library Company of Philadelphia.

In the first full year of operation using the three waterwheels, the Fairmount works supplied 4,844 customers with 1,972,975 gallons of water daily and produced an income of $26,013. This was still well below the $69,269 in expenditures, but soon the expanding system was making money. By 1836 six wheels and pumps delivered an average of 4,178,317 gallons daily and earned $101,266, almost $30,000 above expenses. In addition, the city was successfully upgrading the delivery network. While in 1823 there were fewer than ten miles of iron pipes, now there were almost sixty.[51]

In 1823, with the conversion from steam power to waterpower an accomplished fact, the members of the Watering Committee presented Graff with an ornate silver loving cup. The inscription attested to "their admiration of the taste, judgment, and fidelity, with which he arranged, superintended and assisted in prosecuting to a conclusion the Public

Works at Fair Mount."[52] Graff's future relationship with the committee would have its ups and downs, but after his death in 1847 its members raised a Gothic monument to him—complete with a portrait bust—at the base of Fairmount, close by the works he had directed for decades. On the side of his tomb in Laurel Hill Cemetery, three miles up the Schuylkill from the dam, a view of his waterworks buildings is carved in low relief. Writing in 1842 about Philadelphia's water system, inventor and future U.S. Patent commissioner (as well as historian of technology) Thomas Ewbank offered Graff high praise. "It is impossible to examine these works," Ewbank wrote, "without paying homage to the science and skill displayed in their design and execution; in these respects no hydraulic works in the Union can compete, nor do we believe they are excelled by any in the world."[53]

Looking back on prior decisions and actions—which Philadelphia was doing as soon as 1799—officials explained that they had always acted as effectively as could have been expected, given that virtually every step was an experiment. Of instances where measures had worked out well, they applauded their own wisdom. As for the things that had gone wrong, they reminded others how hard a task—actually, an ever-changing combination of hard tasks—it was to build a waterworks, pointing out that the failures had been instructive and even valuable.

The Department of Water's 1860 history of the works questioned the original refusal to follow Franklin's plan and let gravity bring the Wissahickon's water to Philadelphia's homes and businesses, but it also found a silver lining in all the expenditures on the steam-powered systems that the city then abandoned. The money had not been misspent, since Philadelphia gained many things along the way: "Character and impetus was given to the city; much was done to improve its sanitary condition, and an important feature added to its many attractions as a place of residence." In addition, local mechanics acquired a great deal of useful experience, since "to their pumping engines and the practical knowledge derived from them, may be attributed much of the preeminence which this city has always enjoyed in the constructing of machinery." As a result, "there is scarcely a water works, in the construction of which Philadelphia mechanics have not more largely contributed than any others, and are still contributing; and in this city and vicinity are located the largest works for casting pipe and constructing the various apparatus necessary for water or gas works, in this country." In sum, "it may be seen that the permanent interests of this city have been greatly promoted and advanced by the sums thus expended, although the burden fell heavily upon the citizens in the early history of the works." Uneven as it was, Philadelphia's early water history had taught an important lesson: "No investment made for an abundant supply of water to any city is ever lost, even though it may at first seem to remain unproductive."[54]

Boston: The Long Road to Long Pond

Boston delayed far longer than Philadelphia in deciding to construct a waterworks, but then took advantage of forty years of progress in hydraulic engineering to build a successful system on the first try. In May 1825, Mayor Josiah Quincy appointed a joint committee of the Common Council and the Board of Aldermen, with himself as chair, "to inquire into the practicability, expense and expediency of supplying the city with pure water."[55] What prompted this step was not dread disease but a major fire the previous month that had destroyed several dozen buildings. The joint committee agreed that providing water to Boston was indeed practical, expedient, and worth the expense. Another fire, on November 10, 1825, confirmed the wisdom of this view.

Quincy saw the city's needs for water as going well beyond fighting fire. He declared it "an article of the first necessity" to construct a system that would provide "a sufficient and never failing supply for our city of pure river or pond water" that, in addition to dousing flames, would also serve "all culinary and other domestic purposes." This meant that the water must be "capable of being introduced into every house in the city" and acceptably "soft," that is, low enough in mineral content to dissolve soap and to cook vegetables satisfactorily. This excluded the city's well water, Bostonians' main current source, which was "generally harsh, owing to its being impregnated with various saline substances." This brackishness, he noted, "impairs its excellence as an article of drink, and essentially diminishes its salubrity."[56] Quincy here cited a letter from the eminent physician and Harvard Medical School professor John Collins Warren, who was one of the founders of Massachusetts General Hospital as well as the *New England Journal of Medicine and Surgery* (which in 1828 became the *Boston Medical and Surgical Journal* and, a century after that, the *New England Journal of Medicine*). Warren advised Quincy that "many complaints [of illness] owe their existence to the use of the common spring water of Boston," and that he had seen patients' health improve when they switched to another source.[57]

Over the next two decades, Boston repeatedly seemed to be on the verge of doing something about its water supply, but then retreated. The city commissioned waterworks reports not only in 1825, but also in 1834, 1836, and 1837, and it held referenda on water in 1836 and 1838. After listing several pressing reasons why Boston needed to build a new system, a group of petitioners to the City Council early in 1838 exhorted, "*Let the thing be done*, and done as soon as by any exertion consistent with prudence and reasonable economy, is practicable."[58] Speaking for the members of the Joint Standing Committee on Water, Mayor Samuel A. Eliot expressed "their wish, their hope, their earnest prayer that some decisive action may at length be had upon this long delayed subject, and that they may live to see a sufficient quantity of water provided for every family and individual in the City."[59] But the thing was not done, and decisive action was not to be had.

It is not clear why Boston did not begin a serious discussion of a waterworks until almost thirty years after Philadelphia had done so, or why it continued to drag its feet. Perhaps it was because Boston, while hardly exempt from epidemic disease, never experienced anything as horrifically motivating as yellow fever's assaults on Philadelphia. In addition, a large number of residents believed that they could make do with their present water sources. And there is no question that economic conditions following the Panic of 1837 discouraged a major investment in infrastructure or anything else, however badly it was needed.

By the early 1840s, however, Bostonians concurred they could delay no longer. Included in the wealth of information that pioneering statistician and sanitary expert Lemuel Shattuck packed into his 1845 city census was a survey of how the population of 114,366 was supplied with water, "that indispensable element of health and comfort." Shattuck, who would address the issue of city water many times over the next few years, found that while 5,287 of Boston's 10,370 houses had wells, only 214 of these wells furnished water soft enough to be acceptable for washing, and 1,052 were effectively dry.[60] The city's poorest residents suffered the most difficult circumstances. On the one hand, they were plagued by flooding in their homes due to the absence of adequate drainage, while, on the other, they were hard-pressed to secure the clean water they needed to sustain even a minimal standard of decency.

Although there was now a consensus that Boston had to do something about water, there was still no agreement as yet on the best source for a central water supply or on how it should be funded and managed. Among several options under discussion, there were two leading contenders. The first was to construct a publicly owned system that would bring water from Long Pond, located about nineteen miles west of the city in Framingham and Natick. The other was to entrust Boston's needs to a private company whose main source would be Spot Pond, about ten miles north in Stoneham. Both sources were at a higher elevation than Boston, so no pumping would be necessary, though each would require the construction of an aqueduct to carry its water to the city.

Near the close of 1844, the City Council, backed by yet another report, determined that a city-owned system drawing on Long Pond was the only acceptable choice, mainly because of its larger capacity (the consultants judged the purity of Long and Spot Ponds to be of comparably high quality). Though backers of Spot Pond protested loudly, in a referendum on December 9 the Long Pond and public ownership proposal carried the day by a margin of almost three to one.[61] A few more hurdles remained, however, though none seemed especially high. The legislature would need to pass and the governor would need to sign an act authorizing the city to construct the works, and then Boston voters would have to ratify its terms in another referendum.

During testimony before a special committee of the General Court (as the Massachusetts legislature is called) in February and March 1845, opponents of the Long Pond plan focused on how much the new system would cost, as well as how to pay for it. Citing a pamphlet on the subject he had recently published, Lemuel Shattuck warned that "to bring water from Long Pond would be ruinous to the city." His evidence was the price of New York's publicly owned Croton system, completed in 1842, which doubled the original estimate. Besides, in his informed opinion, Boston did not need the volume of water that Long Pond would supply; the fifty-year-old Boston Aqueduct, supplemented by Spot Pond, would suffice for another twenty years.[62]

Boston's attorneys produced witnesses who stated that the situation was dire and that Long Pond was by far the best solution. Thomas Curtis, a member of the Common Council, recounted episodes of an "underground war" between neighbors when one drilled a new well that drained water from existing ones nearby. A builder named George Cram testified that Bostonians were desperate for water. He pointed to the fact that one well in the city was being used by as many as forty-three families, and that another served people who came from several blocks away because they had no better alternative.[63] The attorneys also cited the enthusiastic endorsement of Long Pond by the voters. The legislature passed and the governor signed what was called the Water Act; a referendum on the charter was scheduled for May 19, 1845.

The intensity and bitterness of the public discussion of water preceding this referendum had no equivalent in Philadelphia or Chicago. Opponents of a public system attacked the Long Pond plan in letters to newspapers, polemical pamphlets, and impassioned broadsides, which those favoring the proposal answered in kind. The lawyer for an anti–Long Pond group accused city officials of acting with "hot haste and a want of due deliberation and thorough investigation."[64] Partisans of both sides held spirited meetings in many venues, including schools and churches, and they separately assembled in Faneuil Hall, where they listened to addresses and passed resolutions, a tradition that went back to the town meetings before Boston was incorporated as a city in 1822.

Notices for these gatherings promised that there would be good speakers and that seats would be reserved for the wives and daughters of voters. Since they could not vote and did not hold public office, women's voices were almost entirely absent from political debates and deliberations relating to water, but these notices acknowledged that water was as vital and central to their existence as it was to men's. Perhaps even more so, given that the onus of housework, including the management of domestic water, fell on them.[65] One wag composed a poem about the dismay and frustration that women felt because they were being inundated with publications advocating one source or another at the same time that they lacked water in which to wash their children's soiled clothes:

> Their houses are deluged with pamphlets untold,
> Of green, blue and gray, such a tale to unfold!
> Of Shakum, and Mystic, Charles River so fair,
> Long Pond, and Spot Pond, and Natick's fresh air.[66]

The tone of most of the debate was dead serious, if often hyperbolic. The call for one meeting described the water act as "the only mode worth consideration, by a community NOW SUFFERING, AND BURNING for the want [of water]."[67] The Boston *Daily Advertiser*—which under the editorship of its owner Nathan Hale (nephew of the illustrious Revolutionary patriot whose name he shared, and brother-in-law of the statesman and Harvard president Edward Everett) zealously promoted the Water Act—admitted that a meeting of citizens who were favorable to bringing in water but who were "either entirely opposed to the principles of the water act, or are in doubt as to the expediency of adopting it," drew one of the largest crowds.[68] The key points in question at this meeting were not whether Boston needed a new water supply, but how the commissioners who would run the waterworks were to be chosen, and whether the act granted them excessive power, independence, and pay. The gathering produced a statement signed by sixteen prominent citizens, who protested "that principles are interwoven into the act, which, as they are contrary to the spirit and practical operation of all our civil institutions, will work unkindly—will produce mischief and engender discontent."[69]

These objections had merit. The Water Act specified that the mayor, aldermen, and Common Council members would form a committee of the whole that would select the commissioners and set their salaries. This meant, critics pointed out, that it was mathematically possible for a commissioner to attain office even if he was not supported by a majority in both chambers—in fact, he could be elected without a single vote from the Board of Aldermen, since it consisted of eight members while the Common Council had forty-eight. If only a bare quorum of thirty-one members participated in the selection of commissioners, sixteen votes would carry the day. Once selected, the commissioners could borrow funds for the project without City Council approval, never mind that of voters. Even the minimum salary in the range prescribed by the legislature ($3,000 to $5,000) was higher than the mayor's ($2,500), and this salary was to be paid, also without City Council endorsement, from the funds the commissioners were authorized to borrow. Further, the act did not define a time limit for the commissioners' term of service; they could be removed for reasons of "incapacity, mismanagement, or unfaithfulness," but only by a three-fourths majority vote by the council. There were no restrictions on the number of staff the commissioners could hire and at what salary.[70]

As the referendum approached, Mayor Thomas Aspinwall Davis ordered that samples be taken from Spot Pond, Jamaica Pond, and the Charles River (which had long been a favored alternative of some Bostonians, in spite of the low quality of its water), as well as Long Pond. Davis had these placed where citizens could view them, so that "the people might know the truth, as to the comparative merits of the water, and to rebut the influence that it has been attempted to create prejudicial to the water of Long Pond, the popular source, whence it is proposed to obtain a supply."[71] At this time, water that looked clear and was free of objectionable odor was usually assumed to be fine for domestic use, especially if its mineral content did not make it hard. There were some who were concerned about the tiny organisms, often called "animalcules," that are visible in virtually all water when examined under a microscope, but physicians were generally able to allay fears that these were abnormal or harmful. The water from Long and Spot Ponds passed inspection, and both were deemed superior to the Charles.

Ten days before the referendum, the self-named Faneuil Hall Committee—consisting of ten leading Bostonians chosen by the people who had met there in the fall of 1844 to "take the general charge and direction of such measures as might be expedient to '*foster and promote the Water Project*'"—issued a pamphlet that reminded voters of their past support for a comprehensive system and appealed to them to stand up and be counted one more time. "The fact is," the committee warned in gender-charged language, "the time has arrived when the people must act. They must gird on their armor, and fight manfully against all attempts to debar them from the accomplishment of a measure so essential to their interests as the one in question."[72] Despite such pleas, the people decided not to act, by the slim margin of 329 votes out of the 7,669 that were cast (at this time Boston's population was about 115,000 people, which indicates how few people participated in such a momentous decision). If only a relative few of those casting ballots against the Water Act were moved by the contention that it granted the commissioners too much compensation and unsupervised authority, that was enough to have made a decisive difference.

It appeared that Bostonians would continue to function as best they could with what water they currently had. Trying to rally a favorable vote, the *Daily Advertiser* had maintained just before the referendum that if the act was defeated, "there would seem to be very little prospect, of any concentration of opinion or effort, in favor of any other measure, for a long time to come, either in the City Council, or on the part of the public."[73] After the votes were counted, the *Daily Evening Transcript* predicted that the water question was now "put at rest for years, or until some dreadful conflagration takes place, to remind the citizens of their utter destitution of this important element, for the extinguishment of fires, to say nothing of the great inconveniences to which thousands of families may be put for the want of it for domestic purposes."[74] Sensing a revived opportunity, the owners of Spot Pond renewed their efforts to persuade the city to contract with them.

The Long Pond proposal did not die, however. Its backers refused to accept the referendum as the last word. On the evening of the day of the vote, a group of supporters met to resolve that it was "the will of the citizens that water shall be introduced into this city at the city's expense, either from Charles River or Long Pond, not withstanding the result—of this day's balloting." They called for Bostonians to assemble "without respect to party" on a ward-by-ward basis in order to keep the cause alive.[75] Within a short time, representatives from each of the city's twelve wards organized the Union Water Convention, open to anyone regardless of other political affiliations, in favor of public water. They declared that the defeat in the referendum should not discourage proponents of the Long Pond plan, since they had lost "by a small vote, and a trifling majority."[76]

At the convention, attendees chose as their leader former mayor Charles Wells, and they urged citizens throughout Boston to form what they called Ward Water Unions in order to promote the cause. At another meeting, which immediately followed a fire that struck South Boston in September, speakers pointed out that this latest misfortune yet again demonstrated the perilous inadequacy of the current water supply.[77] For its part, the City Council's Joint Standing Committee on Water decided to commission one more report, choosing two presumably neutral outside experts to submit their opinion: John B. Jervis, the renowned chief engineer of New York's Croton Aqueduct, and Walter R. Johnson, a professor of chemistry from Philadelphia. The owners of Spot Pond had meanwhile offered the city a minority stake in the company. The company sweetened the deal with attractive terms, but when a delegation visited Spot Pond on September 30, 1845, the pond's low water level cast enough doubt on its adequacy as a supply to effectively doom its chances.[78]

By November the Ward Water Unions claimed to have over four thousand members, a larger number than those who voted down the 1845 Water Act. On the eighteenth of that month, Jervis and Johnson tendered their report. It was long and technical, but its recommendation was easy to understand: "We have no hesitation in stating, as our opinion, that Long Pond is decidedly the most appropriate source to which the City can resort, to obtain an adequate supply of pure and wholesome water for the present and future use of its inhabitants."[79] Within days the City Council unanimously rejected Spot Pond's offer and endorsed the idea of a public waterworks system drawing on Long Pond. The death of Mayor Davis on November 22, shortly before the end of his term, did not break the momentum, since all the candidates to succeed him in the December 8 election also favored Long Pond. It was the council's responsibility to choose a person to finish Davis's term, and its members selected Josiah Quincy Jr., the son of the former mayor. Quincy had already been elected to succeed Davis in January, and so he assumed office a few weeks early. Since the discussion of building a system at public expense began in the administration of his father, this gave the city's water history a nice symmetry. There was little opposition when Boston again petitioned the legislature and governor to authorize the waterworks, and the vote in the April 13, 1846, referendum—4,637 to 348—left no room for doubt that the voters backed the revised legislation, which answered most of the political objections to the 1845 Water Act.[80] The tally suggests that many who voted nay the previous spring stayed home.

The city broke ground at Long Pond on August 20, 1846. Even more than had been the case in Philadelphia when the Fairmount dam was erected, there was some complex legal work to be done before construction could begin. Boston compensated owner W. H. Knight for exclusive rights to the water of Long Pond and nearby Dug Pond; for two water privileges (i.e., rights to take water as a power source) on a nearby stream; and for Knight's "manufacturing establishment," including mills, machinery, fixtures, and worker housing. The commissioners agreed to rent Knight the site for three years and to install a steam engine large enough to power his factory. The city also purchased another pond in nearby Hopkinton for $25,000 to use as a compensating reservoir, and in 1847 alone the water commissioners reached terms with the owners of more than sixty different properties near Long Pond and in the path of the aqueduct. During the following year, they settled with the selectmen of Brookline for encroaching upon local roads, and they consented to erect an aqueduct for Newton after residents there claimed that construction of the underground portion of Boston's aqueduct had dried up their wells.[81]

The Boston system was not as pathbreaking a feat of engineering as was the Philadelphia waterworks, but building it was still a very ambitious and complicated project. The fourteen-mile aqueduct (approximately half the length of New York's Croton Aqueduct) ran from Long

Pond to a holding reservoir just south of Boylston Street in Brookline. From there the water flowed to a pair of distributing reservoirs, one behind the State House on Beacon Hill and the other on Telegraph Hill in South Boston. The unevenness of the terrain the aqueduct crossed required the construction of two bridges, over which water was carried in inverted siphons, and two tunnels. To make possible the building of the latter, laborers had to sink seven vertical shafts along its route. Seven steam engines ran constantly in order to pump water from the excavation, which was bedeviled by patches of quicksand. Before Long Pond water was let into the aqueduct, an inspection of the tunnel portion revealed cracks that had to be repaired.[82]

In April 1849, six months after the water was first introduced into Boston, the City Council extended the life of the original commission another eight months.[83] At the close of the year, with the power granted to it by the state, the council water committee drafted an ordinance establishing the Cochituate Water Board, which would take over management of the newly built works. The board consisted of a commissioner to superintend all activities, an engineer to oversee construction projects, and a water registrar to set rates and administer other financial matters. Each was appointed by the City Council for a one-year term. The council also chose a water comptroller, who was not a member of the board, to keep its books and manage its funds. The water committee was charged with inspecting the works once a year and auditing the board's annual accounts.[84]

Like the Philadelphia system, Boston's works encountered numerous financial difficulties. The first water commissioners calculated building costs at about $2.6 million. As in Philadelphia, the city allowed for a margin of error, deciding that a round $3 million would be a safer estimate.[85] The actual bill by the time Boston officially finished construction at the end of April 1851 was $5,184,984, almost exactly twice the original projection.[86] But Bostonians took to their new water supply far more quickly than had Philadelphians. In its first calendar year of operation, the system delivered an average of more than 10 million gallons a day via 80-plus miles of pipes to over 900 hydrants and more than 12,000 customers. In their last report, issued early in 1850, the original water com-

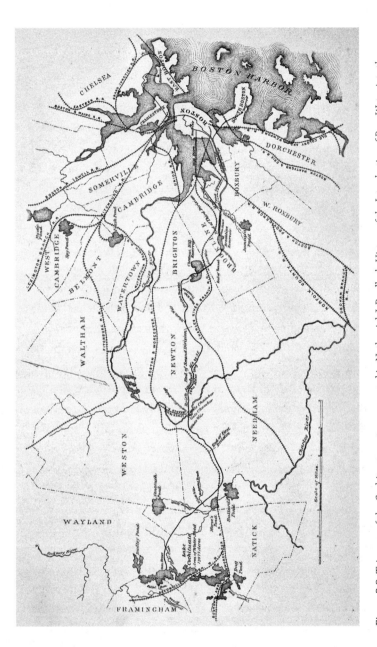

Figure 2.3. This map of the Cochituate system appeared in Nathaniel J. Bradlee's *History of the Introduction of Pure Water into the City of Boston* (1868). It shows the line of the aqueduct connecting Lake Cochituate (*on the left*) to the Brookline Reservoir, located between the Chestnut Hill Reservoir, which was not built until after the Civil War, and Jamaica Pond, which had been used by the Boston Aqueduct Company. From the Brookline Reservoir the water headed to the city. Courtesy of the Massachusetts Historical Society.

missioners expressed their great pleasure in the abundance of the supply and the "manifest improvement in the transparency and purity of the water, since its introduction."[87]

Chicago: The Tunnel Beneath the Lake

One might assume that access to an abundant quantity of good water would be the last problem Chicago would face, given the presence of the Chicago River and Lake Michigan. But there was a catch: as was frequently the case with lake cities, Chicago took its water and dumped its waste in the same place.[88] Since the river flowed into the lake, to pollute the river, as Chicagoans most spectacularly did, was to risk ruining both as healthy water sources.

The young town of Chicago's first small step in constructing a water supply ignored both the river and the lake. On November 10, 1834, the town trustees allocated $95.50 to dig a public well just north of the river's short Main Branch (this and the North and South Branches divide the city into its North, South, and West Sides, then called divisions).[89] It was at this time that Chicago's population began to climb very rapidly, from a few hundred people in 1833 to 4,170 by 1837, when the town was incorporated as a city. The numbers kept rising, approaching 30,000 by 1850, 110,000 a decade later, and 300,000 only ten years after that. Water needs grew even faster, making Lake Michigan—rather than the small, sluggish, and, by this point, filthy Chicago River—the obvious choice for meeting them. As in Philadelphia and Boston, however, there was a very significant gap between the idea of constructing a water system and its successful accomplishment.

The first Chicago waterworks was privately owned. The 1836 charter of the Chicago Hydraulic Company allowed it four years to begin construction. The company needed the entire period because of the same hard times that delayed Boston in the late 1830s, but building finally commenced in 1840, and the works was running two years later. The system consisted of a steam-powered pumping station located at Lake Street and Michigan Avenue, just east of the expanding downtown. Thanks to landfill projects that began later in the century, this intersection is now about a half mile from the lake, but at that point it was very near the shore. The engine drew the water through a 320-foot intake to two holding tanks, both raised high enough for gravity to push their contents through wooden pipes to the second floor of buildings. The steam engine was more than equal to the job—indeed, it had enough surplus power to run an adjoining mill (as Latrobe had vainly hoped to do with excess engine capacity in Philadelphia), but the system had other liabilities. Within a year the lake level dropped unexpectedly, which limited the effectiveness of the intake pipe. And since water near the intake was turbid as well as shallow, it required filtering. Customers of the Chicago Hydraulic Company were also dismayed to find not microscopic animalcules but all-too-visible small fish in their water. They were still better off than the many residents who lived outside the company's modest service area, which covered mainly the South Division, reaching only a few people in the West Division and none in the North. As a result, in 1850 about 80 percent of Chicagoans had to get their water by some other means.[90] This usually meant fetching it from the polluted river or the lake, or purchasing it from watermen.

As Chicago's leading nineteenth-century historian, A. T. Andreas, recollected, "For a large and rapidly growing city this state of affairs was alarming, especially as the general health was perceptibly suffering."[91] The state of Chicago's water moved from being a source of annoyance to one of ridicule, shame, and danger, the last because of the vulnerability of much of the city to fire and to the outbreaks of cholera in 1849 and succeeding years that caused dramatic spikes in mortality.[92] The irony of the situation was obvious. "With an unlimited quantity of the purest water upon the continent of America, in the closest possible proximity," the Tribune complained, "our city is nevertheless but poorly supplied with facilities for an abundant and cheap supply of the article." The paper hoped that the state legislature would soon approve a plan to empower Chicago to build a public system, one worthy of its ambitions, which were as limitless as the lake. "Among the many other subjects which address themselves to the earnest attention of all who take a laudable interest in the welfare and prosperity of our beautiful city," the paper's editors declared, "none are fortified by so many considerations that would be likely to move a man, as that of the cheap and abundant supply of pure water."[93]

This view enjoyed much influential support. In April 1850 a group of citizens, similar to those who had campaigned for a public water supply in Philadelphia and Boston, determined to do likewise in Chicago. Records of the unicameral Chicago Common Council (the name was changed to City Council in 1875) include petitions and remonstrances asking it to provide better water service.[94] As in Philadelphia, if not in Boston, results followed rapidly. The state incorporated the public Chicago City Hydraulic Company on February 15, 1851. Its three original commissioners were appointed, but their successors were to be popularly elected.[95] The commission's first duty was to "examine and consider all matters relative to supplying the city of Chicago with a sufficient quantity of pure and wholesome water, to be taken from [L]ake Michigan, for the use of its inhabitants." The charter empowered the commissioners to "adopt such plan as in their opinion may be the most advantageous for procuring such supply of water," and to "ascertain as nearly as may be, what amount of money may be necessary to carry the same into effect." They hired the capable engineer William J. McAlpine to prepare a plan.[96]

McAlpine's proposal resembled the private system it replaced, though the new works was to be better situated and far larger. He presented two choices for where to locate a new lakefront water intake: at Chicago Avenue (about a half mile north of the Main Branch of the river) or at Twelfth Street (now Roosevelt Road, a little less than two miles south). In either case, Chicago could distribute water from a single raised reservoir near the intake point or one in each division. The commissioners selected the Chicago Avenue site and the three reservoirs option. McAlpine hoped to place the intake six hundred feet from shore in order to be certain that it was always well below the water's surface, but "the boisterous condition of the lake" made it impossible to go out that far.[97] On March 2, 1852, Chicago voters endorsed this design by a majority of over 80 percent. Construction began that summer, and the works were pumping Lake Michigan water by early in 1854.

While acknowledging many unexpected costs, Chicago officials emphasized what a bargain their system was compared to waterworks in

Pittsburgh, Cincinnati, Detroit, Albany, Boston, Buffalo, and New York.[98] Like other new systems, at first the hours of operation were limited— there was no city water available on Sundays except in case of fire—and until the reservoirs were constructed, service could be erratic. In time the pipes (first wood, but soon iron) reached more and more customers, and delivery became more reliable. By 1856, with demand for water approaching the capacity of the high-pressure steam engine in the pumping station, the city decided to install a bigger and more powerful machine without interrupting service, a tricky feat that was accomplished the following year. On January 1, 1858, Waterworks Superintendent Benjamin F. Walker proudly stated, "It is a source of gratification to me to be able to represent the works in a perfect state of repair and their ability to supply the demand made upon them, to fully equal the anticipations of those who designed them."[99] A year after that the commissioners announced, "The Water Works with unimportant exceptions, are throughout in good working order, and as a whole were never in so serviceable condition."[100]

"Serviceable" did not quite mean "satisfactory," however. The city was expanding so fast that it was outstripping both the larger capacity engine and the pace at which pipes could be laid.[101] But the quality of the water was the more distressing problem. First of all, the presence of fish in the pipes persisted. In April 1855 the *Tribune* predicted: "Unless a finer strainer be put upon the reception pipe of the Water Works, there won't be any fish left in the Lake shortly, judging from the rate at which they are being sucked in at present." The paper claimed that "little juvenile white fish" were "alive and kicking in different parts of the city, after they have passed for miles through the pipes and into the houses." It warned readers to be "careful in drinking not to get fish bones in your throat."[102] This was meant to be an amusing exaggeration, but the situation that prompted it was disconcerting and unacceptable. In their report for 1859, the water commissioners admitted this in a section candidly titled "Fish in the Pipe," in which they discussed the problem and possible solutions.[103]

The fish were a minor inconvenience compared with Chicago water's other defects. The river, which by this time was appallingly fouled by slaughterhouses, distilleries, tanneries, tallow renderers, and soap and candle makers, was contaminating the water by the intake. The supply was described as "a villainous compound of decomposed animal and vegetable matter, titurated with sufficient water to give it a semi-fluid consistency."[104] Chicago's new sewerage system—one of the first in the nation, constructed in the mid-1850s to counter flooding and cholera—made matters worse. This system was designed by the most important figure in the city's early water history, Ellis Sylvester Chesbrough. Chicago's sewerage commissioners had lured Chesbrough away from Boston, where he had been in charge of the section of the Cochituate waterworks between Long Pond and the holding reservoir in Brookline from 1846 to 1850, when he became city engineer. His sewers greatly improved Chicago's drainage, but they did so by sending effluent into the lake, both directly and via the river.[105]

Well aware of the worsening situation, the Chicago water commissioners passed a resolution in March 1860 asking Chesbrough to propose how the city might secure clean water. He submitted five alternatives: extending the intake pipe a full mile out into the lake and presumably beyond the reach of contamination, building an intake tunnel of the same distance under the lake's clay bottom, moving the intake about twenty miles north to a point near the lakefront suburb of Winnetka, constructing a system of filtering beds, and erecting a subsiding reservoir to allow the water to clear before it was pumped to users. Other ideas briefly considered included placing the intake five miles up the lakefront by the town of Lakeview (later incorporated into Chicago) or bringing water by aqueduct from Crystal Lake, about fifty miles northwest of the city. Chesbrough said that it was not immediately necessary to take any of these steps, but that the situation needed to be monitored.[106]

In 1861 the newly constituted Board of Public Works assumed responsibility for Chicago's water. The board consolidated the supervision of what had previously been separate administrative areas: sewers, public parks, streets, river and harbor, public buildings, and public lighting, as well as waterworks. Chesbrough became the board's chief engineer—he would be appointed city engineer and then, when the Department of Public Works was created in the mid-1870s, the city's first commissioner of public works. In its first report, the board stated that the fish problem would be readily dealt with—as it was—by a modification of the intake opening, and that it was time to do something about the more serious hazard caused by the "large quantities of blood and other waste and foul substances that find their way into the river" from the city's constantly growing number of manufacturing concerns.[107] While board members insisted that the terrible quality of the supply had been exaggerated, that "ordinarily the water of the city is as good as that furnished to any city," and that "statements that the daily drink of our citizens is the discharge of the sewers, the refuse of the slaughtering establishments, etc." were "wholly untrue, and cannot fail to do harm," they agreed that the current system was no longer viable. A recent freshet—a sudden rise in water level due to heavy rain and rapid thawing—had flushed a large volume of Chicago River water into the lake at a time when the river was in "an extremely offensive state." This had turned the city's supply "very bad" for several days.[108]

Acting on Chesbrough's advice, the board decided that the best alternative was to build a tunnel under the lake to a point not one but two miles out, which members believed was surely farther than river pollution could reach, thanks to the diluting effects of the massive amount of water in the lake. The superiority of this strategy over the other choices lay in the fact that the tunnel seemed to combine "greater directness to the nearest inexhaustible supply of pure water" along with "permanency of structure and ease of maintenance."[109] The board predicted, however, that carrying out this plan was not going to be easy. Some Chicagoans thought that it would be impossible.

They had reason to think so. The physical challenges of building such an enormous aqueduct beneath a lake bed, especially at that time, are hard to overstate, and its construction was one of the most heroic engineering feats in this age of such achievements.[110] The slightly oval tunnel, five feet wide and two inches larger than that in height, was to be big enough to deliver 50 million gallons a day, which was thought to be enough for a population of a million Chicagoans, more than six times the number at the time. Sundays and holidays excepted, construction proceeded around the clock, with crews of miners and masons work-

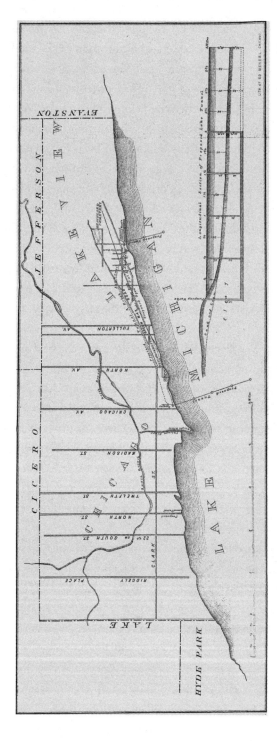

Figure 2.4. This is from the *Second Annual Report of the Board of Public Works to the Common Council of the City of Chicago*, which appeared in 1863, when the city was considering several different alternatives for building a new waterworks. North is to the right. Note how the Chicago River divides the city into its North, West, and South Divisions. As is the case now, the commercial center of the city was just south of the Main Branch of the Chicago River in the South Division. At this time, before Chicago annexed the adjoining towns of Lake View, Jefferson, Hyde Park, Lake, and part of what is marked here as Cicero, the city was far smaller than it is today. Included on the map is a proposed intake tunnel from the shore in Lake View, a rejected option, as well as the tunnel that Chicago did build (twice as long as the one pictured here) at Chicago Avenue. Note also the sectional drawing of the tunnel itself. Courtesy of the Northwestern University Library.

ing in alternating eight-hour shifts. Despite the presence of ventilation shafts, smoke from workers' lamps and evaporating perspiration from their bodies and those of their work mules made it hard to see. At any moment they might hit pockets of gas that could burst into flames or explode. At one point they were blocked by a boulder that was so big they needed to blast it away. The greatest progress made in any one week was ninety-three feet.[111] The tensions of laboring so intensely in such close quarters led to a deadly dispute that broke out shortly after midnight on August 24, 1864, between two foremen, Patrick Hunt and Michael Corry. When Corry, who had recently been fired because of his frequent quarrels with Hunt, came to retrieve his gear, the two men got into another argument. They agreed to return to the shore and settle their differences with their fists, but before they ascended from the tunnel Corry stabbed Hunt to death.[112]

In July 1865, sixteen months after the start of construction, three tugs towed an enormous wooden pentagonal structure, forty feet high and fifty-eight feet wide on each of its sides (requiring 675,000 feet of lumber and 200 tons of metal bolts and fastenings), to the projected terminal point of the two-mile tunnel. The plan was to fill it partially with stones and settle it into place on the lake bottom. The so-called crib was to enclose the intake and provide a multi-level work platform at this point. After a harrowing delay of three days as the unanchored and unwieldy crib was tossed about by a violent storm that threatened to break it apart, it was finally put in position. Digging under the lake soon began shoreward from the intake, so that there were now two sets of crews burrowing toward each other.

Chesbrough had already supervised this kind of construction—also with men working night and day—in building the tunnel portions of the Boston aqueduct, but he had never directed an excavation carried out under an enormous lake. When the two sets of crews met on November 30, 1866, the tunnel sections were within a very manageable 7.5 inches of right on the mark, thanks to the calculations of assistant engineer William Clarke.[113] The tunnel was put into service the following March. A heavy rain that once again drove polluted Chicago River water into the lake near the shore—though not as far as the crib—soon of-

Figure 2.5. With nothing but oil lamps for illumination, crews worked around the clock to dig Chicago's lake tunnel and, as seen here, to line its interior. Visible is one of the mule-drawn rail cars that laborers used to transport materials and themselves. The project also required air shafts for ventilation and machinery to remove the excavated earth. This is one of several illustrations that appeared, among other places, in J. M. Wing and Company's *The Tunnels and Water System of Chicago* (1874). Courtesy of the Northwestern University Library.

fered assurance, at least for the moment, that the city had followed the right course. Chesbrough expressed the widely shared hope that Chicago would "be entirely freed from the necessity of using offensive water for domestic purposes hereafter."[114]

At the same time the city was digging the tunnel, it was constructing a new engine house at the Chicago Avenue location used by the Chi-

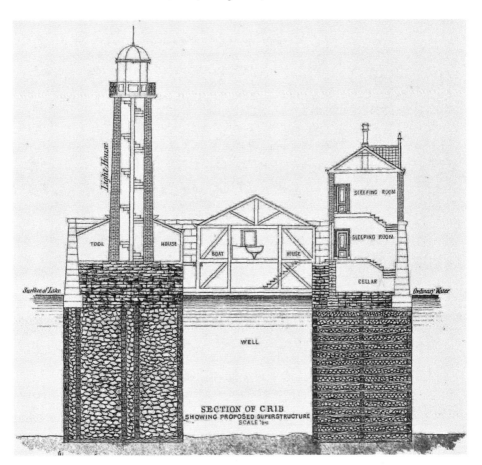

Figure 2.6. This sectional drawing of the crib is from the *Eighth Annual Report of the Board of Public Works to the Common Council of the City of Chicago* (1869). The cutaway view reveals the crib's hollow center, the rocks that were used to settle this wooden structure on the lake bottom, and the living and work spaces used by its resident crew. Courtesy of the Northwestern University Library.

cago City Hydraulic Company.[115] High-powered pumps would enable the Board of Public Works to phase out the reservoirs.[116] Across Michigan Avenue, laborers also began work on a 154-foot tower, which would house a standpipe.[117] The Pumping Station and the Water Tower (the names of these now-iconic buildings are commonly capitalized) were designed by W. W. Boyington, the leading Chicago architect at the time, in the "castellated Gothic" style, with battlemented cornices. They were both

Figure 2.7. This illustration is artist R. W. Wallis's depiction of the crib on a stormy day. The hope that placing Chicago's water intake so far from the shore would protect the city's supply from pollution by the Chicago River proved unfounded, as did the expectation that the new supply would meet the city's needs. From J. M. Wing and Company, *The Tunnels and Water System of Chicago* (1874). Courtesy of the Northwestern University Library.

covered with the fashionable cream-colored limestone known as Athens marble, cut from quarries southwest of the city. This made the Pumping Station in particular resemble a late medieval fortress. Boyington crowned the Water Tower with an iron cupola, from which, as Andreas pointed out, "might be obtained a magnificent view of the lake, the city and surrounding country."[118]

In late March 1867, Chesbrough decided the moment had come to give the lake tunnel a final internal inspection by first filling it with water and then, after emptying it halfway, floating through it in a small rowboat from the crib to the shore.[119] He invited reporters from three Chicago dailies to accompany him. They had second thoughts as they followed their leader—outfitted with a rubber coat, high boots, and a broad-brimmed sou'wester—in being lowered by rope down a dark, damp, and dripping shaft into the tippy boat, which had taken on some water even before they boarded it. According to the correspondent from the *Chicago Republican*, only the concern of each of the reporters that the others might scoop him sustained their mutual resolve. Under Chesbrough's direction, these anxious sailors directed their craft by pushing against the tunnel's interior walls with their hands. Though the water was "as clear as crystal, and sweet to the taste," their trepidation understandably increased as their lamps dimmed and then died. They began

Figure 2.8. In the late 1860s, the Chicago firm of Jevne & Almini produced a series of somewhat idealized lithographic views of the city, including this charming depiction of architect W. W. Boyington's "castellated Gothic" Water Tower and Pumping Station. The two buildings look much the same today, though, thanks to landfill, now both are farther from the lakeshore, and the Water Tower, which was once one of the tallest buildings in the city, is dwarfed by its neighbors. The view is northeast along Chicago Avenue, with Lake Michigan in the distance. Courtesy of the Chicago History Museum, ICHi-06859.

to fear that they would perish along with Chesbrough, who remained inexplicably unperturbed. They struck up "The Star-Spangled Banner" to muster their courage.

They had progressed well toward their destination when the headroom between the boat and the tunnel roof diminished to a point that made it impossible to go farther even if the men lay flat, and so Chesbrough reversed course. About a mile from the crib, and after much feckless bailing, the precarious vessel finally swamped. Forced to wade through this cold, wet, and eerie chamber in total darkness, the men felt their bodies growing numb and found themselves "thinking sorrowfully of the four small obituaries that would appear in the morning papers." The comic near-tragedy ended with the members of this improbable crew, by now cramped and dizzy, making a slippery ascent back up to the deck of the crib. Chesbrough was evidently satisfied by what he had seen. As the reporters resuscitated themselves with hot coffee, they heard a great "rush and roar" as the tunnel was refilled.[120]

On March 25, the city put the tunnel into operation with great public fanfare. The members of the Board of Public Works attested that Chicago now rejoiced "in the object of previous years of labor and anxiety—the supply to the city of the waters of the lake in abundance and absolutely pure."[121] Two years later they bragged that Chicago's water output was now approximately the same volume as Boston's, and that the new works was "hardly equaled anywhere for commodiousness and beauty."[122] A group of wealthy Chicago businessmen rewarded Chesbrough with $11,000 in water bonds—his regular annual salary was $1,500—in gratitude for his prudence, energy, and skill. They described the tunnel, which had been "undertaken in the face of much opposition," as a "triumphant vindication of his wisdom and foresight."[123]

Usage records show how dramatically the new system improved matters. In 1866, when the city's population was just over 200,000 and there were 13,634 taps in service, the average number of residents per tap was 14.5. By 1868, the figure was down to 11.7. In that same time period, the daily supply leaped from 8.7 million to 14.7 million gallons. By 1876, at which time the population had doubled, the number of takers had more than quadrupled, and there were only 7.1 Chicagoans per tap.[124]

Not only did more houses and businesses connect to the supply, but water also reached to more and more places within individual structures,
some equipped with multiple fixtures, though much of this kind of pen-
etration did not take place until later in the century.[125] City boosters celebrated the building of the water system as the latest proof of Chicago's
astonishing energy, ambition, and capability. A contemporary history of
the tunnel crowed that accomplishing a feat like this was not for the
weak of heart, who could not even read about it without shuddering. "If
all the people of Chicago had been of this sort," it contended, "we should
have drank foul water all our days."[126]

Not that the system was problem-free. In order to serve people in
all divisions from a single pumping station, the board had to lay pipe
across the shallow and busy Chicago River. In the course of constructing
a bridge over the North Branch of the river in 1856, pile drivers severed
a submerged pipe, which interrupted the flow of water.[127] In 1854 an anchor from a river-going vessel had destroyed a main at State Street, cutting off the regular supply to the South and West Sides for four months.
The city then forbade the dragging of anchors, but in August 1869 a
lumber barge under tow neglected to obey this law and tore out a main
at Chicago Avenue, which interfered with service to the West Side for
several weeks.[128]

The board seriously underestimated both how quickly the city of Chicago—and its demand for water—would grow and how much the system
would cost to run. In their first report in 1851, the water commissioners
included a detailed table that forecast the city's population, waterworks
expenditures and income, and the water debt for the next twenty-five
years. They erred significantly on the population, which they predicted
would climb to 162,000 by 1875, rather than the approximately 400,000
it actually reached that year. They were even further off when they
said that waterworks expenditures for 1875 would amount to about
$31,000. The figure turned out to be almost $628,000 (to be fair, revenues were also higher than expected).[129] The commissioners were not far
wrong in one important respect, however. They predicted that income
would exceed annual expenses in two years. While it ended up taking
five, this was certainly very good when compared to either Philadelphia

or Boston. What the commissioners did not anticipate at all, however, was that Chicago would need to build an almost entirely new system in the 1860s.

No one could have prophesied the Great Fire of 1871 and the devastation it caused. All things considered, however, the new water supply system was relatively fortunate. Of an estimated $2.2 million in fire-related losses sustained by infrastructure overseen by the Board of Public Works, only $248,910 was due to damage to the waterworks. Although the roof of the engine house collapsed in flames, the pumps halted, and Chicago lost almost $100,000 in wasted water when 15,000 service pipes melted from the intense heat, the Pumping Station's walls and engines remained largely intact. The tunnel was safely under the lake, and although the crib was reportedly threatened by airborne brands carried far out into the lake by the fire's fierce updraft, it was not harmed. Until the machinery could be returned to service and pipes were replaced, Chicagoans made do with water from artesian and conventional wells, ponds in city parks, the lake, and the polluted river.[130] The Water Tower was—miraculously—one of the very few structures in the path of the fire that was spared. Standing solitary and unscathed amid the charred ruins of the North Division, which was otherwise devastated by the fire, it became an enduring emblem of the city's irrepressible spirit.

As difficult and complex was the task of designing and erecting comprehensive water systems in all three cities, the physical effort of construction was only part of the story. Every step along the way involved a similarly challenging consideration of its implications, large and small, for the communities that were building these systems. When city people were talking about water, they were always discussing many other equally critical matters. These included the character of the society they were forming, the alterations that urbanization was working on the natural world, the health of the urban individual and of the city as a whole, and where their metropolis stood in the course of human events. As was the case with the waterworks they were building, the progress of their discussions was neither simple nor direct, but a continuous process in which finding a right answer proved an elusive goal.

3

THE INDIVIDUAL AND
THE COLLECTIVE

Water, Urban Society, and the Public Good

The seventeenth-century founders of Philadelphia and Boston had hoped to establish a society based on a shared high purpose and mutual regard. This vision evolved into the democratic principle that the individual is most rewarded and fulfilled by dedication to the collective, which in turn benefits most by respecting the rights and liberty of each of its members. As cities grew larger and more polyglot, and their social and economic divisions became more distinct, the sense that all residents were united by a common cause and the feeling that every individual should think of the welfare of others in the community became harder to sustain. Different portions of the population were unlikely to know and get along with one another, much less see themselves as knit together by any principle other than self-interest. Yet at the same time they all faced some common challenges—among them the need to provide enough good water to serve everybody—that demanded collective solutions.

Advocates of waterworks in Philadelphia, Boston, and Chicago contended that these systems did not just provide water but, if properly conceived and constructed, would turn the very aspects of urbanization that might seem to undermine community into the means of enriching and fortifying it. However much the growth and diversification of a city's population might have weakened ties among individuals, its size empowered its members to do great new things. The same urban development that made it hard for city people to obtain water on their own and caused them to be strangers to one another provided the credit, labor, and expertise to construct a waterworks that would serve every individual and strengthen the whole.

The practical matter of building waterworks thus also involved a continuing and often contentious attempt to define the public good and how it might best be served. This included a debate over whether something as vital as a waterworks should be publicly or privately owned, which opened into a consideration of the proper purviews of city government and private enterprise. All three cities ultimately decided on public systems, and the celebrations they organized to mark the stages of construction were meant to affirm how effectively these new waterworks integrated the individual and the collective to the benefit of both. But the discussion of city water also exposed doubts about urban democracy. In addition, certain practical questions posed by water, notably how to charge for it and what to do about the individual waste of this essential public resource, revealed the tension between self-interest and the needs of everyone.

Individual and Public Good

There were few more profound moments in the process of urbanization than when individuals switched from the disparate local water sources they had been using and joined with many others in taking water from a central works. One was now relying on others for access to an indispensable commodity.[1] A person who previously obtained water independently decided to connect to the new system in the expectation that time, effort, and possibly money that would otherwise be devoted to getting water would be freed for other purposes. The means of that liberation was the slender service pipe—called a ferule—that tied an individual property into the underground grid. By linking oneself to a central water supply, one irrefutably became an urban person, one of thousands of individuals whose everyday existence required this shared resource. At the same time, the success of the entire undertaking depended on having enough paying customers participate in it. The recognition of this forced many to ponder the significance of what they were doing.

Perhaps the most eloquent voice in the cultural conversation about city water was that of Dr. Walter Channing. Channing was the first professor of obstetrics and medical jurisprudence at Harvard Medical School and, like his colleague John Collins Warren, a founder of the *New England Journal of Medicine and Surgery*. He was the grandson of William Ellery, a signer of the Declaration of Independence, and the brother of both the noted Unitarian clergyman William Ellery Channing and of Edward Tyrrel Channing, Harvard literature professor and editor of the *North American Review*. A major figure in several different reform movements—health, anti-poverty, housing, temperance, and peace—Channing was also a leader in the quest to obtain water for Boston. In his 1844 pamphlet, *A Plea for Pure Water*, he told of a resident who realized that he could get good and plentiful water only with the construction of a common waterworks. By allying his fortunes with those of others whom he did not know, he was able to serve himself far better than he could alone. "Understand what I say," the man explained. "I want [water] for myself, and for my family. I do not want it for the poor, nor for the rich. I cannot bring it here to my door myself. It will cost too much. So, I have joined with others to get the [water] act."[2]

Support for a shared waterworks may have been based in self-interest, but some of those who, like Channing, argued for the construction of citywide systems claimed that such projects possessed a transcendent quality that went far beyond the narrow motives of personal convenience, comfort, and thrift. A waterworks, they held, was at once an expression of the urban collective and a force that united it. At the laying of the cornerstone of the Beacon Hill reservoir in November 1847, Mayor Josiah Quincy Jr. explained why this kind of structure was more

significant than other worthy civic institutions: "The corner stone is laid
of churches of a particular faith,—of monuments of past events of general but not universal interest,—of hospitals, of whose benefits a small
number only will partake,—of colleges, where the fortunate few alone
derive immediate advantage." Water was different; there was "nothing
sectarian, nothing sectional, nothing exclusive about it." Its benefits
were not limited to a particular "party in politics or sect in religion." It
was "an equal blessing to the high and low, the rich and the poor, the just
and the unjust." What Quincy called "the gift of water" served every city
dweller, from "the poor woman, toiling for her children at the wash tub"
to the "proudest beauty in the luxury of her chamber." Its value hardly
stopped there: "Nor will the blessing be confined to man; nothing that
enjoys animal or vegetable life will exist on this peninsula, for centuries,
without sharing its benefit."[3]

The term "public good," and close variations such as "public interest"
and "common good," appeared repeatedly in discussions of waterworks.
When the members of the Philadelphia Watering Committee announced
in 1818 that the system would need an additional loan of $70,000, they
admitted this was "unpleasant" news, but they asked Philadelphians to
keep in mind that such borrowing was "necessary to their comfort and
health." The purpose of the loan, after all, was to achieve something that
"the public good imperiously requires."[4] The choice of the adverb "imperiously" pointed to the fact that in this case—so different from places
and times where and when the construction of a works might be an act
of self-aggrandizement on the part of a monarch—here it proceeded directly from the needs of the people. Bostonians who petitioned for a
new system in 1838 maintained that it was "their unqualified opinion,
that the public good requires the introduction of a supply of pure water
into the city as soon as the proper works can be constructed."[5] Chicago's
first water commissioners stated that in deciding to issue bonds in order to fund their project, they were "but acting for the common good."[6]
Such phrasing conceptualized the urban population as "the public," an
abstract entity made concrete by the indisputable fact that every person
in it required water.

It is helpful to distinguish between *a* public good and *the* public good.

A public good is an object or institution that, unlike Quincy's examples of a hospital or a college, serves everybody. A public good may be anything from a road sign to a standing army. *The* public good is the beneficiary of these separate public goods. It is usually defined in aggregate terms, such as the condition of the population's wealth, health, or safety. Proponents of comprehensive urban water systems posited that these projects were so essential that they were indistinguishable from the public good.[7] This invested the specific public good of water with a very special status. Walter Channing based his arguments for building a waterworks on what he described as "*broader grounds.*" Channing wrote, "I see its necessity in the wide public want. I look for its accomplishment in a wise care for the public good, in generous purposes, in large and true policy." In his view, the value of a community acting together to provide water for itself went far beyond even such unquestionable desiderata as prosperity, health, and fire protection. "I see this project to supply this city with pure water in its moral relations," Channing declared.[8]

A system that furnished a generous supply of clean water represented the animating force behind all the individual interests that brought it into being. Without city water, shimmering in the sunlight at its source and rushing through the mains beneath the streets, there would be no city life—or public good—at all.

Public or Private?

Once a city decided that it was necessary to build a waterworks, the first and most basic question it had to answer was just who would own and administer it. To a certain extent, this involved practical issues, but it became clear almost immediately that it was impossible to think of these issues apart from questions of principle, including whether doing the right thing and doing the most sensible thing were two separate matters.

Philadelphia, Boston, and Chicago were in the substantial majority of larger nineteenth-century American cities that constructed publicly owned systems, several of them after private companies could not or would not dependably furnish a plentiful supply of good water. While over 58 percent of American waterworks in 1860 were privately owned, twelve of the sixteen biggest urban centers—the exceptions were New Orleans, Buffalo, San Francisco, and Providence—had settled on public systems.[9] These numbers make the decision by Philadelphia, Boston, Chicago, and other cities to build their own supplies seem more simple and uncontroversial than it was. There was a long-standing history of public pumps and wells, but constructing a city-owned waterworks that would serve the entire population was a completely different undertaking.[10] Building a central system enormously expanded the size, responsibility, and expense of urban government. A report issued in 1825 by the original Boston City Council Joint Standing Committee on Water understated the situation considerably when it noted that whether the building and management of a waterworks "ought wholly to be left to private enterprise, or wholly effected at the expense of the city, are questions on which there is a diversity of opinion."[11]

Between a historical suspicion of political rulers that went back to the Revolution, a strong libertarian bent, a faith in capitalist enterprise, and a concern about incompetence and dishonesty among public officials, city people were averse to government involvement in city life.[12] Through most of the nineteenth century, public sentiment preferred that private companies provide even widely desired public goods, and the encouragement of the interests of business became identified with the well-being of the American economy and with national progress as a whole.[13] Before Philadelphia and other cities built their waterworks, no municipality had ever taken direct responsibility for a public good of the scale, intricacy, and cost of building and managing a citywide water system. As Boston mayor Thomas Aspinwall Davis observed during the water debates of 1845, a public waterworks was a project of "great magnitude, surpassing anything hitherto entered upon by the City Government."[14] In many respects, it would have been safer and easier for cities to pass on to private investors the financial risks and technical responsibilities that these enormous and complex systems entailed. There was also a strong precedent in London, which was served by several private water companies.

Why, then, did Philadelphia, Boston, Chicago, and most of urbanizing America's leading cities build their own waterworks? Some reasons were practical. The Philadelphia Watering Committee rejected the Delaware and Schuylkill Canal's proposal to provide water because its members concluded that while the canal might become a potentially important new avenue of trade in addition to a supplier of water, it might also be very expensive, take a long time to complete, and serve neither of its purposes well.[15] By mid-century, supporters of public ownership could cite instances in which private companies had failed to do even a satisfactory job. Boston city attorney Richard Fletcher perhaps overstated things a bit when he told the state legislature in 1845, "Every where in this country where [water] supplies had fallen into the hands of private companies it had been a source of disappointment and regret," but his point was well taken.[16] In Chicago, the privately owned Chicago Hydraulic Company was obviously unequal to the task of furnishing a sufficient amount of water to fight fire and meet the demands of domestic and commercial customers in the booming city.[17] London's private companies charged high prices for bad water. Fletcher was one of many who meanwhile cited the success of public projects, notably in Philadelphia and New York, the latter of which had previously suffered from the greed and indifference of the privately owned Manhattan Company.[18]

Some maintained that the purposes and constituencies of private corporations argued against choosing them to furnish city water. The Philadelphia Watering Committee, whose members were men of affairs who firmly believed in business enterprise, remarked on the unsuitability of anything other than public ownership of the water supply. Since by definition the primary purpose of a private corporation was to make a profit, such a corporation would naturally put the goal of earning its stockholders the maximum return on their investment ahead of the public good of Philadelphia. For example, it would concentrate on those parts of the city where income was most likely and skimp on those that were less promising, which conflicted with the committee's responsibility to furnish water to all the residents of Philadelphia. It would be hard for a private company to reconcile the wishes of both the city and of the company's owners, while the Watering Committee's loyalties would, in theory at least, be undivided.[19]

Speaking in 1826, Boston mayor Josiah Quincy Sr. made the same

points when he asserted that Boston "ought to consent to no copart-
nership" in procuring water. "No private capitalists will engage in such
an enterprise without at least a rational expectation of profit," he explained.[20] They would pursue the cheapest water, the best customers,
and the highest price, while the city wanted the best water to be delivered to everyone at the lowest cost. Quincy had very recently written to
Joseph Lewis of the Philadelphia Watering Committee to ask his advice
on the issue of ownership, and Lewis responded at length. After pointing
out the shortcomings of London's water companies, Lewis stated that
the reason for these deficiencies "does not require much consideration
to discover." The problem lay in "the fatal error of suffering interested
individuals to have the supply of an article of the most indispensable
nature and without which health and comfort cannot be enjoyed." When
it came to meeting a city's water needs, the desire for profit, which was
the reason why private companies existed, was beside the point. Lewis
advised, "Expence [sic] is not to be regarded; if a Company can supply
your City, they will expect to profit by it, and this profit might as well
be saved by your Corporation." Lewis added that it would be wrong to
expect a private company to lose money in providing water, since "individuals should not suffer by forwarding of a great public object." And
if a water company did run at a deficit, the city would suffer, since "the
citizens will feel it by a pinched and partial supply."[21]

Quincy agreed. "There are so many ways in which water may be desirable, and in such a variety of quantities, for use, comfort, and pleasure,"
he reasoned, "that it is impossible to provide, by any prospective provi-
sions, in any charter granted to individuals, for all the cases, uses, and
quantities which the ever-increasing wants of a great city in the course
of years may require." In addition, unlike a private owner, the city would
encourage widespread use of water right away "by reducing as fast as
possible, the cost of obtaining it, not only to the poor but to all classes
of the community." Finally, the right to dig up existing roadways, which
a water company needed to do in order to lay and maintain pipes, "ought
never to be intrusted to private hands, who through cupidity or regard
to false economy, may have an interest not to execute the works upon a
sufficiently extensive scale, with permanent materials, thereby increas-

ing the inconvenience and expense which the exercise of the power of breaking up the streets necessarily induces."[22]

The 1839 report of Boston's Joint Standing Committee on Water raised similar points. A private company would not act "without a pretty sure prospect of profit" based on the certainty of being given a monopoly, which in turn would make the city dependent upon it. If the company suspected that the city might later assume control when its monopoly rights expired, it would build a provisional system, which could cost taxpayers more in the long run than if they took the responsibility of constructing a durable waterworks. And, as Quincy had also explained, "the work is of such a character, and involves so many interests connected with the streets, the fire department, and various other branches of the public service, that if entrusted to private hands, there would be constant danger of unpleasant collisions, and injurious controversies with the City authorities."[23]

That water was so indisputably a public good prompted many backers of public ownership to invoke Channing's "broader grounds." In a self-respecting community devoted to the public good, they argued, there were certain things the people must do for themselves, and providing water was one of them. This assertion was heard often in Boston, where the advocates of a public system faced strong opposition. The Joint Standing Committee on Water declared in 1838 that it could "entertain no doubt that it is best for the city to execute the work by its own agents."[24] The public good was so dependent on water that ownership of the latter needed to be kept from private hands. Channing took the same position. City water must never "be given in charge to any man, or body of men" other than the city as a whole, he insisted. And if there was a profit to be made, "let the city make it," which would multiply the benefits of public ownership to the public good. Any excess revenue would go "from it, from its government, back to the people again, in new forms of usefulness and of blessing; with full interest added, to make its agency more important and more useful."[25]

Boston city attorney Richard Fletcher maintained in 1845 that the principle at hand "was one of no ordinary importance." Only rarely did an issue of such "deeper, or more serious interest" confront lawmak-

ers. The water question "did not relate to pecuniary speculation or enterprise, but to the comfort, health and lives of a large portion of the
people of the Commonwealth," and it arose "not occasionally, or at intervals," but affected "every living being, every hour and every minute."
The need of humans for water could not be approached "as the factitious
and artificial wants of society were." Charles H. Warren, who joined with
Fletcher in presenting to the General Court Boston's request to build a
publicly owned waterworks, went a step further, coming close to demon-
izing private entrepreneurs. The city's appeal for permission to form its
own company was far more worthy than that of "speculating individuals seeking to accumulate wealth to themselves out of the wants of the
many," since it was presented by "the City Government acting for and
under the instruction of 110,000 people, a constituency forming one-
seventh part of the whole population of Massachusetts." The city government's sole motive was to serve the public good, which he defined in
terms of "security of property, the preservation of health, the promotion of cleanliness, and the extension of comfort." Bostonians, whom he
melodramatically characterized as "beggars for a cup of cold water," were
not asking "for powers to trample on any rights of individuals or of the
public, but simply for leave to do what should be necessary for the supply
of a great public want."

Unlike a works owned by private company, Fletcher maintained, a
public system would take it as its duty to furnish water generously, for
the common good. He shared Channing's contention that a public system would be dedicated to the proposition that "water should be like the
air we breathe, everywhere, in doors and out of doors, ministering to
the comfort, cleanliness and health of every living thing."[26] A public system would be beyond partisanship. The city's proposed plan for a public
waterworks was "not the project of merchants alone, or mechanics alone,
or of any class alone, but all had united in what they considered the common good of all." This meticulously thought-out project, planned with
the utmost concern for both individuals and the collective, reconciled
them into one glorious entity, the people, whom it would both serve and
symbolize.

Advocates of private ownership, especially proprietors of existing or projected water companies, vigorously attacked the arguments for public control. They angrily questioned the assumption that there was an inherent opposition between private ownership and the public good. After their proposal to provide water for Philadelphia was rejected, Delaware and Schuylkill Canal investors denied the Philadelphia Watering Committee's charge that they were "acting from interested motives, separate from the community." They insisted that they were motivated by "the most disinterested principles," which included a willingness to set a limit on their profits. It was insulting to assert "that *personal interest* will overrule *public good*, in a *trust* for the community." The Watering Committee's repeated attempts "to stun the public ear, in a perpetual *peal*," about the dangers of private interest were unjust.

The canal owners then went on the offensive, dismissing the notion that public ownership was by definition more virtuous than private. The Watering Committee, they claimed, was trying to construct an argument for public ownership out of a vaporous and insulting notion called disinterested public spirit, while experience had shown the many tangible public goods that private enterprise had provided. "But what could *public spirit* alone, however disinterested, have accomplished in the great works of general improvement throughout the state?" they asked with calibrated sarcasm. Private entrepreneurs had made America what it was. They dug its canals, raised its crops, manufactured its goods, and advanced its scientific knowledge and medical practice. They laid roads "through rugged and pathless mountains, dreary swamps and wildernesses," and were about to erect a mighty bridge across the Schuylkill. Private ownership was the foundation of the public good. Switching to a different kind of appeal, the company added that awarding it the project would also serve the public good by reducing the cost of public aid for "numbers of poor," since building the canal would create jobs.[27] More than three decades later, when the Schuylkill Navigation Company added a lock at Fairmount over the city's protests, it likewise maintained that its expansion benefited the community. After all, its 1836 report to stockholders claimed, "The interests of the Company have always been inseparably united with the public good; so that in promoting the one, they have of course advanced the other."[28]

During the Boston water debates of 1844–45, proponents of private ownership made their case by emphasizing how they would in fact more fully benefit the public and do a better job than would a public company. Unlike backers of privatizing public services today, they did not sing the virtues of free-market competition as such, but they did say that private enterprise could serve the community better, and at a lower cost, than could the government. The author of the pamphlet *Hints to the Honest Taxpayers of Boston*, who called himself "Temperance," charged that the Long Pond plan, which he derisively dubbed "a long job," had "neither prudence nor economy" in its favor. Another pamphleteer, who took the name "Prudence," contended that by contracting with a private company, the city "will avoid an everlasting pecuniary embarrassment," and "the citizens will happily escape the fiery ordeal of the tax gatherer." "Prudence" also said that a private company would construct a Spot Pond aqueduct in half the time it would take the city to complete a Long Pond system.[29] In a letter to the *Daily Evening Transcript*, "B" made the same prediction, adding that publicly funding such a large project invited incompetence and corruption. "B" cited railroads as a major enterprise that proved the superiority of private corporations in getting great things done.[30]

The construction of publicly owned systems in Boston and Chicago badly hurt private companies that previously had supplied water by putting them out of business and reducing the value of their property, forcing them to settle for a buyout at a low price.[31] Owners of these companies objected not just to the financial losses they suffered, but also to what they saw as a betrayal. After Boston's new public system opened in 1848, the Boston Aqueduct Company's largest stockholder, Lucius Manlius Sargent, wrote a series of long and irate open letters to the *Daily Evening Transcript*, which he subsequently collected into a book, expressing his resentment that the city resorted to the strained notion of "municipal progress" to victimize civic-minded entrepreneurs like him. Such "progress" in the name of the public good violated "high moral principles." Sargent accused the city of abandoning those businessmen who devotedly served Boston, often at a financial loss to themselves. He described his personal commitment to "the prosperity of this highly-favored city, in which I was born and reared" as "deep rooted and sincere." His company deserved far better than the "injustice, illiberality, and even meanness" with which it was being treated, "which has no parallel here, since Boston was founded."[32]

The Chicago Hydraulic Company arguably had more legitimate reasons for complaint than did the Delaware and Schuylkill Canal or the Boston Aqueduct Company, since it had stepped in at a critical and financially vulnerable early moment, "when," as it said, "the city was small and its credit weak," to ensure Chicago's growth. Its stockholders claimed that they were loyal citizens who had directed their corporation's income not to their own financial well-being but to "the astonishing expansion of the city" that "has of late outrun all calculations." It was a cruel irony that this very expansion had led directly to the creation of the public company that would displace the valiant private one. The reward for unselfishly advancing the public good was public ingratitude. In spite of this insult and injury, the Chicago Hydraulic Company's owners were willing to "retire out of the way," though simple fairness demanded that they receive a very modest payment: "our capital expended, without interest, *i.e.*, the bare cost of our works." If Chicago offered anything less, they said, it would undermine the principles of a just democratic society by resorting to "plain *repudiation and confiscation*, such as DESPOTISM seldom dares." In this concept of civic duty, Chicago's refusal to pay a fair price expressed contempt for individual legal rights, sound policy, and the city's credibility, all of which were part of the foundation of the public good in a democratic society.[33]

In all three cities and many others, however, the defense of private ownership could not overcome the opinion that public control was the best choice, though several factors complicate this generalization. In Philadelphia and Boston, the leading private contenders offered plans deemed inferior to the public ones, regardless of who built and ran the works. In Chicago, the job of delivering water to everyone was too big for the existing company to handle, and there were no private alternatives available. The champions of public ownership also effectively made the case that what was at stake was not just the delivery of water but the public's collective character and integrity.

With the May 1845 referendum looming, backers of Boston's Long Pond plan intensified their critique of the evils of private ownership of water and their praise for the virtues of public ownership. It was "vastly better and more consonant to the feelings and wishes of the people . . . to have and exercise full control, instead of depending upon a mercenary private corporation," announced the Faneuil Hall Committee. If the city yielded to "the querulous and interested objections of the few" because of their influence and efforts, this would be a certain indication "that Boston, with all her proud aspirations for high character and consistency, of noble and judicious enterprise, is doomed, for a long period to come, to stand the laughing-stock, the disgraceful spectacle, of a lack of that public spirit which most other large cities manifest."[34]

This warning pointedly alluded to Puritan leader John Winthrop's advice to the first settlers of the Massachusetts Bay Colony that if they were to survive and prosper, each individual must look out for every other. Winthrop's message contained a warning and an exhortation. While he cautioned his fellow Puritans that forgetting their sacred duty to each other would cause God "to withdraw His present help from us," so that "we shall be made a story and a by-word through the world," he reminded them that dedication to acting together as one would "make us a praise and glory."[35]

Reservoirs of Pride

The planners of urban water systems consciously emphasized the symbolic dimensions of these projects as emblems of civic achievement that expressed the community's commitment to the public good. They demonstrated this in the design of waterworks, which they built to embody their city's heroic sense of itself and what it aspired to be. Not content to let the splendor of their waterworks speak for itself, cities also organized civic ceremonies whose purpose was to make explicit the exalted meanings of these projects as expressions of the high aims and achievements of the urban community.

The designers of the waterworks paid a great deal of attention to, and spent precious city funds on, the appearance of these otherwise utilitarian structures. In its early years, the Philadelphia Watering Committee became most enthusiastic when it described Latrobe's engine house in Centre Square, in spite of the fact that the aesthetics of the structure had virtually nothing to do with how well the machinery within it worked, and the cost of this building was one of the things that drove the price of the works well beyond the original estimate. Latrobe had originally proposed a modest building, but the otherwise highly critical committee members had only praise for the much more splendid one he erected. In their eyes as well as Latrobe's, the engine house's visual appeal possessed a different kind of utility, one that was well worth the price. The elegant water temple would bring Philadelphians together, by both attracting them to gather around it and engendering justifiable pride in their city. Late in 1803 the committee reported that it had spent almost $54,000 on the engine house (the estimate for the plainer one had been $15,000) and close to $2,000 more for landscaping and fencing in Centre Square.[36] As the committee explained, given the "conspicuous situation" of the building, it was important that it be "ornamental, as well as useful to the city." It would have been a significant error "to place a homely mass of building," members continued, "in the best situated square belonging to the citizens of Philadelphia."[37] The engine house, after all, stood for the entire waterworks, which demonstrated the sanctity of Philadelphia's uncompromising commitment to the public good.[38]

The Watering Committee continued to invest in the "ornamental" improvement of the Centre Square works. In 1809 it placed a statuary fountain next to the engine house. Titled *Water Nymph and Bittern* (alternately referred to as *The Allegory of the Schuylkill, The Spirit of the Schuylkill*, and, incorrectly, *Leda and the Swan*) and carved out of pine that was then painted white, the statue depicted a nubile maiden in a diaphanous gown, which clung to her body as if it were actual cloth dampened by the slender jet of real water that sprayed a dozen feet or more into the air from the regally upturned bill of the bittern perched on her shoulder. Said to be the earliest public fountain statue in America, *Water Nymph and Bittern* was the work of America's first important native-born sculptor, William Rush. In the small world of this relatively large city, Rush was also a member of the Watering Committee, and Nancy Vanuxem,

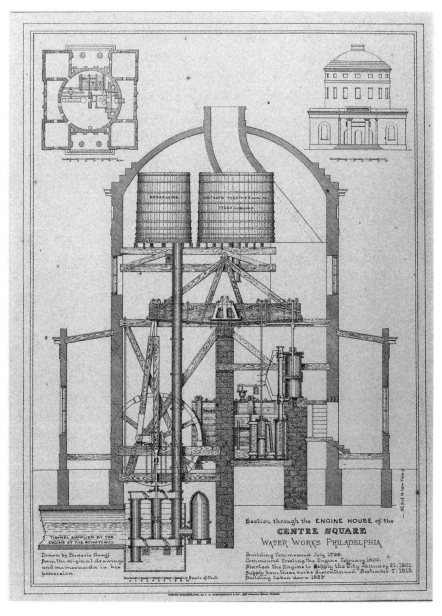

Figure 3.1. This section of Latrobe's Centre Square engine house was drawn by Frederic Graff, son of Frederick Graff, who succeeded his father as superintendent of the works. Note the engine and the two storage tanks, which were elegantly housed and hidden in Latrobe's building, a drawing of which is on the upper right. Courtesy of the Historical and Interpretive Collections of the Franklin Institute, Inc., Philadelphia, PA.

Figure 3.2. The original pine statue of the *Water Nymph and Bittern*, exposed to the elements for decades, rotted away, and all that remains is the head. This bronze copy of the entire figure was cast in 1872 and is now on loan from Fairmount Park to the Philadelphia Museum of Art. Courtesy of the Commissioners of Fairmount Park.

daughter of Rush's fellow committee member James Vanuxem, posed for the nymph.[39]

As noted, when the works moved to Fairmount, Superintendent Frederick Graff expanded on the cultural statement made by Latrobe in Centre Square with the stately assembly of neoclassical buildings that Graff designed for the new location. Other cities followed suit in emphasizing the decorative value of their works. Boston's water commissioners constructed the stone-masonry Beacon Hill Reservoir "with great care and labor." They dressed the principal façade along Derne Street with imposing Romanesque arches, commemorative tablets, and a graceful cornice. Descriptions focused on its "unity and grandeur," which reflected well on the city that built it.[40]

The first Chicago water commissioners specified that their city's reservoirs would be "neat and tasteful, and well adapted to the uses for which they are designed, combining ornament and strength." The commissioners were certain "that the small expenditure which will be required in making the buildings to be used as reservoirs, pleasant to the eye, will not be regretted." They executed the original 1850s engine house on Chicago Avenue in "the modern Italian style."[41] System engineer and future mayor DeWitt C. Cregier praised the 1860s Water Tower as being "the most imposing feature among the whole mass of buildings comprising the works," and "without a doubt the most substantial and elaborate structure of the kind on this continent." The building itself eloquently testified to the greatness of Chicago as a civic collective. "The perplexities, anxiety and risk arising during the prosecution of the work cannot well be conveyed in words," Cregier explained, "but as all that was proposed has been successfully accomplished, the spectator will observe *wisdom, strength* and *beauty* everywhere prominent in the designs and execution of the work." A sentence later he added, again employing italics, "To the citizens of Chicago the water works of our city will stand as one of the enduring monuments which so fully characterize their *energy, enterprise,* and *liberality.*"[42]

Since numerous waterworks structures were visually arresting, prominently located, and, as such, specifically intended to serve as inspiring representations of the urban communities that created them, it is not surprising that they became appealing subjects for visual artists. This was particularly true in Philadelphia. Scenes of Centre Square and, far more so, views of the works at Fairmount appeared in dozens of paintings, drawings, illustrations, lithographs, and even commemorative dinnerware and decorative vases.[43] The most revealing of these depictions as a statement of the importance of waterworks to city life is *View of Centre Square on the Fourth of July* (1812, Pennsylvania Academy of the Fine Arts), a small oil painting by John Lewis Krimmel. Krimmel, one of the first important genre painters in the United States, had come to Philadelphia in 1809 from the southern German duchy of Württemberg, where he was born twenty-three years earlier. Württemberg also gave Philadelphia Lewis Wernwag, the engineer who designed the dam at Fairmount and what was called the "Colossus," his 340-foot single-span bridge across the Schuylkill just downriver from the site of the waterworks.

The "Colossus" was nearing completion when Krimmel displayed his *View of Centre Square* at the Pennsylvania Academy of the Fine Arts in May 1812, as one of the works in the second exhibition of the city's Society of Artists. Sixty local painters, engravers, sculptors, and architects had organized the society in 1810, electing Benjamin Latrobe their vice president. The academy itself was founded only in 1805, and Latrobe designed its first building. The Society of Artists and the Pennsylvania Academy of the Fine Arts were thus, like the waterworks, recent additions to Philadelphia life. These institutions contributed to the development of a vibrant urban culture, one that attracted talented people such as Krimmel, Wernwag, and Latrobe to Philadelphia. Krimmel saw the Centre Square waterworks as playing a starring role in the life of the city.

The painting is an ambitious work for a young painter, portraying over fifty individuals, who congregate informally in front of Latrobe's Centre Square engine house. Two pairs of men frame the more than twenty foreground figures of different ages, genders, races, religions, and classes. The pair on the far left are fashionably attired, while the pair on the extreme right are dressed in less stylish clothing. The attention of all four is arrested by Rush's *Water Nymph and Bittern*, which some had judged shocking when it first went on display. If the shapely nymph prompted disapproval at the time, none came from Krimmel.[44] The way she attracts stares adds a light comic touch to this genre painting. Krimmel placed a young Quaker family in the center foreground, with the fountain just behind them. The father raises his right index finger to chastise his restless young son, while his wife, her arms clasped around her husband's right elbow, turns her bonneted head to consider the statue. A top-hatted young man in the left foreground is purchasing what may well be an al-

Figure 3.3. In John Lewis Krimmel's 1812 painting of Centre Square on the Fourth of July, the scene is busy but orderly. Philadelphians, seen here as a congenial and varied mix of people, gather near sculptor William Rush's *Water Nymph and Bittern* fountain statue, behind which are Latrobe's engine house and the Lombardy poplars planted by the city's Watering Committee. Note the Quaker family in the foreground and, behind them and slightly to the left, the fashionably dressed African Americans by the fence encircling the statue. Courtesy of the Pennsylvania Academy of the Fine Arts, Philadelphia. Pennsylvania Academy purchase (from the estate of Paul Beck Jr.).

coholic beverage from an elderly woman vending it from a small table, under which an African American child is surreptitiously sampling her inventory. This is a another gentle dash of humor that does not disrupt the geniality of city life as Krimmel presented it.

View of Centre Square on the Fourth of July was well received. "There are few people (if any) who visit the Academy," noted a review in the literary magazine the *Port Folio* (another impressive recent addition to Philadelphia and American cultural life), "who are not perfectly acquainted with the scene of which this is so familiar and pleasing a presentation."[45] Krimmel here declared that the place and time to see the city at its best was at this significant gathering place on this day of celebration of independence, freedom, and unity. He depicted Philadelphia as the genteel embodiment of the idea of unpretentious democratic civility. The Centre Square waterworks plays a central role in the painting. Despite its position behind the foreground figures, Latrobe's white temple, with the smoke of the steam engine spiraling from its oculus, dominates the scene both formally and thematically. It does so by dint of its size but also because of what it represents: a technological and aesthetic achievement that stands for the city as a whole, which Krimmel presents as a convergence of industry and art that provides the basis of a sound society. The snowy naiad, who embodies the natural purity of the water, captivates the men and women who choose to congregate by the fountain on this special occasion. The engine house pumping life-giving water beneath Philadelphia's streets is at once the cause and symbol of the concord it hosts.

Although at the time that Krimmel created this work Centre Square was still west of the densely settled part of the city, the improvements carried out by the Watering Committee had made it, as the painting documents, a popular place of public recreation, especially on the Fourth. In 1803 a parade of militiamen accompanied by drums and fifes proceeded westward from the downtown intersection of Third and Chestnut Streets to the square, where they fired a seventeen-gun salute. The marchers then joined cavalry and infantry troops in discharging "a *feu de joie* [i.e., another celebratory firing of rifles] of three rounds."[46] Such demonstrations were common in Philadelphia and other antebellum cities, as governments and civic groups frequently staged carefully orchestrated banquets, parades, and public events that expressed pride in the local population and the nation.[47]

The completion of a waterworks by and for the people was an occasion that shouted out for a mass public demonstration. In spite of the fact that the Centre Square works first began operation in the dead of winter, Philadelphia greeted the flow of its water with a spirited if relatively modest celebration. According to one eyewitness, "On this pleasing occasion the Mayor and the members of the two Councils attended at the Centre Square. The water was delivered in considerable quantities, and the whole experiment succeeded to expectation." He pointed out that spectators included visitors from other places, who "gaped with astonishment as at the tenth wonder of the world." This observer felt sure that these visitors would "speedily return home to communicate the marvelous tidings." While the start of water service was "particularly gratifying" to the public officials who "through varied and multiplied difficulties, persevered to the completion of an object of the first magnitude, both as it respects the health and convenience of the city," it was also "a joyful circumstance to the citizens at large."[48]

Boston mounted a magnificent series of celebrations at key moments in the construction of its waterworks. It took each of such occasions as an opportunity to commend itself for what it was accomplishing. The first of these events was the groundbreaking on August 20, 1846, at Long Pond, to which a special Worcester Railroad train conveyed a party of about 250. On arrival, the group strode to the beat of Flagg's Brass Band to a spot fifty feet from the water's edge. After completing a few remarks, Mayor Josiah Quincy Jr. was given a special spade with which to initiate the work. He doffed his coat, and as the band struck up "Hail Columbia," Quincy hoisted the first clods of earth that made way for the aqueduct. To the crowd's great approval, he then called upon seventy-nine-year-old former president (and distant relative) John Quincy Adams, now a member of the United States House of Representatives, to take a turn. Described by one reporter as "rather feeble in health, though his eye was bright and his spirits seemed buoyant," Adams accepted the invitation, earning "the most vociferous applause," and the band performed "Adams and Liberty," written almost a half century earlier in honor of his father.

The first mayor Quincy, himself seventy-four and recently retired after sixteen years as president of Harvard, then tried his hand, to the accompaniment of "Yankee Doodle" and more cheers.

The group retired to a nearby pavilion to dine on a generous lunch and listen to a series of speeches. In his short address, Adams praised the project as one that reflected and advanced civic unity. Calling the groundbreaking "one of the happiest days of my life," he confessed his relief to be away from Washington and the contention over the Mexican War, to which he was deeply opposed. He rejoiced to be present at this scene "of quiet, of calmness, of peace, and of endeavors to promote the general welfare of man." The round of toasts that preceded the return trip to Boston similarly focused on how this project advanced the public good. Among those who lifted a glass and spoke a few words was Dr. Walter Channing. In the final toast, Henry Williams, who had joined with Channing during the water debates to sway public opinion toward the Long Pond proposal, saluted "the people—They are the arbiters of great public measures."[49]

The groundbreaking and the procession from City Hall that preceded the laying of the cornerstone of the Beacon Hill Reservoir were a mere prelude to the colossal celebration of the actual introduction of Long Pond water into Boston on October 25, 1848. The day's events closely resembled the public spectacle New York had staged six years earlier to commemorate the opening of its Croton waterworks.[50] As in that city, Boston's celebration began with an early morning hundred-gun salute and an exuberant cacophony of church bells. School and business were canceled, and an enormous throng of residents and people from out of town, including two thousand factory girls from Lowell, assembled to witness a parade with so many marchers that it took over two hours for the entire procession to pass any given point on the route. The parade comprised public officials and dignitaries from near and far, clergymen, Harvard students, firemen, representatives of several trades, and members of benevolent organizations, fraternal orders, and temperance societies.[51] They moved variously on foot, on horseback, and in carriages, escorted by bands and military companies. Many in the parade wore colorful costumes and carried banners, placards, and symbols that declared their calling or cause. Laborers who had constructed the waterworks strode proudly behind a wheeled platform groaning under the weight of a section of water main. It took "seven superb black horses" to pull the platform. The marchers made several turns and passed under temporary arches as they weaved their way through the commercial downtown and then onward to Boston Common. Quotations on the arches proclaimed the lofty meaning of the day. Some of them were passages from Shakespeare: "There will be a world of water shed" (*Henry IV, Part 1*), and "Our best water brought [in] conduits hither" (*Coriolanus*).[52]

The official ceremonies took place by the amoeba-shaped Frog Pond on the Common, which was jammed with parade participants and spectators. The Handel and Haydn Society led the celebrants in a hymn of thanks to God. In his invocation, the Reverend Daniel Sharp continued the theme of thanksgiving, expressing gratitude "that there has been this wise forethought and this provident care, for all classes amongst us."[53] A chorus of schoolchildren sang "My Name Is Water," with lyrics by poet, editor, and future Harvard professor James Russell Lowell, then only twenty-nine. The occasion inspired other "water music," including the lively *Cochituate Grand Quick Step*, composed by George Schnapp and "respectfully dedicated to Mayor, Alderman [*sic*] & Common Council."[54]

In his prepared remarks (the speeches were abridged to complete the ceremonies by sunset), *Daily Advertiser* owner (and now water commissioner) Nathan Hale described the project as "a great public work which is of equal benefit to every citizen."[55] Hale's newspaper would expound further on this theme a few days later, when it asked: "Who does not this moment feel happier and better for that out-gushing, and up-gushing water! Every body is concerned in it. Every body has contributed to it. It is alike the gift and the possession of all."[56] In his speech, Mayor Quincy proclaimed that the arrival of the water was a utopian moment of collective harmony. "Every sect and every party, every age and every calling have this day forgotten their distinctions and their differences, and remember only that they are children of a common parent who are to receive a common blessing," he told the multitude. "We have come to congratulate one another, in what this element is to do for health and

Figure 3.4. We see here the scale and pomp of the parade that preceded the ceremonies on Boston Common marking the introduction of Cochituate aqueduct water. One of the many floats in the procession, here moving through Tremont Street, is a fully rigged ship, which is about to pass beneath a Moorish arch on which is inscribed, "THERE WILL BE A WORLD OF WATER SHED," from Shakespeare's *Henry IV, Part I.* Onlookers crowd the streets and the second-floor balcony of the flag-festooned Boston Museum, whose owner, Moses Kimball, had the arch built and served as chief marshal of one of the divisions in the parade. Courtesy of the Boston Athenæum.

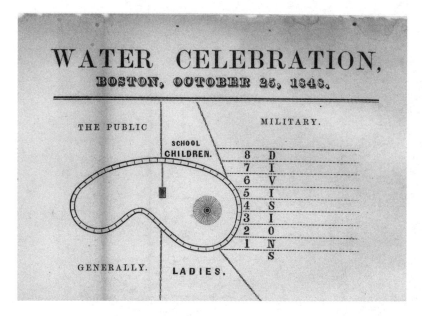

Figure 3.5. This diagram specifies how participants and spectators were to be arranged around the Frog Pond for the great celebration of Boston's new water supply. "Military" refers to the military escort that led the parade through Boston's streets and onto the Common, and "Divisions" refers to the divisions of marchers. The small rectangle in the Frog Pond was the speakers' stand, and the circular symbol next to it represents the fountain through which Cochituate water gushed at the culmination of the ceremonies. Courtesy of the Massachusetts Historical Society.

purity, temperance and safety, and to rejoice that these blessings are secured to those who shall inhabit this peninsula forever."[57]

As the sun dipped in the western horizon, Quincy asked the crowd "if it were their pleasure, that the water should now be introduced" into a new fountain in the Frog Pond that was fed by the aqueduct. The response was a resounding "aye!" As the valve was slowly opened at Quincy's signal, "the water began to rise, in a strong column, increasing rapidly in height."[58] The *Daily Evening Transcript* vividly described the onlookers' shared feeling of tension and release: "A moment's pause, and there was a gush of rusty-looking water, small and doubtful at first, then spreading, and gathering strength, then rising with beautiful gradations higher and higher, until it towered up a strong, magnificent column of at least seventy feet in height, flashing and foaming in the last crimson rays of the setting sun!" According to the *Transcript*, "The spectacle so far surpassed public expectation that every one seemed taken by surprise. The height and volume of the jet were much greater than had been anticipated." Some spectators waded into the pond as it filled with Cochituate water. Amidst all the other shouts of delight, "nine tumultuous cheers arose," and the Handel and Haydn Society sang a chorus from Felix Mendelssohn's *Oratorio of Elijah*, composed two years earlier. The words of the oratorio began: "Thanks be to God! He laveth the thirsty land."[59]

As the last light of this glorious day "tinged the summit of the watery column," bells sounded, cannons fired, fireworks filled the sky, and a circle of Bengal lights illuminated the pond with their bright blue flame. The gathered multitude pulsed with "intense interest and excitement, which it is impossible to describe, but which no one can forget." The release of the water evoked powerful emotional responses: "Some laughed aloud—the men swung their hats and shouted—and some even wept."[60] As night fell, the thousands of tapers that Bostonians lit in the windows of public and private buildings emitted a magical glow. The fountain played freely all the following day, and crowds continued to flow to the Common, including many children, thanks to Mayor Quincy's impromptu announcement at the conclusion of festivities the previous evening that school would once again be suspended on the morrow.[61]

In its inclusiveness and complexity, the day was a masterpiece of coordinated planning and effort. Estimates of participants ran between 50,000 and 100,000 people. Considering that the population of Boston in 1848 was about 130,000, the turnout for city water was extraordinary. As a poem published in the *Advertiser*, titled "Water Ode for Oct. 25, 1848," expressed it:

> What brings out all this *flowing* crowd,
> This grand array, this flaunting proud?
> And why these echoing thunders loud?
>   Thy flow, thy flow, COCHITUATE WATER![62]

Figure 3.6. This lithograph depicts the enormous crowd gathered on the Common at the celebration of not only the introduction of Cochituate water, but also of the city that constructed this great public work. We here witness the magic moment when the water rose high into the evening air from the fountain in the Frog Pond. Courtesy of the Boston Athenæum.

The official account of the Boston water celebration emphasized how orderly it had been, free of confusion, drunkenness, and inappropriate behavior.[63] "The universal disposition for order," not external force, it explained, "kept every individual in his place, and carried into effect the prescribed regulations." As a result, "decorum prevailed in every part of the city," so that the occasion "was passed in social enjoyment, with the exhibition of a universal kindly feeling."[64] The parade and the ceremonies that followed were a display of cooperative social engagement, a field day of intended signification, a milestone in American urban democracy. The celebration was an achievement analogous to the waterworks themselves, and both the works and the festivities reinforced the idea of Boston as a triumphant urban collective.

Among the floats was one atop which printers operated presses, striking off and distributing copies of an original song, "For the Merry-Making on Water Day," to the spectators who lined the parade route. The first verse announced:

> Away, away, with care to-day!
> There's naught but joy before us;
> A gladsome shout from the mass goes out,
> And we will join the chorus.

It was Cochituate water that brought "the mass" to this wonderful moment:

> Its journey passed, 'tis here at last,
> And hailed with acclamation;
> And every tongue shall swell the song,
> Whate'er its rank or station.

"Unseen it comes to all our homes / To cheer the high and lowly," the lyrics continued. All should therefore join "a universal chorus" of praise and thanks. The water had dissolved class differences and melted all of "Boston's sons and daughters" into a cohesive community.[65] As the parade streamed through their city, the printers thus simultaneously participated in it, exercised their trade, and spread word of the celebration as it was happening to the tens of thousands who were watching. The flow of words, ideas, and actions was dramatized by the parade, the arrival of city water, the fountain, and Boston as a productive confluence of the will and energy of its people.

Chicago also marked the building of its great engineering projects with public celebrations, including the festivities that accompanied the construction of the lake tunnel, the Pumping Station, and the Water Tower. At the groundbreaking for the tunnel on March 17, 1864, Mayor Francis Sherman delivered a brief address before swinging a pick into the ground.[66] Many dignitaries were aboard on July 24, 1865, when the tugboat *A. B. Ward* helped tow the crib to its position two miles out.[67] The Chicago Board of Public Works staged a much larger proceeding for the placement of the commemorative last stone in the tunnel on December 6, 1866.[68] The participants consisted of two groups, who entered at either end of the tunnel and then traveled under the lake in mule-drawn railcars (the same means that workers who built the tunnel had used) to their meeting point. This was near a small gap in the tunnel lining in which Mayor John B. Rice was to set the final stone. Before he did so, Rice announced that he was performing this act not as an individual but as the representative of the city as a whole. After he had put the stone in place, the entire group rode eastward in the cars and ascended to the crib, where members of the Chicago Bohemian Club greeted them with a seven-gun salute. In the course of more speechifying, the mayor again stressed the great value of this undertaking as a public good that benefited all citizens. The tunnel would furnish "every inhabitant of the city of Chicago now and for years to come with pure, sweet water; and a supply in excess of the demand, sufficient for a million of inhabitants more." A militia company stationed on land announced to Chicagoans that the momentous deed was done by firing a round of artillery, and flags were raised downtown.[69]

On Monday, March 25, 1867, water from the tunnel finally coursed into the city, and Chicago laid the ceremonial cornerstone of the Water Tower. In spite of chilly weather, a large number of residents skipped work to turn out for the accompanying parade. The city's Masons, in

Figure 3.7. Chicagoans gathered by the partially built Water Tower on March 25, 1867, for ceremonies marking the opening of the new lake tunnel, which included laying the cornerstone of the tower. Building materials on the construction site made access difficult, but neither this inconvenience nor cold weather could qualify the great enthusiasm that surrounded the arrival of Lake Michigan water through the much-improved system. Courtesy of the Chicago History Museum, ICHi-64424.

full regalia, played a starring role. Prominent in their ranks was DeWitt Cregier, who in addition to being the city's water engineer was deputy grand master of the Grand Lodge. The parade made its way from city hall across the Main Branch of the river to the North Division and the site of the Water Tower and Pumping Station. The *Tribune* wrote, "The long body, brilliant with its diversity of colors and its emblems of authority, skill, and power, moved at a free pace throughout the streets, lined with thousands of people on the walks and corners, in doorways and windows, on roofs and fences, all wearing an exhilaration of countenance which showed that the great events which the display indicated were of vital interest and happy import to all." "Free pace" was an exaggeration. The *Evening Journal* noted that the marchers, faced with the likelihood of sinking knee-deep into Chicago's unpaved and muddy streets, "tramped silently but desperately" through some stretches and then took to the wooden sidewalks.

Once finally arrived, they reassembled by the tower-in-progress, now only twenty-seven feet high, where an estimated twenty thousand people awaited. Masonic grand master Jerome Gorin praised all who made the day possible, including "the truly public spirited people of this metropolis of the West."[70] As in Boston, the purpose of the day was to demonstrate that city water had given city people a wonderful reason to be glad that they had joined the even grander parade that was city life.

Whose Decision? Whose City?

But just who should lead that parade? The discussions of city water in terms of the public good prompted heated disagreements over which residents should have a say in determining public policy. The question was not whether the city needed water, but who was qualified to decide how to provide it and, beyond that, to whom the public good should be entrusted. City water thus became a forum for intense debates about the nature—and value—of urban democracy.

This was especially true in Boston. Lucius Manlius Sargent, the Boston Aqueduct investor who complained about how unfairly Boston had treated him by devaluing his business, attacked the political process, including the series of referenda, by which the decision to build a public system had been made. A confirmed social conservative, Sargent doubted whether most members of the urban public possessed the ability to judge what was actually in their interest. He attributed the victory of the Long Pond plan to greedy and opportunistic schemers with their own personal agendas, who had cleverly duped the all-too-tractable voters by inventing a fictive opposition between "the people" and "private interests," thus undermining the commercial spirit that was necessary to a free and thriving society. Sargent was incensed by what he felt was the unseemly pride with which some councilmen boasted that by winning approval for a city-owned system they had "*killed*" his company, which had faithfully delivered water to its customers for five decades. These supposedly "*disinterested*" politicians had wrongly smeared Boston Aqueduct's owners by calling them "*selfish monopolists.*" "By the aid of the press, and of street-corner talks, and popular harangues, [Long Pond proponents had] taught the populace to believe these things, and thus the very name of the *Boston Aqueduct* became a by-word and a reproach." The result—as terrible for Boston as it was for his fellow Boston Aqueduct shareholders—was that his company was "doomed—a sacrifice to the popular will," which "was shaped and actuated, by the agitators," who were themselves acting selfishly and against the public good.[71]

Sargent's commentary can be dismissed as a solitary rant by a disgruntled property holder with a caustic disposition, but he was not alone in questioning whether, in a large metropolis, "the people" could be trusted to determine the public good in regard to water or any other issue of consequence. In 1844 the backers of Spot Pond reissued a sixteen-page pamphlet titled *Thoughts about Water*, originally published in 1838, which they believed was now even more apropos. Its pseudonymous author, who used the name "A Selfish Taxpayer," asserted that the water question should not be settled "in the foolish and intemperate spirit" of those participants in the debate over ownership whose aim was to "coerc[e] entire boards of council to act contrarily to their consciences, by the force of an *instructing* power presumed to reside in an excited and multitudinous body." He called the tactics of some public water advocates "the very maximum of jacobinical absurdity," thus comparing these measures to the barbarous acts committed in the name of "the people" by the radical Jacobins in France only fifty years earlier.[72]

As matters came to a head in the spring of 1845, supporters of private ownership again charged that public support in favor of the Long Pond plan was the result of demagoguery rather than of careful reasoning. One spoke of the actions taken by some citizens and elected officials as "characterized by hot haste and a want of due deliberation and thorough investigation." There was "no sufficient evidence that the public necessity or interests require that authority should be granted to carry the proposed plan into execution."[73] Others attacked the 1844 Faneuil Hall gatherings in favor of the Long Pond plan as subversive of the public good, in spite of the fact that these were attended by many highly respected citizens and open to anyone who wished to attend.

The debate over who would control city water often revealed a deep suspicion of majority rule in general and of urban democracy in particular. Alderman Henry B. Rogers called for the careful weighing of all the facts and positions on the water question, adding that "all attempts to check inquiry, or to carry a particular measure, involving multifarious details and nice calculations, by creating a popular excitement, by noisy speeches, at Faneuil Hall or elsewhere, should at once be put down by the respectability and sobriety of the community." Rogers said that he was "fully satisfied" that most Bostonians, "including among them our men of competence, and our men of wealth," were willing to "accede cheerfully" to any plan that served "the good of the inhabitants of the city."[74] But his reference to competence and wealth suggested that such qualities were required of those who were to determine what served the public best. Spot Pond supporter Jonathan Preston held that the water issue "properly requires more time and more facts than are at command by the majority of our citizens. It cannot be properly investigated in a popular assembly in Faneuil Hall." He maintained that he himself had at heart the best interests of "the mechanical and industrial classes of this city" to whom the water issue was arguably more important than to any other group in Boston.[75] Both men claimed that they were saving all Bostonians from the tyranny of an uninformed majority.[76]

Alderman Rogers apparently saw no conflict of interest arising from the fact that he was one of the several dozen prominent Bostonians who owned stock in Spot Pond. Other propertied residents with no such personal stake also argued, however, that they should have a stronger say on the matter of Boston water than people who did not own land. Their point was that the distribution of power to decide such questions should be based on economic rather than political fairness. They contended that since urban financing depended most heavily on real estate taxes, their views should take precedence in the decision on how to proceed with such an enormous and potentially risky expenditure as a publicly owned waterworks.[77] Attorney Elias Hasket Derby, testifying against the Long Pond proposal before the Massachusetts legislature in 1845, maintained that referring the matter to the voters was anything but right and proper, given that nearly 15,000 of the 19,000 adult male Bostonians eligible to vote paid into the public coffers little other than the poll tax, and so they would not be subject to any rise in levies on land that the construction of a waterworks would require. Derby said that he "represented a large proportion of the wealth and property of the city, men who anticipated a failure if this great work was commenced, and who if it resulted in the manner a similar work had in New York [where the actual cost had doubled the estimate], would be obliged to pay the interest on the outlay, which they believed would add 50 per cent. to their taxes."[78]

Defenders of the Long Pond plan characterized the opposition as self-interested sore losers who attacked the process by which the city had settled the matter because they did not like the result. The strongest argument for going ahead with the project was not just that it would benefit everyone in Boston but also because it enjoyed such widespread support, including from the wealthy. According to the presentation made by attorneys Charles Warren and Richard Fletcher at the 1845 state hearings, this support was truly remarkable precisely because it transcended the many differences, including those of party and financial standing, that usually divided the urban population. The people had made a clear decision after careful and reasoned reflection: "This was not an extreme case of mob violence, or thoughtless radicalism to be guarded against, for it was one of the pleasant circumstances with regard to this matter, that it was the result of the movement of no particular class." There had not been any "clamor of poor against rich or rich against poor." The opposition was a tiny minority who "had had their day of opposition and had failed." They might be unhappy with the outcome, but the multiple votes and other procedures by which it had been reached were unimpeachable. "This cry that the people 'did not understand,'" Fletcher stated, "was one which was a reproach to the institutions under which we live, founded on the principle that the people do understand their own affairs, and act understandingly."[79]

A few opponents of the publicly owned system took the opposite tack, arguing that the Long Pond proposal did not reflect the popular will at all. They protested that the December 9, 1844, referendum result supporting public ownership was not the opinion of a majority. True, a high percentage of those who voted were in favor of the proposal, but they constituted far less than a majority of those eligible to vote.[80] Others tried to stifle public water sentiment by deeming the long campaign for public water an exercise in vanity by politicians, notably the Quincys. "It is quite natural, that the Mayor of a city, having one eye upon the public weal should turn the other occasionally upon his own glory," the pamphlet *Thoughts about Water* charged. "One gentleman associates his name with a splendid market-house [i.e., Quincy Market, which was built during Josiah Quincy Sr.'s mayoralty and named after him], and another would be the founder of an aqueduct."[81]

The most politically effective argument against the Long Pond plan focused not on the Quincys or any other elected officials but on the water commissioners who were to run the works. The nub of the problem, this argument contended, was that the original Water Act (i.e., the one that was defeated in May 1845, as opposed to the one that was finally approved the following year) asked voters to authorize the creation of a tiny oligarchy that would not have to answer to the popularly elected City Council, let alone the electorate as a whole.

The 1845 pamphlet *How Shall We Vote on the Water Act?* charged that the proposed legislation should be more accurately named "either 'an act to annihilate the Board of Alderman;' or 'an act to create a power in the city greater than the city itself;' or 'an act to provide good offices and good salaries;' or 'an act for the benefit of commissioners, rather than the people.'" It noted, in another reference to Jacobins and the French Revolution, "that republicans, when clothed with despotic power, become the worst of despots," and that this legislation placed before the people "a concentration, and a perpetuation of power, constituting a monopoly, 'double-distilled,' anti-republican, and highly dangerous." With this in mind, "We should now make a vigorous attempt to destroy this corporation,—this monster, which has been introduced among us,—before it becomes a permanent resident; and before it obtains power over the money in our pockets, and the very water we may drink."[82] The creators of a stridently anti-public-ownership broadside exclaimed in large letters, "Boston to the Rescue!!" They then called on voters "opposed to the creation of a close Corporation of Water Commissioners, with high salaries" to "repair to the Polls" and "vote AGAINST THE WATER ACT, By depositing a Ballot marked 'NAY!!'"[83]

Like the debate over whether a public or private company should own and run a city's waterworks, the question of exactly who would manage a publicly owned company prompted a discussion of the nature of urban government. It is hard to assess the sincerity of those who argued against the Long Pond proposal on the basis of political principle, since there is the possibility that many of them were more deeply motivated by the fact that they or their associates had a financial stake in defeating it. But it is also true that water committees and commissions, once they were in place, did possess considerable power and enormous financial resources. The water commissions in Boston and Chicago aroused special concern since, whether their members were appointed or elected, they were semi-autonomous governing bodies outside the regular government, and this was effectively the case as well with the Philadelphia Watering Committee, which controlled a large budget of its own.[84] Defenders of the first Water Act said that the way to avoid problems with commissioners was to appoint men of integrity. *How Shall We Vote on the Water Act?* answered this argument by reminding readers that history had demonstrated time and again that power corrupts anyone who wields it, and that the act lacked adequate safeguards against this. "[The Water Act] does not provide sufficient checks against those numerous opportunities for small peculation," the pamphlet read, "which

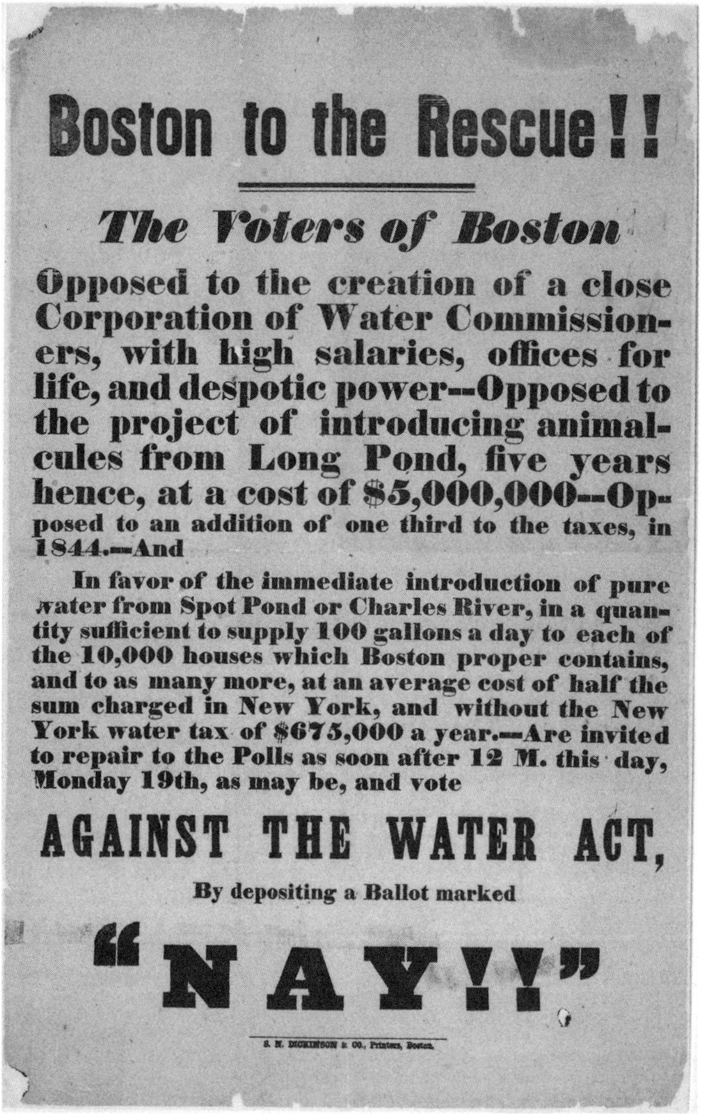

Boston to the Rescue!!

The Voters of Boston

Opposed to the creation of a close Corporation of Water Commissioners, with high salaries, offices for life, and despotic power—Opposed to the project of introducing animalcules from Long Pond, five years hence, at a cost of $5,000,000—Opposed to an addition of one third to the taxes, in 1844.—And

In favor of the immediate introduction of pure water from Spot Pond or Charles River, in a quantity sufficient to supply 100 gallons a day to each of the 10,000 houses which Boston proper contains, and to as many more, at an average cost of half the sum charged in New York, and without the New York water tax of $675,000 a year.—Are invited to repair to the Polls as soon after 12 M. this day, Monday 19th, as may be, and vote

AGAINST THE WATER ACT,

By depositing a Ballot marked

"NAY!!"

S. N. DICKINSON & CO., Printers, Boston.

Figure 3.8. This broadside captures the high feeling of the Boston water debates of 1845. Employing double exclamation points, it urges voters to "rescue" the city by voting against the 1845 Water Act authorizing a public system drawing on Long Pond. It then attacks the high pay and "despotic" power of the proposed water commissioners, as well as the cost of the system, and goes on to make a case for a privately owned system taking water from Spot Pond or the Charles River. Courtesy of the Massachusetts Historical Society.

the peculiar features of this act allows, but which it provides no means of detecting."[85] Others worried that the prospect of possessing so much authority over so much money might attract not the best members of society but people eager to exploit the position for personal financial advantage, with no regard at all for the public good.

Defenders of Philadelphia's public system rejected the charge that its administrators were unresponsive to the needs and wishes of the people. The main criticism of the Watering Committee in its earliest years was not that it was too powerful but that, since the system was working so poorly, it had failed to fulfill a public trust. In its 1799 report, the committee made a point of emphasizing how devoted it was to serving the best interests of Philadelphia. Yes, the $50,000 in unanticipated new property taxes recently levied to cover unanticipated expenses was an unexpected shock, but this was a necessary cost that no one could have predicted, especially given how unprecedented the new system was. The committee felt hurt that some of the same citizens who beseeched their elected representatives to build a works were now savaging those officials who accepted the responsibility. Given the cause for which they were so selflessly working, "without any pecuniary recompence [sic], or the most distant idea of private emolument," committee members asked, "have they not the most cogent and reasonable claims—the most undoubted and legitimate right, to expect the countenance and support of every man who has any just regard to the welfare and prosperity of our city?"

The closing paragraph of the report employed a striking choice of words in posing one last rhetorical question:

> —is it generous—is it just—can it be believed—that, after having complied with the wishes of their constituents, by embarking in a plan maturely digested, and economically and perseveringly pursued, [committee members] should at last be left to struggle with embarrassments for want of efficient aid, and a cheerful and ready cooperation on the part of their Fellow Citizens?[86]

The Centre Square waterworks, then, was not merely an ingenious application of modern technology to meet an urgent need, impressive as it was on these terms. It was the objectification of the fundamental ideals—disinterested service, legitimate rights, prosperity, justice, patriotism, generosity, economy, perseverance, and cooperation—essential to a thriving democratic urban community. The best way for any loyal citizen to demonstrate belief in the city as a community was to endorse the committee's efforts.[87]

But after the defeat of the first Water Act in Boston, a spirited critic of the Long Pond plan claimed that this outcome represented the triumph of democracy, referring back to the American Revolution in exulting that the failure of the proposal represented the best tradition of the plain people of Boston resisting those who would oppress them. The author of a letter to the *Daily Advertiser* who defiantly signed himself "Rabble" declared, "It does my heart good to see that as the spirit of liberty was not extinguished by the British tea in 1775, so the fire of patriotism, to say nothing of numerous other fires, has not been drenched by the water act of 1845."[88] Even the organizers of the Union Water Convention, who successfully revived the Long Pond proposal after the defeat of the 1845 Water Act, conceded that it was "very generally admitted, that the most efficient adverse influence, was the conviction on the minds of many citizens that the general plan of the Act was inconsistent with the spirit and genius of our institutions, and imposed more restraint upon the Municipal powers of our city, than the nature of the case, and a due self-respect justified."[89] At the annual early June commemoration of the Ancient and Honorable Artillery Company of Massachusetts, a public ritual that dated back to 1638, one of the first toasts was to "the City of Boston—As jealous of despotic power in 1845, when it comes in the garb of pure water, as it was in 1775, when it came disguised in the form of tea."[90]

Water for Rent

Once a waterworks was built, practical questions of how water was to be delivered and what it would cost flowed directly into debates about public and private good. To start, councilmen and water commissioners needed to set the prices that individual residential and business users paid for water that they used. In a publicly owned system, this meant deciding how to charge members of the public for water that, in theory at least, they already owned. To distribute water without charge was never a serious option because running a works had substantial continuing costs, in addition to the funds needed to pay back the interest and principal on construction loans. At the same time, however, no humane community could deny an individual access to water even if that person could not afford to pay for it. In short, figuring out fees for water, while it entailed financial considerations that were challenging for technical reasons, also posed more abstract questions of individual and collective obligations and rights.

Good fiscal policy dictated that charges be set at a level that would generate enough revenue to cover operating expenses, carry out necessary maintenance and repairs, help redeem water bonds, and fund improvements and expansion. Every official involved in planning waterworks shared the hope that Chicago's water commissioners expressed in their original report that the waterworks "may ultimately be regarded as a fruitful source of revenue."[91] Under pressure to cover costs, they tried to make sure that "ultimately" arrived sooner rather than later. If a water system was losing money, the city would have to increase charges, authorize new or higher taxes, borrow more money, cut costs, or some combination of these.

Sound fiscal policy might be bad public policy, however. Deciding how to charge for water meant choosing between two alternatives. The first was to treat the distribution of water like any commercial transaction in which one party purchased something from another. The second alternative was to price water not as an ordinary commodity but as an extraordinary public good that was absolutely essential to the well-being of the community, whose health and prosperity depended on all of its members receiving and using good water. The first way of doing water business conceived of city people as a multitude of private individuals and businesses separate from one another and from the collective that was the city. The second understood them as constituents of an interdependent society. Those who spoke most eloquently for a public system based their arguments on the latter view. Having refused to entrust its water to private individuals in search of profit (though companies usually agreed to some limits), a city that built a public works was under pressure to furnish water at a minimum cost to those who connected their property to the system and at no cost to those who could not pay for it.

The pricing of water demonstrated how hard it sometimes was to reconcile the individual and the public good, especially in the formative period of Philadelphia's, Boston's, and Chicago's water systems, when each city hoped to make its waterworks financially self-sufficient. There was no question that operating such a large venture required a "stream" of revenue. When they could, cities generally tried to adhere to a private model in regard to services overseen by the government, such as the regulation of commerce. The most straightforward way to do this was through defined fees for such things as licenses and for the use of facilities that the city supervised, notably markets. As for water, cities billed for it in three ways: taxes to pay for installing pipes under the streets, a onetime fee to link a property to the system, and annual usage charges (usually paid in two six-month installments), which were commonly referred to as water rents. The fee to connect to the system was straightforward. But the other two were far more problematic both in practice and principle.

The laying of service pipes posed a particular conundrum in terms of the relationship between the individual and the public good. One way to pay for water pipes in any neighborhood was through special assessments. A special assessment is a very local tax that covers the cost of a specific infrastructural improvement that benefits nearby property, such as opening a street or constructing a sidewalk.[92] The common procedure is that a group of property holders on a block or similarly defined area petitions the city for a particular desired improvement, consenting to divide the costs according to some formula. The public good is involved to the extent that the city is expected to keep the welfare of the population as a whole in mind when deciding whether or not to approve the petition, and if the designated officials or body approves it, the city makes sure that the improvement meets its specifications either by doing the work itself or supervising a private contractor.

Although he did not use the term "special assessment," Benjamin Latrobe proposed this kind of tax as a way to fund the laying of mains
"through all the crouded [sic] parts of the city." Latrobe suggested that
wealthier Philadelphians who owned real estate "would pay in proportion
to their fronts" while "the poor would be slightly affected," since "the expence [sic] in fact would fall upon the landlord" (though this might well
be passed on in the rent). The value to the individual who paid this tax
was clear. "It would be in fact the *purchase money* of health and convenience, and occur only once," Latrobe wrote. "Every new house would pay
its share, as it was built, and thereby contribute to the future repairs."[93]

Latrobe overlooked several complications that make the installation
of water pipes resistant to the special assessment system. First of all, it
is very impractical to place pipes in the ground piecemeal; unlike streets
or sidewalks, water mains (and other networked services and utilities,
including sewers, gas, and electricity) cannot efficiently and economi-
cally skip over one neighborhood on the way to another.[94] In addition,
charging for mains and distribution pipes by special assessment would
have followed a practice opposed by critics of private companies, since it
would almost certainly mean that wealthier neighborhoods, where people were most able to pay, would receive service ahead of others. Of the
three cities examined here, only Chicago relied on special assessments at
all in laying water pipes, and it did so to a far lesser extent than it did for
other improvements.

This did not mean, however, that the administrators of public waterworks placed pipes without regard to the income they would produce.
For all the civic-minded rhetoric of their backers, these systems, while
eschewing special assessments, commonly themselves gave priority to
sections of the city that seemed to promise the most revenue. This revenue sometimes included funds needed for building the waterworks in
the first place. In order to improve the local market for bonds it sold to
erect the original Centre Square system, the Philadelphia Watering Committee offered purchasers of these bonds (who were called "subscribers") two special benefits: once the water began to flow, they would be
"entitled, in the first instance, to a preference of supply to one dwelling
house for each share so subscribed," and they would "enjoy the use and

convenience of said supply for the full term of three years, free of any charge of water rent."[95] Preferential treatment in return for lending the city money to construct a waterworks that would eventually deliver water to everyone was different from a special assessment for a very local improvement, but it similarly offered more prosperous citizens access to a public good ahead of others. In effect, this meant that they received the benefit of a special assessment without the cost, since they were lending the city money that would be repaid, and with interest. Of the first sixty-three private houses fitted with water service, thirty-six were owned by subscribers to the water loan.[96]

The linking of water service to the prospect of income was even tighter in Chicago. The city grew so fast that demand for water far outpaced its ability to install pipes. The desire to maximize cash flow led the Board of Public Works to decide which areas to serve first in the same way that advocates of a public supply had claimed private companies would act. In its 1864 report, its members stated, "To avoid entailing a heavy, unproductive debt on the Water Works, the Board is compelled to limit the laying of the pipes to such places as are in most urgent need of the water and where, also, the buildings needing water are most numerous and will yield the greatest amount of water tax."[97] These same buildings might also be in areas where potential property losses to fire might be the highest, but they certainly did not include the homes of poor people, most of whom still had to fetch their water, though now they could fetch it from free public pumps fed by the system.[98]

The board was aware of the implications of the choice that it was making. It acknowledged in 1865 that water was "one of those necessities which cannot be dispensed with," and that "it has not seemed to us proper to cease extending the pipes merely on account of the unusual cost of the work," but it also admitted that it decided to put pipes "only where there would be a sufficient income to pay the interest on its cost when laid."[99] The board reaffirmed this policy in its 1870 report, though it added that installing mains at a slight loss in certain areas was a good idea, since "experience shows that the laying of the pipes is speedily followed by the erection of new buildings and such streets soon become self-supporting."[100] Urban expansion would go where water led, to the benefit of the overall prosperity of the city. In this way, the water system could be said to serve the public good as a whole, but it still meant that the board had its eye firmly fixed on which areas would provide the most "fruitful" income.

In the developmental phase of new water systems, which stretched over years, the decision of where to lay pipes raised other thorny questions about what the city and the individual could expect and demand from one another. Should property owners accept the proposition—one that would never arise if the company were private—that it was their responsibility and even in their interest to pay for water they could not currently receive because pipes had not yet been laid in their neighborhood? In this case, the argument that they should pay rested on the view of the city as a collective, not that the water should go first where there was the greatest likelihood of income from water rents. The rationale was that although water service for the time being reached only certain portions of the city, this still benefited everyone. After all, improved fire protection and sanitation anywhere made the entire city safer and healthier since both fire and disease could spread quickly. Besides, the water would only reach everywhere if there was sufficient current revenue to apply to new construction costs or to back additional loans.

Residents and businessmen in East Boston, located across the Mystic River from the city's North End, objected to such thinking. Because of the logistical challenges of reaching this section of the city, East Boston would not be serviced from Long Pond until a few years after water was widely available on the Shawmut Peninsula. During the water debates of early 1845, its residents had persuaded the legislature to pass an amendment to the first Water Act that exempted them from having to contribute to the cost of building the original system.[101] Pleading against East Boston's exemption, Boston city attorney Charles Warren delivered a civics lecture on the indivisibility of the public good. If people in this section of the city really believed that they should not pay, Warren stated, they displayed "a poor conception of the duties of a good citizen." Their reasoning could be invoked against any city tax, no matter how worthy its purpose. "It would be a monstrous doctrine to maintain that every part of land in the city must be individually benefited by every public expense or the owner was not liable for its tax," he warned, implicitly condemning a system of funding based on special assessments. By the same principle, people without children could refuse to pay the portion of their taxes devoted to public schools, which would go against the principle that the education of every child was a public good and a common responsibility. "The remonstrants seemed to be forgetful," Warren held, "that what was for the benefit of the corporation [of Boston] is for the benefit of all." He added that Boston taxpayers en masse, not just the people of East Boston, currently shared the cost of street lighting, paving, and fire protection in that part of the city.[102]

Fletcher contended that the objections of East Boston were "founded upon a false principle" that was inimical to the idea of urban community. To accept this principle would undermine the viability of the democratic experiment as an attempt to blend the one and the many for the good of both. East Boston's insistence on highly selective and strictly non-redistributive taxation defied "the principle of social life and the municipal compact, without which," he reminded legislators, "we could have neither streets, school-houses, fire apparatus or any of the like results of civilized combination of means." People who cared about the public good knew that "what was for the benefit of one was for the benefit of all."[103]

Individuals who did not want to take available city water because they preferred to stay with private wells presented a related question. Should they be charged for water nonetheless, since the pipes placed in their neighborhood enhanced their property value, fire protection, and sanitation, whether or not they connected to the water? In 1802 a member of the Philadelphia Common Council explained why he and other members were opposed to allocating general municipal funds to pay for logs to be used as pipes. While they did not wish to interfere with the construction of "these valuable works," they believed that the project should be finished "on just principles, to wit, at the expence [sic] of those who are to receive the benefit—exempting those who are not."[104] They lost 10–9 in a vote on the matter, which revealed how closely opinion was divided.

In 1839, seven years before Boston finally decided to build the Cochituate system, some in that city who were satisfied with their current water supply sought the same kind of exemption in the event that their city did erect a waterworks. Mayor Samuel A. Eliot told these people that they were being shortsighted even on selfish grounds, let alone in terms of the idea of citizenship that Warren and Fletcher would advance. Urbanization was on the march, and these residents who depended on wells deluded themselves if they thought that they could continue to do so for very long. Boston's growth was already compelling residents to dig more and deeper wells, "either to find a new spring, or to prevent a neighbor from drawing off the water." The wisest course for everyone was to acknowledge that the individual and public goods were becoming more and more indistinguishable. "No man can safely say, therefore, that he is supplied, and sees no occasion for paying a tax to save his neighbor the expense he has incurred in digging a well," Eliot advised. It was time for this man to abandon narrow-minded self-reliance for his own good. At the very moment a person was declaring that his own supply would last indefinitely, "his neighbor may be drying his well by the digging of another, which would have been rendered unnecessary by the introduction of an abundant supply from sources which might be relied on."[105]

Six years later Charles Warren repeated this point, charging that anyone who was opposed to the city building a waterworks because he or she currently had a reliable well "was entirely forgetful of the duties of a good neighbor, and showed a selfishness which was unworthy of the privileges of the community in which he lived." In any case, obedience to these duties was good insurance. Like Eliot, Warren told such individuals "that if they now felt no compunction for such selfishness, they might soon feel it, when as repeatedly happened, from the digging of new wells, their own sources of supply became worthless."[106] Thinking about the entire community was not only the right way but also the shrewd way to approach city life.

The original charter of Chicago's public water company excluded from assessment those properties that were supplied by a private source, but four years later the commissioners applied to the state legislature "for authority to assess all buildings on the line of the pipes which can be supplied with water, regarding it just and equitable that the owners of property on the line of the water pipe should be assessed, their property being protected from fire, and enhanced in value." This measure would serve the individual good, broadly conceived, since it would enable the commissioners to lower individual water charges "to such rates as will secure to all the comfort and convenience of pure and wholesome water," which in turn would advance the public good of providing water for more and more people, a development "whose salutary influence on the city has been so frequently demonstrated."[107]

Assessing property owners for water service whether they took it or not—and thus strongly encouraging them to sign up—would also perhaps help deal with another difficulty relating to the public good. Many poorer city residents were denied convenient access to water by their landlords' refusal to hook up to the system. The proprietor of a restaurant or hotel might seize the opportunity to take city water since doing so made good business sense, but the owner of a run-down rental property would very likely decide to forgo this privilege—or, more precisely, forgo it for his tenants—as costing more money or trouble than it was worth. As a result, the occupants of his building had to carry water from the public pumps or get it from any other source they might find, good or bad. If landlords were paying for water whether or not they connected their property to the system, they might have a stronger incentive to connect.[108]

Cities also wanted to enroll as many residents as possible since this would reduce the need for publicly accessible hydrants from which individuals could draw water, which system administrators disliked for several reasons. In a private letter written in 1803 to an acquaintance in Baltimore who was seeking advice on how to construct that city's system, Benjamin Latrobe strongly recommended that public hydrants should be avoided, since they often broke down, and, if they did work, they discouraged people from subscribing.[109] The original 1851 charter of the public Chicago City Hydraulic Company empowered its commissioners to bill "the person or persons occupying, or owning, any house or other building situated in the vicinity of any public hydrant," whether or not the house or building was directly connected to the network of water pipes. The original purpose of this was to discourage people who could pay for water from taking it from a hydrant for free.

A half century after Latrobe had complained of the problems that free water had raised in Philadelphia, the Chicago Board of Water Commissioners grumbled that the city's approximately thirty public hydrants "have embarrassed the collection of the Water Rents from parties having the water in their buildings in the vicinity of them." In addition, "Great difficulty has also been found in keeping them in repair; many of them have been broke down by malicious persons, or abused so as to be rendered useless." Others had frozen up in winter.[110] Two years later, Superintendent Benjamin F. Walker characterized public hydrants as "a necessary evil." The new ones that had been installed during the recent cholera outbreak to provide water both to poor people and to residents who did not have domestic service were "rather troublesome to keep in repair, and no profit to the works."[111] Engineers and inventors, notable among them Chicago's DeWitt Cregier, gradually improved hydrant designs, but for many years cities resorted to more primitive solutions aimed at preventing freezing, such as packing hay, bark, or manure around the base of hydrants to insulate them.[112] Chicago set fines for vandalizing hydrants and phased them out by the mid-1860s, contending that they were no longer necessary.[113]

Determining what the water rent should be for individual users might seem a simpler task than deciding in what order to lay pipes and how to cover the expense of doing so. It made sense that residents should pay for the water they received and that charges to customers be as low as possible. But this still did not settle the question of what exactly those rates should be. It was very hard to calculate how much water individuals and businesses would use once a supposedly limitless supply of good water was at their fingertips, and it was comparably difficult to know what would be the cost of delivery. Even after the systems were operating, it proved impossible to say just how much water most individual customers actually did consume, since there were very few water meters at all until mid-century, and at this time they were practical only for large commercial customers. As the Cochituate Water Board stated in 1857, the greatest objection "to the *universal* use of the meter" was that installation throughout the city would require an outlay of from $20,000 to $25,000, which was more money than they would save. That was not the only problem. "There is also difficulty in finding a reliable instrument at any price," the board explained.[114] Meters became more common by the late 1860s, but just for very big consumers. In 1874 the Chicago Board of Public Works concluded that while the only fair way to charge was by gallons consumed, the expense of meters meant that "their general application [i.e., to most homes and small commercial users] cannot be resorted to for years."[115]

Without a way to measure individual use, cities set rates based on estimates of how much water customers in different categories would consume. The original Philadelphia works established a uniform base price of $5 a year for private homes connected to the system with a single service pipe of up to a half-inch in diameter.[116] The charge for "manufactories" was $24 annually for those with a one-inch pipe, $53 for an inch-and-a-half pipe. Prices charged to businesses depended not only on the size of the service pipe but also on how water-intensive a particular customer was assumed to be. The bills of curriers rose $5 per additional half inch; dyers and hatters, $10; soap and candle makers, $15; distilleries and sugar refiners, $20.[117] Breweries, stables, and banks likewise paid prescribed charges.

The growth and diversification of Philadelphia can be traced in the annual tabulation of water rates. With each year, the number of customers and the categories in which the Watering Committee placed them increased. In 1809 the prison and a bathhouse were each charged $30 a year; Pennsylvania Hospital, $50; and both the almshouse and house of employment, $100.[118] The city would soon begin imposing an extra fee on residences for every horse they kept and for special plumbing amenities, such as bathtubs. In 1826 the merchant firm of Thomas Pym Cope, Latrobe's onetime antagonist, paid $10 for the right to supply its packet ships with water. The highest-paying customer, at $200, was a sugar refiner that operated a steam engine.[119] Four years later, the water bill for six morocco houses (which processed soft morocco leather) was $35 each.[120] Soon there were more distinctions in charges for different-size dwellings, including the multi-family tenements that were becoming more common. The rate schedule for 1859 indicates that the basic water rent for houses remained $5 (half that sum for small homes), with a premium for every bathtub, water closet, urinal, and washbasin. Hotel

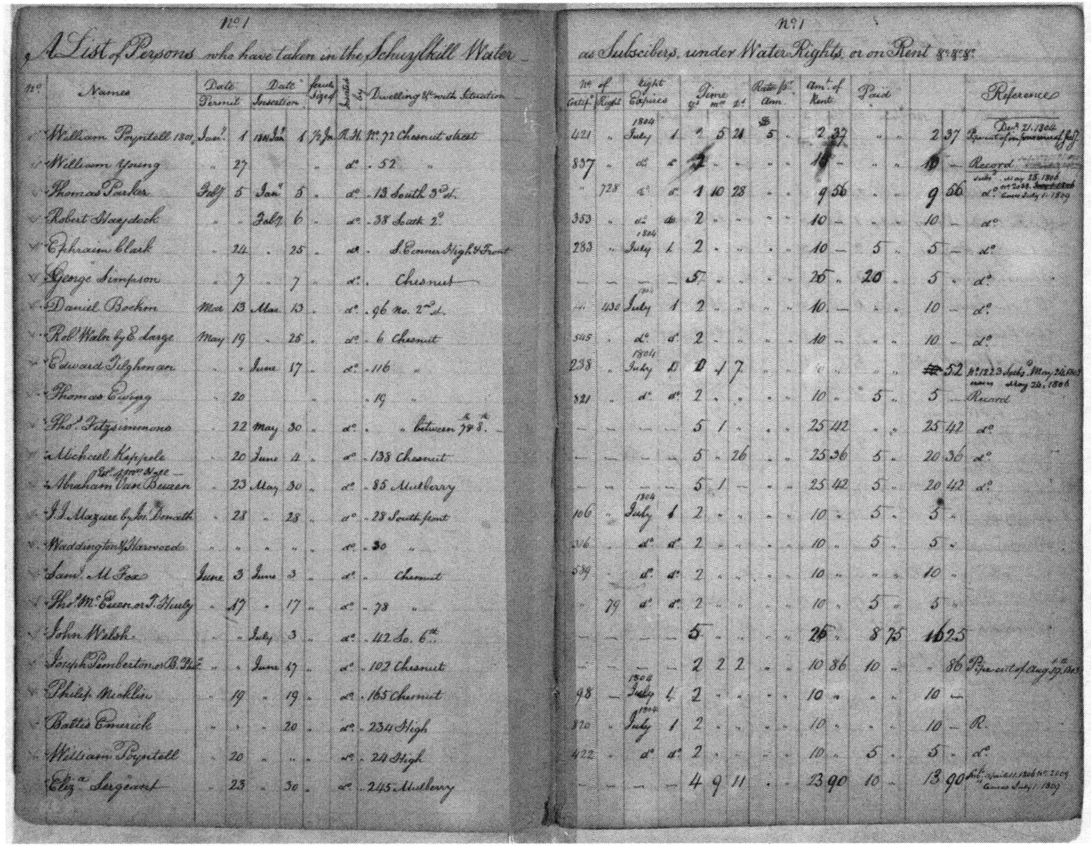

Figure 3.9. This handwritten Watering Committee account book presents "A List of Persons who have taken in the Schuylkill Water as Subscribers, under Water Rights, or on Rent." In order to improve the market for the water bonds that funded the Centre Square system, the committee offered three years of free water service to "subscribers" of the loan. Others paid the annual water rent, which for most customers was $5 per year. Courtesy of the Historical Society of Pennsylvania.

water fees also reflected the plumbing conveniences they offered, while stables were assessed per stall, bathhouses per tub, schools per 100 children, steam engines per horsepower (with different rates for high- and low-pressure engines), and packet ships per 100 gallons of water.[121]

Boston's evolving water rates reveal its continuing struggle to bring revenues more closely in line with operating costs while also maintaining equity. After some discussion, the city established a detailed schedule of rates in 1850. The base charge was $5 for a single-family dwelling assessed at up to $1,000. From there rates rose with the value of the structure and the number of families living within it. For each additional family, the annual charge increased by $2; and for every $1,000 of valuation, it climbed by $1. As in Philadelphia, hotels and large boardinghouses were billed per bed, stores per plumbing fixture, stables per horse, public baths per tub, printers per press (not including steam presses, which were charged per horsepower), railroads per locomotive and passenger station, builders per cask of lime or cement. If one wanted a hose up to five-eighths of an inch in diameter, that cost $3 a year.[122] A team of inspectors monitored consumers. Shortfalls in income prompted a discussion of raising rates, as well as regulating or even banning certain kinds of uses, notably hoses, which were sometimes left running continuously, whether for convenience or to avoid freezing. It did not help revenues that many water takers in all three cities were delinquent in their payments.

Charitable and nonprofit institutions that served the public good raised another issue when they asked for free water or reduced water rents. In 1850 the Children's Friend Society of Boston, which cared for destitute and dependent youth, requested that it receive water at no cost. The City Council referred the matter to the city solicitor, who held that while the society was a very worthy organization, Boston's charter forbade giving a particular group public resources without compensation. "If in a general view of the subject, the water is to be regarded as valuable property," he explained, "the City Council have no more right to give it away, than they have a right to make a donation of the City lands, or of the money in the City treasury." To do so would transfer public property to a private organization, albeit one of an "admirable character." If the council could donate city property to an organization because that organization had limited funds or was "doing great good in the community," by the same principle its members could give water service gratis to a poor or good man. "In short, the same reasoning would authorize the City Council to vote the water free at once," and stop the flow of income to the fund that was expressly designated as security against the money the city borrowed to build the system.[123]

Thirteen years later, the trustees of the Massachusetts General Hospital petitioned for a significant discount on the hospital's water bill, which had recently risen to over $600 a year. The trustees' central argument closely resembled that of the Children's Friend Society: the hospital merited a lower rate because it served the public good, which made it different from most users, especially commercial ones for whom water provided a "pecuniary benefit." The trustees maintained that the water rent schedule should make special provision for "hospitals and other charitable institutions," whose purposes "justify a very low and moderate rate of charge." After all, Massachusetts General Hospital provided the best medical and surgical treatment "to persons who have *little or no means*"; fully three-fourths of patients in recent years had received free care, which saved the city far more than what the hospital paid for water. The trustees conceded that the City Council was legally bound under the conditions on which the water bonds had been authorized to demand payment from all who connected to the system, but they felt that Boston could still reduce water rents for public service organizations like theirs.[124]

In making clear that they were not asking to be given city water for free, the trustees revealed their awareness of how the city had answered the Children's Friend Society's appeal. But Massachusetts General Hospital fared no better. The City Council Joint Standing Committee on Water stated that "the beneficent character of the Institution" and "the great public benefits it confers, especially in the cases of gratuitous relief to the sick poor," compelled its members to take the trustees' appeal very seriously. Convinced that the case the hospital presented was "as strong as any which is ever likely to arise," the committee again asked the city solicitor for his opinion. He took barely a page to declare that the petition "cannot consistently with the legal duties of the City Council be granted." As laudable and valuable was the work of the hospital, to reduce its water rent would be tantamount to transferring public funds directly to its support. And even if the council wanted to do this, he explained, a tiny fraction of the public could stop it from doing so. The legislation authorizing the waterworks required Boston to set water rates at a level that would provide funds sufficient to pay the interest on the water debt, and if it failed to adhere to this condition, a petition signed by a mere one hundred citizens could ask the Massachusetts Supreme Court to step in to make certain this goal was met.[125]

Officials occasionally commented in their reports on how hard it was to set rates. In 1875 the Chicago Board of Public Works served a city whose population had surged over 20 percent in the preceding four years, from around 325,000 to about 400,000, this in spite of the fact that in 1871 Chicago had suffered a disastrous fire that had leveled a third of its property and left an equivalent fraction of its residents homeless. Board members seemed to be asking for sympathy when they explained the challenge of predicting usage and costs in a place of "large magnitude and variety" that was continuously under alteration. The calculation of water rates, they explained, "involves constant and faithful labor on account of the peculiarly complicated character of the tax."[126]

Private Waste of the Public Good

The most exasperating problem that urban water officials faced—one that demonstrated individual disregard for the public good and seriously threatened the effort to supply every citizen—was the colossal amount of water wasted by users. Some waste is unavoidable in any waterworks system, but a great deal is always attributable to human disregard for the consequences of using far more water than one truly requires, especially if these consequences are not immediately evident. To get individual users to avoid waste involves inculcating in them a sense of responsibility to see beyond their needs to those of the public good. This has always been difficult, rarely more so than when city people, many of whom once had to haul their own water (or pay someone else to do so), suddenly had what appeared to be an unending supply piped directly to them.

It was only logical to expect that consumption would rise sharply once users connected to the system. In the first report it issued after the original Centre Square works opened in 1801, the Philadelphia Watering Committee acknowledged that "the great and increasing value of the [Schuylkill] water, for culinary and other purposes, will induce a great demand for it, whenever the distribution shall be more widely extended, and be in a more perfect state."[127] Boston mayor Theodore Lyman similarly predicted in 1834 that "one thing is quite certain, that the moment a supply of soft and pure water [is] obtained, a much greater use [will] be made of water than is now the case."[128] Still, no one came close to estimating accurately just how thirsty city people would be. Average per capita consumption leaped very quickly from a few gallons a day to somewhere between fifty and a hundred gallons, and soon even much higher.

Evidence of how much water was wasted was on view everywhere. Visiting Philadelphia during a tour of the United States in the late 1830s, British naval officer turned author Frederick Marryat was struck by how profligate local residents were with Fairmount water. Since the city was so "admirably supplied," he remarked, "every house has it laid on from the attic to the basement; and all day long they wash windows, door, marble step, and pavements in front of the houses." Philadelphians "have so much water, that they can afford to be very liberal to passers-by," he added wryly. "One minute you have a shower-bath from a negress, who is throwing water at the windows on the first floor; and the next you have to hop over a stream across the pavement, occasioned by some black fellow, who, rather than go for a broom to sweep away any small portion of dust collected before his master's door, brings out the leather hose, attached to the hydrants, as they term them here, and fizzes away with it till the stream has forced dust into the gutter."[129]

Up to a point, cities wished to encourage residents to use water generously, on the assumption that this would be good for the collective. Even the poor should have water in "wasted profusion," Walter Channing argued.[130] "Profusion" would not be "wasted" at all in his opinion because increased use of water was so clearly beneficial to everyone in so many ways. All systems have a finite capacity, however, which is determined by their design and how well they function, regardless of the size of their source. The Schuylkill River, Lake Michigan, and even Long Pond contained far more water than their respective cities needed, but the amount the local works could deliver in a given period was limited.

Despite Channing's comment, it was essential to instill an awareness in individual users that while there was plenty of water for all essential needs, there was not enough to squander. A few weeks after Boston's new system opened in 1848, a contributor to the *Daily Advertiser* dismissed "the popular feeling of distrust in regard to the quantity because it is to come from a pond, and not [as in Philadelphia] a river." This source would provide enough "for the present generation, and for the succeeding ones, until the city shall have more than doubled its present population." And when that day arrived, the citizenry would once again "provide for the occasion by uniting some of those many streams which nature has placed in a condition to meet the emergency." Nevertheless, he warned, "vast as is this accession to our means, there is only enough for use and nothing for waste." To use water without being sensitive to this was tantamount to robbing the urban collective, and to no one's benefit. The user who "sets his pipes to run uselessly and wastefully, is actually stealing a portion of the common property, that not enriches him, and makes his fellow citizen poor indeed." The author conceded that "our habits need reformation in this respect, for in regard to all God's favors, we Americans are the greatest prodigals on earth."[131]

All too true. Once the new systems opened, delighted users immediately forgot old habits of husbanding their supply and luxuriated in the abundance. The irresistible convenience of having unlimited running water right at hand drowned out any calls for conservation. All the personal incentives were on the side of heedless waste. Having good water available at the turn of a tap encouraged excessive use, since the water renter no longer needed to exert the effort required to fetch it. The flat rates that systems charged meant that the more water one consumed, the better the bargain.[132] And the introduction of meters for heavy industrial users could in fact raise consumption since cities sometimes dropped the cost per gallon when business customers passed a certain level. On the one hand, this supported the growth of the local economy, but newer and larger kinds of factories required enormous amounts of water for cleaning, diluting, processing, lubricating, and cooling. In Boston the rate could fall from as high as six cents per hundred gallons to two cents. The Cochituate Water Board noted that some customers "have run large quantities to waste, for the purpose of being charged a lower rate." Graduated charges also discriminated against small manufacturers and complicated the work of the board, which recommended a uniform charge of three cents per hundred gallons.[133]

Leakage added to the problem, as did vandalism and insufficiently skilled plumbers. "The easy access to [hydrants] by boys and mischievous persons renders them difficult to be kept in repair," the Philadelphia Watering Committee reported almost immediately after Latrobe's original waterworks opened. One difficulty led to another. In the winter, the water that puddled up around leaky hydrants froze into large and dangerously slippery patches of ice.[134] Late in 1814, Philadelphia Watering Committee member James Vanuxem, whose daughter had posed for William Rush's fountain statue in Centre Square, wrote to Mayor Robert Wharton advising him to warn citizens about this peril. He also told Wharton to inform Philadelphians that it was unacceptable to "solve" this problem by "the practice of leaving, through the night during the inclemency of the winter, the cocks of the Hydrants open," since to do so was both hazardous and wasteful.[135]

For all their vaunted New England frugality, the people of Boston were spendthrifts when it came to water, in spite of the fact that the ordinance establishing the Cochituate Water Board included several regulations meant to prevent "all unnecessary waste of water." It prohibited paying customers from sharing their water with non-renters and authorized the water registrar to "enter the premises of any water taker, to examine the quantity used, and the manner of use."[136] These measures proved no deterrent to freewheeling Bostonians. Board president Thomas Wetmore (perhaps the most aptly named person, along with Mayor Charles Wells, involved in Boston's water history) informed Mayor Benjamin Seaver in

1852 that per capita usage was more than sixty gallons a day, a jump of better than ten gallons a person from the prior year and twice as much as had been planned for.[137] Wetmore was alarmed that the massive demand made it hard "to supply the necessary wants of the City hereafter, with our present means, for more than a very limited period." He was angry at "the reckless wastefulness which now prevails," which he attributed to a demand "entirely incommensurate" with the number of customers. While every kind of water taker was guilty, Wetmore singled out hotels that ran water "unnecessarily and illegally," as well as livery stables and other businesses that left hoses open continuously. In their defense, some users would leave water flowing in cold weather to avoid the same problem of freezing that beset hydrants. Wetmore urged stiffer penalties for "wanton and illegal waste."[138] The city decided that this measure would be hard to administer, though it passed a law in 1861 that made "corrupting" the water or damaging the works a crime for which the perpetrator could draw a fine up to $1,000 and a prison sentence.[139]

Water closets and urinals, whose increased production and sale accompanied the availability of running water, were another significant source of waste. Many of these fixtures leaked, while some designs required that water be flowing through them constantly.[140] Boston banned not only open hoses in livery stables but also a water-greedy forerunner of the modern toilet called the hopper closet. The city hoped to improve the quality and reduce the wastefulness of improperly installed or functioning fixtures by another form of regulation, the licensing of plumbers.

Mayor Seaver took Bostonians to task for "the *reckless, and . . . continually increasing wastefulness* in the use of the water, which seems to prevail almost universally." Seaver reminded residents that the city had constructed the works with a capacity based on its estimate of water needs at thirty gallons per person per day. This figure would have been on target "if any *ordinary discretion were exercised in the use of water.*" How to remedy Bostonians' water addiction was a matter "for grave consideration" by city officials. Unless the waste ceased, either the supply to the worst offenders would need to be stopped, or "*another main must be laid to the Brookline Reservoir at a great expense.*"[141]

Two years later, in 1854, Wetmore was still beside himself about "the exorbitant consumption of water in the City."[142] Three years after that, the board (of which Wetmore was no longer a member) reported that the "evil" of waste, "instead of being abated, has greatly and uniformly increased." The board then restated and expanded the advice on civic responsibility that the *Daily Advertiser* had proffered when Cochituate water first became available. Wasteful usage revealed the absence among individual Bostonians of "a wholesome and effective public sentiment." The waterworks, like any public good, was not intended solely for the benefit of each user as a separate person. Rather, "the *public*, the masses collectively, have an interest in them, perhaps equal if not paramount to that of individuals." While an individual might make do with very little water, this was not the goal, since it was "for the interest, the health and comfort of his neighbors that he should use it freely." In other words, "Every inhabitant of the city, besides the interest he has in the matter of consumption on his own premises, has also a direct interest in the consumption of his neighbors, and indirectly in that on the premises of every other citizen." This same interest, however, dictated that no one draw from the public supply without good reason, since "extravagant consumption" due to "sheer carelessness and waste" threatened to bring the whole waterworks system down, and with it community well-being.[143]

Waste confounded Chicago's water officials, too. When the public system began operation in 1854, the Board of Public Works recalled fifteen years later, "there were but few water takers, and having no reservoir, the water was allowed to run to waste through numerous fire hydrants, in order to keep the small engine running." The board immediately added, in parentheses and italics, "(*It is not so now.*)"[144] The city passed a law in 1855 governing water consumption that resembled the Boston ordinance of five years previous. Chicagoans were not to leave water running in any tap, washbasin, water closet, bath, or urinal "when not in actual use." The city could send an inspector at any time to check for waste, and it could cut off water service if the inspection discovered a violation.[145] Five years later the water commissioners candidly admitted that they themselves lost a great deal of water by leaving hydrants running almost constantly during the winter to keep them from freezing, cracking, and

leaking.[146] In 1869 the Chicago Board of Public Works forbade customers "to connect with the Water Works any Water Closets which make it
possible for the water to be left running when not in use." It raised the
unpopular possibility of replacing flush toilets with earth closets, which
composted excreta rather than washing them away, and thus promised
to save water.[147]

Chicago responded to the hydrant problem with technical refine-
ments, but human indifference was a far tougher challenge, especially
once the lake tunnel was completed in 1867. Per capita water consumption, which had leveled off at less than 50 gallons a day, suddenly jumped,
and it kept on rising. The Board of Public Works was obliged to report in
1868 both good news and bad: the improved quality of the city's water
was "a subject of general congratulation with the citizens," but the "extraordinary increase" in average daily consumption—from 8,681,536
gallons to 11,560,730, or more than 33 percent—meant that demand
would soon exceed the capacity of the brand-new system.[148] Two years
later, City Engineer Chesbrough stated that the "rapid increase in the
consumption of water continues to be not only enormous, but, considering the pumping power at present available, truly alarming."[149] The average daily per capita consumption over the previous year was up to 70 gallons. It climbed as high as 80 in the summer, with the unfulfilled demand
perhaps 20 gallons more. A chart of water use in the 1878 annual report
of the Department of Public Works indicated that consumption ran considerably higher than in either Boston or Philadelphia, peaking at almost
150 gallons per capita per day in August 1878.[150]

Appeals to individual users that they should consider the needs of
others failed to convince the worst offenders. In Boston, the Cochituate
Water Board sadly concluded in 1865 that "requests made from time to
time to the water-takers, not to permit the waste of water upon their
premises, have generally resulted in a stinted use of it by those who were
sufficiently economical before, and have had but little if any effect upon
those who carelessly or wilfully wasted it."[151] Even when Chicago's supply
was inadequate to meet universal demand, plenty of water still made it
to customers closer to the Pumping Station, who thus did not experience
the consequences of any shortage and had no motive to limit their wa-

ter consumption. The flow would slow and stop in outlying areas, where people had already learned to be prudent in their water use.

Chicago residents who were left dry complained of the inequity of the situation. "Is there any law or regulation of the Board of Public Works," one inquired at the beginning of the summer of 1870, "to prevent parties from washing sidewalks, and deluging their front yards, and washing their barns during a heated term like the present, while others cannot get water enough for family use?" In an anecdote that recalled Marryat's description of how freely water flowed in Philadelphia, the aggrieved Chicagoan told of witnessing thirteen people on Wabash Avenue near downtown spraying water on their yards and two washing their barns. Meanwhile, in his neighborhood it was "almost impossible to get water enough for daily necessities."[152] What statutes did exist to prevent the misuse of water were not effectively enforced, partly because it was very hard to do so without metering.

Chesbrough condemned as "a reckless waste" another common practice, that of letting water run so that it was always fresh and cool to the taste, and he lamented the leaks caused by faulty or broken plumbing fixtures owned by those who were too indifferent or poor to fix them. He echoed the opinion expressed earlier by the Cochituate Water Board when it condemned the waste of water as a betrayal of the urban community by selfish individuals. Chesbrough pointed out that excessive water use by some left families and businesses "deprived of an absolute necessity, and exposed, in case of fire, to immense loss." He added, "When it is considered that all waste of water causes not only an increase in annual expenditures, but necessitates enlargements and extensions of the water works earlier than would otherwise would have been necessary, it will be seen that every citizen hereinafter will have a direct interest in preventing this waste."[153] The *Tribune* agreed. "We should be glad to believe that moral suasion is a sufficient remedy, but we are afraid it is not," the paper commented, calling for the passage and enforcement of "stringent regulations."[154]

By 1875 the Chicago Board of Public Works used similar language in criticizing water wastrels. The "monstrous waste of water," it declared, was "a reflection upon the moral sense of those who are guilty of it."

Individuals seemed to forget the collective to which they owed the water as they indulged themselves. The board sadly observed that it seemed "impossible for many people to understand that water, unlike air, is not a free commodity" but "municipal property." It was "no more a privilege to waste water than to steal the groceries of the merchant or the fabrics of the dry goods dealer." Notwithstanding all the exhortations, warnings, and vigilance by managers of the water system, "many citizens persist in shamelessly and inexcusably wasting that to which they have neither a legal nor a moral right."[155]

The wealthy were, if anything, the worst offenders. "The greatest recklessness prevails along the most elegant avenues and in the sections occupied almost exclusively by those who are presumed to have educated consciences," the board stated. The businessman who "declaims against the dearth of water in the upper floors of his store or manufactory," and who "censures the city authorities" for this inconvenience, refused to acknowledge that the fault lay in his failure to manage his own household staff responsibly. Meanwhile, "the poor living in remote streets which are reached by small service pipes only, are subjected in the weather when cold water is their greatest boon, to actual suffering for the quantity for which they pay, and which is essential to health, cleanliness and comfort, while it is being stolen from them and from the city by the conscienceless servants of the rich."[156]

Since warnings and most existing ordinances failed to keep people from wasting water, the authorities took additional steps. One was to restrict what were defined as non-essential water-intensive uses. For example, they curtailed the hours during which public and private fountains could play. In 1850 Boston rejected petitions for the installation of drinking fountains. To monitor private consumption generally, it hired nighttime as well as daytime inspectors. But water boards and public works departments were forced to conclude that neither legal regulations, appeals to civic duty, nor quasi-sermons on the immorality of waste were effective deterrents. They would fight selfishness with self-interest by raising rates. Water takers, especially large commercial ones, protested every threatened or actual increase as unfair and stifling to business, and, therefore, not good for the community. Boston's water administrators replied that the original purpose of building the works was "to furnish the mass of the people water, at the lowest possible rate, and it is very apparent that, if we sell it to manufacturers and others, who use it for the money they can make from it, at less than cost, the loss must fall upon the general consumer."[157]

Breaking Ranks

The construction of the Philadelphia, Boston, and Chicago water systems were great accomplishments of civic unity, but as the challenges of rates and waste indicated, the arrival of city water was just as likely to reveal the continuing tension between the needs and desires of the individual and the public good, especially in an urban collective that was fragmented and fractious.

Another Krimmel waterworks painting provides revealing evidence of this. The artist returned to Centre Square on July 4, 1819, to see if he might again find a subject there.[158] The resulting watercolor, *Fourth of July Celebration in Centre Square* (1819, Historical Society of Pennsylvania), is a more complex work than the oil he had executed seven years before. Now fully a hundred separate figures are arranged in nine different visual vignettes, all part of an Independence Day observation suffused with patriotism recently reenergized by the War of 1812.[159] The differences between this painting and the earlier one show Krimmel's more advanced skill, but also that he was perhaps more faithfully reporting the raucous and contested nature of public life in Philadelphia.

By 1819 Centre Square had developed a reputation as a place where heavy drinking and other vices intruded on the Fourth. A complaint in the July 4, 1821, *Daily Advertiser* noted that the square had "too often on this day been disreputably distinguished." The paper singled out "gambling establishments [that] abound there in the open day, to which apprentice-boys and others are enticed." Instead of learning to associate democracy and independence with civility, these impressionable youths were being "initiated in the wretched school of gambling, and may possibly, at a future period, trace their ruin to the deviations at Centre Square." Two years later, Mayor Wharton issued a proclamation decrying "scenes

Figure 3.10. John Lewis Krimmel depicted Centre Square on Independence Day of 1819 as far more crowded and raucous than it was in his 1812 painting. The celebrants are loud (note in the foreground the fiddler and the two boys—one with a toy cannon, the other with a pistol) and demonstrative. The proper Quaker family and the well-dressed African Americans of the earlier painting have no equivalent here, Rush's *Water Nymph and Bittern* is also not visible, and Latrobe's engine house is well in the background. Courtesy of the Historical Society of Pennsylvania.

of debauchery, gambling, and drunkenness, with many other acts of excess and riots which annually take place on the 4th of July." He cited as especially objectionable "the booths, tents, and other unlawful restaurants on the public streets and grounds of the city," which he banned. Wharton hoped that the soldiers who paraded in the square would be able to "obtain refreshments from sources less impure" after the day's formal ceremonies, and that they would "see the propriety of banishing from our city limits causes of such ruinous effect to the morals and future usefulness to the rising generation."[160]

Although there is no depiction of gambling or outright debauchery in Krimmel's 1819 painting, we do see the much-criticized booths and tents, in one of which several high-spirited celebrants are eating and drinking while a fiddler saws away. Other activity is also far more boisterous than that depicted in the earlier painting. Many figures gesture in a broad and energetic manner. One of two boys in the foreground fires a toy gun as the other sets off a miniature cannon. And members of the several different groups in the painting have little to do with one another. A military parade marches through the square, ignored by the many revelers who are absorbed in their own noisy doings. Gone are the gentility and social harmony of the first painting. There is no equivalent to the sober Quaker family or to the three well-dressed African Americans who appeared in the 1812 work.

In addition, the engine house is now farther in the background and, as a result, much smaller. Smoke no longer rises from its dome, attesting to the fact that by this time the engine was out of service, since the works had moved to Fairmount. Latrobe's marble water temple was now reduced to use as a very expensive storage shed; it would soon be dismantled, its columns cannibalized for the construction of the Congregational Unitarian Church at Tenth and Locust Streets.[161] Just as the engine house no longer contains the throbbing machine that once pumped refreshing Schuylkill water to Philadelphia, it does not frame the human action in front of it as it previously did; the nobly proportioned building is an artifact of a more orderly Philadelphia that no longer exists, if it ever did in the first place.[162]

Even the performances of unity in civic celebrations of new water systems could become exercises in unintended irony. In Boston, which witnessed the largest and most organized of these observances in any of the three cities, the orderly parade of citizens marching through the city streets and into the great social concourse on the Common belied serious divisions. The parade was in danger of falling apart before it ever assembled. Some participants carped at the choice of the route, while native-born firefighters balked at being placed behind Irish and other immigrant organizations. Organizers were able to resolve the different disputes, so that the *Daily Evening Transcript* could report with some relief two days before the event, "All the great difficulties have been got over; the masses 'fraternise' most harmoniously; and the strife now is, who shall do most to make the celebration a glorious one."[163] But the celebration took place less than two weeks before the presidential election of 1848, and the competing Whig and Free Soil Parties both tried to politicize the public gathering to their advantage. Their respective local champions, Daniel Webster and Charles Sumner, were on hand not just to be present at the great moment of municipal accomplishment, but also to address their constituencies.[164] Each party planned a separate torchlight procession, the Whigs the evening before the great day (though it was postponed by rain until October 30), the Free Soilers a few hours after the water celebration ended.

In Boston and Philadelphia, the discussions of water itself did not always break down along political divides, since no party was against having good water. More common was the second-guessing of the kind that the Philadelphia Watering Committee experienced in the trying early years of the works. In Chicago, however, party figured heavily. The ceremonies surrounding the completion of the tunnel and the dedication of the Water Tower were a rare moment of harmony in a city where the seemingly apolitical task of providing an adequate supply of water had become embattled. Shortly after the new publicly owned Chicago City Hydraulic Company began its work, the *Chicago Tribune* sharply criticized members of the Common Council for questioning the ability and integrity of the original commissioners appointed in 1851. The *Tribune* accused United States congressman and future mayor John Wentworth, then a Democrat (he joined the newly formed Republican Party in 1857), of masterminding the complaints. Wentworth, who happened to be owner-publisher of the rival *Chicago Democrat*, countered that he spoke for several members of the council as well as himself when he declared, "We believe in scrutinizing as closely as possible the *official* conduct of all men." He added, "If there is anything wrong, public good will grow out of that scrutiny; and if all is right, no harm can grow out of it."[165]

The city's water board became even more politicized beginning in 1854, when, as specified by the Chicago City Hydraulic Company's corporate charter, its members were chosen by direct election. Now the *Tribune* tried to guarantee that no mere partisan or self-interested "politician" but a true public servant—that is, someone *it* judged to be a true public servant—would be elected. The editors were concerned that a mere politician might win because his backers were better at getting *their* people, rather than *the* people, out to the polls. The best representatives of *the* people were, of course, the candidates that the *Tribune* favored. It urged its readers—who, in this period of unapologetically partisan journalism, followed the paper because they agreed with its politics—to be sure to vote. The editors appealed to their audience's sense of the public good. "Apathy and negligence, where so much is at stake," they stated, "cannot be otherwise than fatal to the interests of the city, and criminal in the citizen."[166] Evidently urban democracy would succeed only if the right people acted to protect the public from the wrong ones.[167]

It was one thing, then, to promote the building of a waterworks as indubitably essential to the public good, to insist that it be publicly owned, to house it in an ennobling structure, and celebrate its opening as a triumph of the urban collective, and yet another to build and manage a system in a way that served and pleased everyone. The early waterworks histories of Philadelphia, Boston, and Chicago demonstrated that it is frequently hard to assess how *a* particular public good represents or benefits *the* public good, and that this assessment is always open to dispute, especially in a democratic society.

4

NATURE AND ART

*Water and the Reconciliation of the Natural
and the Urban*

At the same time that Philadelphia, Boston, and Chicago faced the issues
that water raised about the changing relationship of the individual to the
collective, they also found themselves looking for the best way to understand another major aspect of urban development: the transformation
of landscape into cityscape. Cities are by definition built environments
where humans have altered the natural world to suit their needs. Waterworks are particularly powerful expressions of the human desire and
ability to overcome, discipline, and even dominate nature.[1] As Benjamin
Latrobe explained, a system not only must move enormous amounts of
water from its source to users, but in doing so it also "must not be liable to interruption from ice or freshe[t]s, but be equally useful in the
severest winter, and in the wettest summer." His steam engines would
keep the Centre Square holding tanks "perpetually full, being of a power
sufficient to supply every possible demand of the city." They would in
effect liberate the people they served from nature, since their operation
"depend[ed] not on the variable seasons, nor on the natural advantages

of situation,—but solely on the *option* of man."[2] Philadelphia would soon build a dam that diverted the Schuylkill River, while Boston constructed one that increased the capacity of Long Pond, which its engineers also fitted with a gatehouse that regulated the flow of the pond's water into the aqueduct—a fabricated river—they erected.[3] Chicago would dig a two-mile tunnel through the clay bottom of Lake Michigan.

Once a new waterworks system was in place, city people quickly adjusted to the displacement and reconfiguration of nature. They now treated water as a human-produced commodity whose source was a mechanical faucet. They gave little thought to the pumps and pipes that delivered it, and even less to its natural source.[4] Most commentaries on new urban water systems envisioned the commodification of water in very positive terms, often asserting that the appropriation of this natural resource was nature's intention, however paradoxical this might seem.[5] They viewed waterworks and the cities they served as realizations of God's plan for the natural creation. Even as city people gloried in subjecting nature to their will, however, they revealed an often strained and sometimes almost desperate eagerness to maintain, observe, and honor a vital connection to the natural world as something separate from and above the one that humans made.[6]

Going with the Flow

Champions of infrastructure projects sometimes maintained that nature was in fact on their side. This was certainly the case when they spoke of ones that directed water to serve human needs. In 1818 state senator and Philadelphian Samuel Breck, concerned that the proposed Erie Canal, which would connect the Hudson River to the Great Lakes, would give New York a major competitive edge over his city, urged his fellow Pennsylvanians to build a competing canal. Otherwise they would be foolishly throwing away the gift of the many promising waterways that nature had bestowed upon their state and letting New Yorkers leave them far behind. "Nature has done her share," Breck stated, and the moment had come to "let art complete it."[7] The dichotomy of "nature" and "art" was a common trope in this period; Breck anticipated by eighteen years the most famous articulation of it, by Ralph Waldo Emerson in *Nature.* Emerson defined nature as encompassing "essences unchanged by man; space, the air, the river, the leaf," while art was "the mixture of [man's] will with these same things." Among the examples Emerson gave for the latter was a canal.[8]

Bostonians also believed that to take advantage of nature's generosity was to fulfill an obligation inscribed in the creation. In 1824 Mayor Josiah Quincy Sr. assured the City Council that "the prognostics of [Boston's] future greatness are written on the face of nature, too legibly, and too indelibly, to be mistaken." Given all the natural resources with which Bostonians had been endowed, it was incumbent upon them to "be willing to meet wise expenditures and temporary sacrifices, and thus to cooperate with nature and providence in their apparent tendencies to promote their greatness and prosperity."[9] Others went to great lengths to demonstrate, or at least to claim, that something much more powerful and sacred than human need lay behind the building of their city's waterworks. The rhetoric of some supporters of the Long Pond system combined their Puritan forebears' conception of themselves as a new chosen people with a Unitarian belief in a benevolent God and an unfallen nature accessible by reason, adding a touch of Transcendentalist mysticism for good measure. The Faneuil Hall Committee pointed out that engineer Loammi Baldwin Jr., who prepared the 1834 water report that recommended Long Pond, "was so impressed with the excellence of the source and the purity of the water," that he was known to declare "*'that God in his Providence'*" had created this source "*'purposely for the supply of the Peninsula of Boston!'*"[10]

Once the selection of Long Pond was sealed twelve years later, spokesmen for the city proclaimed on multiple occasions that Boston was responding to nature's will. During the groundbreaking ceremony, Water Commissioner Nathan Hale declared that the pond's liquid reserves "had been laid up in their present situation from the beginning of the world, in the quiet and seclusion of the country, retired from the neighborhood of a large business population, and exposed to no impurities." The aqueduct would now merely "divert [the waters of Long Pond] from their retirement" and activate them to their long-intended use, "the nourish-

ment and health of the vast population of a great city." The topography of
Massachusetts, Hale contended, revealed that nature had formed Long
Pond in the knowledge that Boston would need it. The pond's elevation
was higher than that of the city, so "the waters would flow [to Boston]
by their own gravitating force," which sanctified the whole undertaking
as being in accord with natural law, and thus profoundly appropriate and
right. Hale here came close to restating the passage from Emerson's assertion in *Nature*, "The axioms of physics translate the laws of ethics."[11]

To regard the tapping of Long Pond as an act of gracious deference to
purposeful natural topography ignored the extent to which impound-
ing this body of water and constructing a nearly twenty-mile aqueduct
entailed a major human incursion into the Massachusetts countryside.
Hale's observation that "nothing was necessary . . . but to remove the
obstacles presented by intervening high lands" dismissed any thought
that these obstacles might be read as nature's resistance to the project.[12]
He also overlooked the fact that a titanic amount of manual labor would
be required before "gravitating force" could do its work, and that this
labor could not start until the city purchased the pond and negotiated
legal agreements with the owners of property and representatives of the
other towns through which the aqueduct would pass.

In his remarks on the same occasion, Mayor Josiah Quincy Jr. likewise
claimed that nothing short of divine will lay behind the waterworks, and
he thanked the Author of nature for his gift. Quincy noted that it was
impossible to stand where he and the other celebrants were gathered,
"gazing first upon the calm, serene surface of the lake below, and then
into the blue expanse above, without having our own minds directed up-
wards to the seat of that wonderful wisdom which, seeing the requirement that would come in the end, provided the means of its fulfillment
long before the human race existed on this globe." The Almighty—whose
"beautiful law which decrees that wherever water exists, it shall always
and ever flow as high as its source"—had invested the landscape with
the means to supply "our natural wants" on whatever scale was required.
Quincy made it seem as if Boston were merely enacting God's good intentions for the city. It was to "His beneficence" that humans owed their
opportunity to construct an aqueduct that would "assuage the sufferings

and satisfy the cravings of thousands and thousands now many miles from this spot." And it was with God's "permission" that Quincy and his listeners were all here assembled "to declare that the waters of yonder lake, now so placidly reposing beneath us, shall ere many months be over, rise from the bosom of our own beautiful Common, in fountains of brilliant hues, affording the means of luxury to the rich, of cleanliness, comfort, and health to the poor."[13]

Quincy attended to one more piece of business that reinforced this idea. He claimed that the water commissioners had discovered that the local "aborigines" had called present-day Long Pond "Lake Cochituate," and this name would now be officially revived. Then, to "great laughter and applause," he "translated" the word "Cochituate" to mean "an ample supply of pure, soft water, of a sufficient elevation to carry it into the city of Boston at a moderate expense." This witty bit of linguistic legerdemain implied that the area's displaced Native American population consciously assented long before the fact to the conversion of this body of water into the life source of what would one day be the region's major metropolis.[14] Quincy did not mention that what made the aqueduct necessary was the destruction of the ample natural supply that had attracted Native Americans and then English settlers to the Shawmut Peninsula.

While Quincy doubtless intended the rechristening of the pond to leaven an already happy moment with humor, he and others were serious in seeing the Cochituate waterworks project as a reaffirmation of what the city's founders believed was their covenant with nature's God. In his 1846 Thanksgiving Day sermon, delivered three months after the groundbreaking, Congregationalist minister Nehemiah Adams cited Revelation 14:7 in stating the hope that the waterworks project "may help to turn our adoring and grateful thoughts . . . to Him 'that made heaven, and earth, and the sea, and the fountains of waters.'" (Adams omitted any reference to the first part of this verse, in which an angel warns humankind to "Fear God," since "the hour of his judgment is come.") Adams's message was of God's continuing generosity toward a worthy Boston. His main text—and the source of the title of his sermon, "The Song of the Well"—was Numbers 21:17: "Then Israel sang this song, spring up, o well; sing yet unto it." As Adams explained, the passage describes how God fulfilled his promise that He would provide water to the thirsty Israelites on their long journey home from slavery in Egypt. "God gave them streams in the wilderness and refreshed his heritage when it was weary," Adams remarked.[15]

Now Bostonians found themselves about to receive a similar gift. While Adams repeated Mayor Quincy's point that the "fountains" of Cochituate water had been known to "the aborigines," he said the pond still qualified as untouched nature since there was no visible human trace upon it. Never mind that there had been an immense amount of alteration of the area by the descendants of white settlers. They had erected a dam on Long Pond, whose water had most recently been used to power William Knight's carpet factory.[16] To acknowledge this would have weakened the contention that the moment for which this pond had been created had only now arrived. As Adams would have it, "For centuries the embosomed water has reflected the image of the heavens from its eye ever lifted upward, as though it were waiting for the purpose of God, in its creation, to appear."[17] Adams even interpreted Boston's long and frequently divisive postponement of building a waterworks as evidence of divine intention. "So in the delay of introducing the waters of a lake into a city," Adams stated, "we know not how necessary it may have been that the secret springs should have had [a] longer time to strengthen themselves, and wear for themselves larger openings into the lake; or what slow, chemical changes it was necessary should be accomplished in the waters, before they were in the highest degree fitted for the use of the great city." Divine will was unerring, and "we shall always find, that, if we are not slothful, though we may be disappointed and hindered in the execution of our plans, God's time is always the best time."[18]

The 1848 water celebration continued this imaginative framing of the construction of the Cochituate water system as an act that was, in Josiah Quincy Sr.'s words, "written on the face of nature." The biblical inscriptions adorning the speakers' stand on the Common interpreted the day as the fulfillment of scriptural prophecy, helpfully citing chapter and verse: "The Lord Spake. Gather the people together and I will give them Water: Numb. xxi. 16"; "We have found water, Gen. xxvi. 32"; "The water is ours: Gen. xxvi. 20"; "Ye shall serve the Lord your God, and He

shall bless thy bread and thy water: Exod. xxiii. 25"; and "Jesus saith, Fill
the water pots with water: John ii. 7."[19]

James Russell Lowell's "Ode" to water, sung that day by local schoolchildren, alluded to Greek mythology rather than the Bible to make the
same point, that nature was giving itself to Boston rather than being
seized by the city. Cochituate water is the speaking voice of the poem,
describing itself in the first of eight sestets as the "crystal vintage, from
of yore / Stored in old Earth's selectest bin." After stating that it has
very productively "toiled and drudged" for Massachusetts industry
("Throbbed in her Engine's iron veins, / Twirled myriad spindles for
her gains"), "Water" proclaims that the new aqueduct through which it
speeds to Boston represents something loftier:

> So free myself, to-day, elate
> I come from far o'er hill and mead,
> And here, Cochituate's Envoy, wait
> To be your blithesome Ganymede,
> And brim your cups with nectar true
> That never will make slaves of you.[20]

This self-characterization of "Water" as "Cochituate's Envoy" and as a
"blithesome Ganymede"—the beautiful shepherd boy abducted by Zeus
to Olympus, where he became cupbearer to the gods—asserted yet again
that by building the waterworks Bostonians were accepting rather than
appropriating a divine gift.

Nathan Hale used his speech at the celebration on the Common to
continue this theme. "Sprung from an elevated source, and devoted to
the service of man, in the supply of his domestic wants," he explained, the
water from the lake "is taught [by the aqueduct] to mount again to the
same elevation, and to maintain its level, until it shall have accomplished
all the purposes to which it is destined." Hale thanked "that Almighty
Being, who holds the waters in the hollow of his hand, and pours them
out for the use of his children."[21] Speaking after Hale, Mayor Josiah
Quincy Jr. folded the interpretation of the waterworks as the realization of nature and God's purpose into a tribute to John Quincy Adams,

who had passed away earlier in the year at the age of eighty-one. Had Adams lived to this day, Quincy told the multitude, this great occasion would have reminded the great statesman of "the Hill in Horeb," where, as chronicled in Exodus, God demonstrated his presence among the children of Israel when water poured from a rock after Moses smote it.[22]

Perhaps influenced by the Transcendentalist view that the miraculous lay in everyday nature rather than extraordinary events, Quincy implied that there was an important difference between the two awesome moments. At Mount Horeb, God's wondrous work disrupted the way things normally were in nature. What He had done for Boston was always immanent in the purity and abundance naturally stored up long ago in Lake Cochituate. Quincy stated that "[God's] wisdom had, in every land, prepared among the hills the receptacles of water," and "His Power had given fixed and permanent laws to the most unstable of elements." Here, in Boston, "His Goodness enabled his intelligent offspring to understand and avail themselves of its use."[23]

By this logic, the powerful flow of ambitious and talented newcomers to Boston was as natural and in keeping with God's wishes as was building the aqueduct that carried the water from Lake Cochituate to the city. This idea was most eloquently expressed by physician and man of letters Dr. Oliver Wendell Holmes when he observed that Boston "drains a large water-shed of its intellect, and will not itself be drained."[24] At the 1846 groundbreaking, Mayor Quincy cited the "magnificent stores" and the "dwellings, which might almost be called palaces," that had been constructed by many a country boy who "entered the city with only a quarter of a dollar in his pocket and all his wardrobe in the knapsack on his back." Such country boys would continue to arrive, Quincy confidently predicted, thanks to the fact that Boston would now have enough water to attract an even larger population of ambitious individuals.[25]

Lemuel Shattuck and fellow statistician Dr. Jesse Chickering, who also surveyed the population of antebellum Boston, verified Holmes's assertion.[26] Shattuck observed that "a large majority of the active businessmen of Boston" were "men who have come to this metropolis to seek their fortunes, bringing with them the economical habits, the industry, the energy, and the perseverance, which are peculiar to the people of interior towns of New England."[27] But Chickering's assessment contained a warning about this trend. He might have been talking about the new water supply when he asserted that it was essential to the well-being of Boston that the city never exhaust the supply of such newcomers. Only the steady arrival of people from smaller settlements and rural areas could enable it to thrive. Chickering directly compared this human flow to water. "The stream continues flowing on to supply the waste," he wrote, "and by the abundance of the flow, a multitude of the adventurers remain to do the labor, to fill up the places, and to give a direction to the affairs of the heterogeneous community."

While the allusion to "waste" implied that there was a downside to the "flow" for some people, Chickering found another trend more disturbing. His data told him that those people he regarded as descended from the best New England stock were lately moving not into Boston but out to western states, and to younger cities like Chicago, taking with them "the hardy enterprise, the industrious habits, the intelligence and the institutions to be found in those parts of the country they have left, and transplanted them in the new states which they have adopted for their future homes."[28] If these other less developed places appeared more promising to such people, Boston's intellectual and economic vitality would go into decline.

Chicago boosters happily agreed with Chickering, staking their own claims to nature as an ally. They relentlessly bragged of their city's natural assets and ignored what limitations could not be denied—or denied them anyway. In their eyes, nature had ensured that Chicago's success was a foregone conclusion. In 1846, two years before the Illinois and Michigan Canal was completed, lawyer and local historian Henry Brown told the Chicago Lyceum, "Canal or no Canal, Chicago will advance." Only a great calamity or the "consummate folly, depravity or impudence of her people" could slow the city's majestic ascent. This was because Chicago's location made it an unstoppable economic force. "So long as yonder inland seas [i.e., the Great Lakes] bear on their surface the wealth of every clime," he intoned, "so long as yonder prairies bloom with verdure, and the 'cattle upon the thousand hills' shall graze their herbage, and so long as yonder interminable fields shall wave with their golden harvest, an effort to blot Chicago from existence, or to depress her rising consequence, would be like an attempt to quench the stars." Chicago's "position is commanding and her progress sure."[29]

Real-estate promoter John Stephen Wright, who was second to none in heralding Chicago's prospects, gloated in agreement when he cited a Cincinnati paper's concern that "trade and travel naturally flow east and west on lines of communication" to Chicago. He similarly quoted a St. Louis paper that sadly admitted that "the tide of commerce of the Grand River Valley of Missouri is now tending towards Chicago, and that St. Louis is fast losing the trade of its own State." Wright announced that the "Hub" of "our National Wheel of Commerce" was unquestionably his city, which was, he said—employing the familiar dichotomy—"pivoted by Nature and by Art." Wright agreed with those who held that natural advantages only mattered if people knew what to do with them. "NATURE" had "laid from the Atlantic Ocean into the heart of the Continent, this chain of rivers and lakes, over a thousand miles of the grandest navigation on the globe," and "ART" had "perfected" nature's work with canals and railways. Building a place as magnificent as Chicago required generous helpings of both. "Nature never makes a city," Wright asserted. "No human institution is more artificial, success depending upon a conjunction of causes, which, however liberally bestowed by nature, lie dormant until operated by human effort and ingenuity." Without this natural endowment, in turn, human effort would face a mighty challenge. If a city did prosper, as Chicago was doing, this proved that nature was its ally. "Art," he explained, "would have a difficult task in localities neglected of nature, and easy upon sites she favored." His city's unequaled pace of growth "would therefore be strong *prima facie* evidence that nature favored Chicago."

"As in every natural object which astonishes us for its beauty, its ingenuity, its perfect adaptation to its purpose," Wright explained, "it is more unreasonable to suppose it the creation of accident, than of an intelligent Creator; so in this union of many free and independent wills, effecting these great purposes as with one mind, one soul, an over-ruling Power must govern." Making use of what nature had provided was a matter of sacred duty. "We cannot examine into the why and wherefore of our growth," he wrote, "without becoming reverently impressed with the truth that this is not man's work alone."[30] Wright here integrated a belief in technological progress into the period's powerful current of post-millennial thinking. City building was a form of perfectionism, and Chicago above all other places was God and nature's chosen place.

No wonder, then, the excitement when Chicago seemed to prove this by bringing the cool, crystal waters of Lake Michigan to its people. At the ceremonial placement of the last stone in the lake tunnel, Mayor John B. Rice congratulated Chicagoans "that they have the pure refreshing water of our mighty lake," which, along with the prairie winds and fertile fields, "tend to make Chicago the most favored of cities." Rice finished with a tribute to the wisdom and will with which his city developed the features that made it so "favored." "Hail! Chicago, metropolis of the great West," he exclaimed, describing the city as a community that was "vast in her resources, fortunate in her citizens, whose genius, industry and integrity secure to us the use of all those advantages and blessings which are vouchsafed to us by the Creator and Dispenser of all the things which we have."[31]

A reporter for the *Chicago Republican* who covered the completion of the tunnel (which a headline writer described as "The Wedding of the Lake and the Land") wrote that "more truly than ever before, is Chicago entitled to her name of the Garden City. For water is the condition of the Garden." His use of the capital *G* signaled that this was not just any garden. "In the description of Eden," he explained, "half of the space is taken up by a statement of the circling streams that irrigated its meads and diffused everywhere their grateful coolness."[32] Chicago was not quite the second coming of paradise, but thanks to the water the tunnel supplied from the unsullied reaches of Lake Michigan, it was ready to take its place as a center of a great productive harmony between humans and their Creator.

Eloquent Skeptics

The new waterworks that Philadelphia, Boston, and Chicago erected vividly exemplified how urbanization involved the construction of what has been called "second nature" upon the "first nature" that was the unaltered creation.[33] The distinction between first and second nature is similar to the one that Samuel Breck, Ralph Waldo Emerson, and John Stephen Wright, among others, discerned between "nature" and "art." Emerson said that the sum total of all the impositions of man's will on the natural world was insignificant—"a little chipping, baking, patching, and washing"—so that "in an impression so grand as that of the world on the human mind, they do not vary the result."[34] The rise of large cities troubled this kind of assessment, causing some to wonder whether urbanization and industrialization violated rather than fulfilled nature's purpose.[35]

Amidst antebellum America's enthusiasm for urban and technological development, there were skeptics who were concerned about the consequences of this process, which they saw as a destructive and alienating violation of nature, not a fulfillment of its intentions. Henry David Thoreau believed that the "chipping" was deeply sinister. He despised what he considered the ignorant and dangerous arrogance behind the national passion for internal improvements, which he called, with his characteristic fondness for wordplay, "external and superficial." If people examined with a "true gaze" the profoundly false and inessential world they were building, he insisted, it "would all go to pieces."[36]

Writing about his experience at Walden, where he lived at the same time that the Boston water debates were at their most intense, Thoreau appealed to readers to reconsider all the false steps that humans had taken in constructing their world, pointing in particular to several major European and American cities, to which he added his own hometown. "Let us settle ourselves, and work and wedge our feet downward through the mud and slush of opinion, and prejudice, and tradition, and delusion, and appearance, that alluvion which covers the globe, through Paris and London, through New York and Boston and Concord, through church and state, through poetry and philosophy and religion," he wrote, "till we come to a hard bottom and rocks in place, which we can call *reality*, and say, This is, and no mistake." Only having once reached this natural philosophical bedrock, "a *point d'appui*, below freshet and frost and fire," would a person of insight discover "a place where you might found a wall or a state, or set a lamp-post safely, or perhaps a gauge, not a Nilometer, but a Realometer, that future ages might know how deep a freshet of shams and appearances had gathered from time to time."[37]

Thoreau was outraged by local residents' ambitions, large and small, to treat nature as if its only purpose were to serve human material desires. In several instances he focused on water. He protested, for example, against the impudence with which the owner of another pond not far from Walden had renamed it after himself. "What right had the unclean and stupid farmer, whose farm abutted on this sky water, whose shores he has ruthlessly laid bare, to give his name to it?" (One can only imagine what he would have thought of Josiah Quincy Jr.'s self-serving translation of "Cochituate.") It would have been far better, in Thoreau's view, to call the pond after the fish, fowl, and other wildlife that lived in or by it, or "some wild man or child the thread of whose history is interwoven with its own." This natural treasure should certainly not be named for "him who could show no title to it but the deed which a like-minded neighbor or legislature gave him," who would reduce nature to its instrumental and cash value. Having heard of a proposal to use Walden as the source of a local waterworks system, Thoreau was outraged. The villagers, he wrote, "who scarcely know where [Walden] lies, instead of going to the pond to bathe or drink, are thinking to bring its water, which should be as sacred as the Ganges at least, to the village in a pipe, to wash their dishes with!—to earn their Walden by the turning of a cock or drawing of a plug!"[38]

Thoreau may have taken some small comfort from his knowledge that it was not possible to exclude, transform, or manage nature with anything approaching total control. Nature was not nearly as willing a partner in human affairs as others would like to believe. As cities piled levels of built environment every which way upon each other, nature frequently refused to cooperate in its own submission. In Philadelphia, freshets swamped buildings along the rivers, and the construction of dams drove fish away.[39] Chicagoans had something of the opposite problem when fish swam out of their water faucets, gifting them with more natural bounty than they wanted.

One of the first Americans to realize in its full profundity that altering nature invited unintended and unwanted consequences was attorney, editor, farmer, mill owner, politician, linguist, and diplomat (he was the American minister to Turkey and then, for over twenty years, to Italy) George Perkins Marsh. This extraordinary polymath is credited, among other things, with being America's first major ecological thinker. Marsh's *Man and Nature; or, Physical Geography as Modified by Human Action*, published in 1864, was the product of decades of wide travel, close observation, and deep reflection. Marsh concluded that "the destructive agency of man becomes more and more energetic and unsparing as he advances in civilization, until the impoverishment, with which his exhaustion of the natural resources of the soil is threatening him, at last awakens him to the necessity of preserving what is left, if not of restoring what has been wantonly wasted." He decried "the prodigality and the thriftlessness [that] former generations have imposed upon their successors."[40] The deliberate changes humans inflicted on nature were bad enough, but these were "insignificant in comparison with the contingent and unsought results which have flowed from them."[41]

Marsh's stated purpose was "to point out the dangers of imprudence and the necessity of caution in all operations which, on a large scale, interfere with the spontaneous arrangements of the organic or the inorganic world." Rome may have fallen because of internal tyranny and external enemies, he allowed, but another key factor was its "ignorant disregard of the laws of nature" in its heedless mismanagement of all the natural resources at its command. The historical lesson was that humans must not disrupt the natural processes that sustain the creation. They needed to make sure that they conducted what he called their "transforming operations" in such a way "as not unnecessarily to derange and destroy what, in too many cases, it is beyond the power of man to rectify or restore." Marsh advanced the strikingly modern recommendation that the soundest policy by far was "to become a co-worker with nature in the reconstruction of the damaged fabric which the negligence or the wantonness of former lodgers has rendered untenantable."[42]

To fail to cooperate with nature would bring grave results. "The ravages committed by man subvert the relations and destroy the balance which nature had established between her organized and her inorganic creations," Marsh advised, "and she avenges herself upon the intruder, by letting loose upon her defaced provinces destructive energies hitherto kept in check by organic forces destined to be his best auxiliaries, but which he has unwisely dispersed and driven from the field of action." Like Thoreau, Marsh questioned the age's enthusiasm for "progress," which, he believed, invited disasters that were anything but "natural" since they were precipitated by human action. In his chapter on "The Waters," Marsh specifically criticized leading early nineteenth-century French health official A. J. B. Parent-Duchatelet for maintaining that permitting Paris's sewers to empty into the Seine would not measurably affect the quality of the river's water, on which the city depended.[43]

The environmental damage brought by rapid urbanization in Philadelphia, Boston, and Chicago confirmed Marsh's warnings. The 1875 annual report of the Philadelphia Water Department sounded very different from the speeches at the water celebrations when it candidly blamed city people for the water woes that required them to build systems that had to reach farther and farther out. Instead of working with nature to make the most of its beneficence, the city had poisoned its benefactor. Between the layers of clay and gneiss on which Philadelphia's built environment rested, nature had placed a stratum of "*water-bearing* gravel and sand." This was "the natural filter, the subterranean water-way, and the underground reservoir, whence issued the numerous springs and brooks that furnished to the early settlers and their immediate descendants, an abundant supply of sparkling, pure, and wholesome water." By covering the ground with streets made of impermeable materials, Philadelphians may have facilitated the "march of civilization," but now rain could no longer "moisten and purify the earth, nor foster and replenish the springs and wells." In addition, cesspools were "foolishly, if not wickedly, *dug through the clay*," infecting natural stores of clean water, "making them vehicles to convey and diffuse the germs of disease and death" while "robbing the soil of its proper and natural nourishment." City people who acted so irresponsibly could not do so with impunity: "Nature, though slow in action yet sure in effect, will not permit the violation of its laws without exacting the penalty."[44]

Chicago waterworks designer and city engineer Ellis S. Chesbrough and two other experts hired by Boston in the 1870s to evaluate its sewers reported that much of the city's large stretches of land that had been created by dumping earth from elsewhere into marshy areas were "filled with material more or less unsuited to the purpose, and the evils of such a condition have been increased from the periodical floodings with water or sewage to which these places have been subject."[45] Perhaps Chesbrough had learned something from having allowed the sewers he built in Chicago starting in the 1850s to disgorge into the river and the lake. To be fair, even at the time, Chesbrough knew this was a flawed choice and expressed some of the same concerns as Marsh had done in his criticism of Parent-Duchatelet. Chesbrough's fellow Chicagoan and public health champion Dr. John H. Rauch regretted that those who founded cities, including his own, paid too little attention to the basic question of whether the natural setting they selected was suitable as a place to live. They cared too much for "trade and barter," as opposed to "healthfulness of surroundings," with the result that they later had to confront "deficiencies in regard to water supply, to soil, to atmosphere, to location and topography."[46]

Looking Backward

One very popular mode of questioning the human transformation of nature by building cities had little to do with such serious philosophical, scientific, environmental, and sanitary objections. In this case, city people did not criticize city life for all the damage it had done but instead expressed a longing for a simpler and better time when they lived closer to nature. It would have been more remarkable had they not done so, since pastoral nostalgia is timeless.[47] But the quickening pace of industrialization and urbanization in the nineteenth century intensified reflections on what had been lost amid all the changes. The period that witnessed the advent of city water, with all its accompanying alterations of the natural landscape, prompted in some city people a yearning to return, at least imaginatively, to the countryside that they had left behind. To a notable degree, this return directly involved water.

This was especially true in the United States, which many claimed enjoyed a special relationship with nature, and whose burgeoning cities were heavily populated by native- and foreign-born newcomers with a rural background. While a multitude of country people from America and Europe chose to cast their lot in urban places, the strong presence of nostalgia in the literary record suggests that many felt at least twinges of regret. To a significant degree, however, this was a form of psychological accommodation, not resistance. Taking an imaginative refuge in the past, away from the routines and pressures of city life, was quite different from being willing to go back, even if that were possible.

Often the site of nostalgic retreat was quite specifically a watery one. Oliver Wendell Holmes Sr. mused, "All of us have been thirsty thousands of times, and felt, with Pindar, that water was the best of things." Holmes, who was born in Cambridge in 1809 and could recall the days when it was a rural village, then shared a personal reminiscence: "I alone, as I think, of all mankind, remember one particular pailful of water, flavored with the white-pine of which the pail was made."[48] Holmes was not nearly as alone as he thought. By the time he was reminiscing about the pine pail, the rustic bucket of pure water as the quintessence of an unrecapturable country life was a familiar convention.

The prototypical example is Samuel Woodworth's celebrated poem and song lyric, "The Old Oaken Bucket" (also titled "The Bucket"), written in 1817. One of the most popular American writers of the first half of the nineteenth century, Woodworth was a native of Scituate, Massachusetts, but he was living on busy Duane Street in lower Manhattan when he wrote the poem. He said that he was inspired by a glass of New York's unappetizing local water, which he drank after walking home from work. Woodworth supposedly exclaimed to his wife, " 'That is very refreshing; but how much more refreshing would it be to take a good long draught, this warm day, from the old oaken bucket I left hanging in my father's well, at home!' "[49] That bucket of cool sweet water represented an earlier way of life that he and so many other city people had left.

In his poem, Woodworth associated the bucket with "the scenes of my childhood." The scenes are also those of the nation's childhood, a world defined by a nearby millpond, graced with a bridge and cataract, as well as "the orchard, the meadow, [and] the deep-tangled wildwood." The "moss-covered" bucket struck Woodworth as "the source of an exquisite pleasure, / The purest and sweetest that nature can yield." He described it as an overflowing "emblem of truth," a phrase that looked forward to James Russell Lowell's characterization of Cochituate water as "nectar true." Both poets thus implicitly associated the urban with the artificial and false. Woodworth conceded that the busy city, "far removed from the loved habitation," was the epicenter of national progress and the place he had chosen to pursue his career, but it could never weaken the hold on his heart of the old bucket and all that it stood for. He confessed:

> The tear of regret will intrusively swell,
>   As fancy reverts to my father's plantation,
>   And sighs for the bucket that hangs in the well![50]

Fancy might revert, but it is important to note that Woodworth also limned many of the alterations that had taken place since his childhood. While Woodworth wrote several other poems in the same vein as "The Old Bucket"—his collected works include, under the broader heading of "Pastoral Poems," an evocation of "My Father's Farm"—he was an enterprising professional who was equally capable of striking quite a different cultural chord in verse that praised the mastery of water and the transformation of the countryside. Woodworth penned a lengthy ode on the completion of the Erie Canal in 1825, which firmly established the national preeminence of New York and—along with the construction of the Illinois and Michigan Canal—impelled the astonishing rise of Chicago in the decades ahead. "'Tis done! Proud Art o'er Nature has prevailed!" Woodworth proclaimed of this engineering triumph.

In another poem about the Erie Canal, Woodworth applauded nature's surrender to what he called "the daring surreption" of human engineering. Woodworth claimed that the canal's construction united Americans in heretofore distant and competing places, who now put aside their localism in what he described figuratively as a kind of family reunion. Thanks to the canal, the Hudson River and Lake Erie—and the commercial interests they promoted—now embraced "like sisters." Blurring the distinction between the constructed and the natural,

Woodworth's simile credited the canal with turning the heretofore "unrelated" river and lake into siblings. That Woodworth saw urbanization in positive terms, despite his yearning for the old bucket, is also apparent in his poem "New York," which hailed his adopted home. He called it a "happy city" where "the arts convene / And busy commerce animates the scene."[51]

The different sides in the Boston water debates sometimes tapped into rural nostalgia in praising the source they favored. The author of *Voice in the City; to Water Drinkers, a Recitative Poem*, who called himself "Bostonius," composed this work in 1839 in support of choosing Spot Pond as the city's water source. The poem pictured the environs of the pond as abounding in the purity, beauty, and wonder only to be found in nature. In spite of his characterization of Boston as his "lov'd city," "Bostonius" praised Spot Pond for being free of the undesirable features of any urban center: "No filthy population broods around; / No lurking nuisance near thy verge is found." By Spot Pond's leafy shores, as "Bostonius" would have it, weary refugees from the city find the pastoral ideal of a refined sylvan retreat. It is a perfectly balanced world of both human sophistication and natural simplicity. "Bland amusement reigns" as "fancy dictates, or as wit ordains," while "tuneful birds, in fascinating glee / Warble their wildly wildering melody." In Spot Pond's environs, "Sweet nymphs and swains in Sunday clothes are seen," soothed by their surroundings. And here nature truly exists to serve them. If these cultivated and handsomely attired nymphs and swains choose to fish, they hook "the silvery pike" "from the coolly bottom of the lake" on "the first attempt."[52] The idea that they might draw what promoters of both Spot and Long Ponds called "pure country water" from an Arcadian retreat made such sources all the more appealing to city people yearning to stay in touch with the best that nature had to offer.[53]

A New and Better Nature

That such a retreat might exist just a few miles away was not enough. Believing that palpable and positive affective contact with the natural was essential to a healthy human spirit, city people took significant steps to reserve a place for nature in their midst, even if they had to fabricate it themselves. They did so by constructing "natural" retreats of various kinds, in which water was often an important presence.

William Penn's original plan for Philadelphia was hardly a city at all, but rather an orderly collection of landed estates. In 1681 Penn instructed the commissioners he sent ahead to America to establish a settlement of 10,000 acres that would be divided into 100 sections of 100 acres each. Each First Purchaser of a 5,000-acre piece of land in the hinterland of the Pennsylvania colony would receive one of these parcels. While a single section might be divided among multiple owners, the resulting lots would still be large. Penn also encouraged settlers to situate their houses in the middle of their property, "so that there may be ground on each side for Gardens or Orchards, or feilds [sic], that it may be a green Country Towne, wch will never be burnt and allways wholsome [sic]."[54] The term "Country Towne" suggested the possibility of combining the qualities connoted by each of the two words.

The impracticalities of this plan soon became apparent to the commissioners and to Penn, who revised it even before he arrived in October of 1682 for a stay of almost two years. Philadelphia would be much smaller—a 1,200-acre rectangle between the Schuylkill and the Delaware—than originally projected.[55] It would also have America's first comprehensive grid. Long straight streets, intersecting at right angles, would form approximately 200 rectangular blocks of nearly uniform size. This layout immediately, unmistakably, and irreversibly imposed order on the natural landscape far more thoroughly than would have the earlier plan. The Cartesian regularity of Philadelphia is the salient visual feature of the plat drawn in 1683 by Penn's surveyor Thomas Holme in his *Portraiture of the City of Philadelphia*.[56] Penn's naming of the streets, however, continued the attempt to balance elements of the natural creation with those devised by humans. He named most of the east–west streets after trees (e.g., Walnut, Chestnut, Filbert, Locust, Spruce, Pine). He used numbers, a human construct, to identify streets that ran north–south.[57]

Penn also specified that there would be an eight-acre square in each of the quadrants, with a fifth square of ten acres in the middle of Philadelphia, where the city's two major thoroughfares, High (Market) and

Broad Streets, intersected (see fig. 2.1). Each of these squares was reserved for a small park. Penn's squares, placed like the dots on the five-side of a die, were locked into the grid, but they provided a variation on its monotony and were meant to afford all residents of the future city access to a green space. Before he returned to England, Penn recommended that a Quaker meeting house and some other structures be built by the central square, and he contributed lumber for the purpose. A wooden building was erected, soon to be replaced by a brick one, but by 1702 this was taken down and rebuilt in the more settled eastern end of the city.[58] It was only when Latrobe located his water temple in Centre Square a century later that Penn's vision of a cultivated reserve in the geographic heart of the city was realized.

"The Center-square has been partially levelled and planted with trees, with the view of affording an agreeable and useful place of recreation to the citizens," the Philadelphia Watering Committee announced in 1801.[59] The landscaping, which cost almost $1,200, was as artificial in its way as Latrobe's engine house and Rush's fountain statue. The trees were a non-native species, Lombardy poplars, favored for their beauty and shade. Two years later the committee reported spending just over $3,400 more, about double the water rents for the year, on additional improvements, including further leveling and more trees.[60] And its members soon added another non-natural creation, the fountain statue of the *Water Nymph and Bittern*. We see the poplars flanking the engine house and filling the upper right of Krimmel's first painting of the Fourth of July in Centre Square. This and other depictions of the engine house and statue, situated by a grove of greenery, conveyed an impression of Philadelphia as a place where the human-made and the natural were in pleasing balance. When the Watering Committee relocated the works to Fairmount, it turned what was already an attractive natural site into a splendidly appointed park. The buildings that Frederick Graff constructed here resembled in their isolation and elegance a neoclassical folly one might find on an eighteenth-century country estate. The city laid sod, constructed gravel walks, and built stairways to enable visitors to ascend the hillside, which it planted with shrubbery and vines.[61]

Sixteen years after he created *Water Nymph and Bittern*, William Rush

Figure 4.1. This print, like that of the street plan of Philadelphia (fig. 2.1), was published by William Birch in 1800, the year before Philadelphia's first waterworks was in operation. It places Benjamin Latrobe's neoclassical engine house in a pastoral setting that seems far from the more settled areas of the city, which were only blocks away. Note the well-dressed mother and her children, who enjoy the groomed countryside around the stately water temple. In a later version, Birch added William Rush's *Water Nymph and Bittern* fountain statue. Courtesy of the Library Company of Philadelphia.

completed two more waterworks sculptures. He carved them from Spanish cedar and, as he had done with the earlier statue, painted them white. They were mounted over the entrances to the wheelhouse (the statues there today are reproductions—the originals are on loan to the Philadelphia Museum of Art from the Commissioners of Fairmount Park). One, titled *Allegory of the Schuylkill River in Its Improved State* (also called *The Schuylkill Chained*), depicts an elderly bearded male figure lying on his left side. He struggles to raise his right hand above his head, which is difficult because his wrists are shackled together. A bald eagle perches by

Figure 4.2A. One of the two statues carved by William Rush in 1825 to adorn the Fair-mount works mill house, this wooden figure represents the Schuylkill in its "improved state," that is, as controlled by humans. A reproduction is on the restored waterworks buildings today, while the original and that of its female counterpart, *The Schuylkill Freed*, are, like the bronze reproduction of Rush's *Water Nymph and Bittern*, on loan to the Phila-delphia Museum of Art. Courtesy of the Commissioners of Fairmount Park.

Figure 4.2B. Titled *Allegory of the Waterworks, or The Schuylkill Freed*, this statue personi-fies the Schuylkill as generously cooperating with the will and needs of man. The water-wheel by the comely figure's raised left knee and the water pipe behind her represent the Fairmount waterworks. Courtesy of the Commissioners of Fairmount Park.

his feet, its wings spread. The sculpture represents the successful subjugation of nature, here gendered male, by human will. In his history of the city, Thompson Westcott read the figure as standing for the river in its "neutralized" condition, "no longer running uncontrolled, but flowing gently from dam to dam, and passing through artificial canals by locks and gates." Westcott interpreted the eagle as untamed nature, "about to abandon the banks of the Schuylkill, in consequence of the busy scene which Art is introducing."[62]

The second sculpture, *Allegory of the Waterworks* (or *The Schuylkill Freed*), personifies the Schuylkill as a beautiful young woman. She sits gracefully erect, her upper body above the water as the lower portion of her gown merges with the coursing river. While the figure is female, she is a symbol of nature's power, if also its beauty and generosity; her upright pose and the positioning of her arms resemble the representation of the manly and muscular Tiber River that since the sixteenth century has been in the Piazza del Campidoglio on Rome's Capitoline Hill. She rests her back against what looks like a water main, next to an upturned pipe that gushes into an overflowing urn. Most telling is that her left hand is gracefully raised, as if in benediction, over a waterwheel of the kind that was lifting Schuylkill water to the Fairmount reservoir and, from there, to thirsty, dirty, industrious, and grateful Philadelphia. Both the title of *The Schuylkill Freed* and the statue's pose strongly suggest that nature is a willing participant in the building of the works. The unruly male Schuylkill may have been subdued, but only in order to serve the high purpose of watering the city.

Philadelphia's water system was the reason for the creation of Fairmount Park. Fairmount is the nation's largest (9,200 acres) urban park, stretching along both banks of the Schuylkill north of the waterworks. It was officially created in 1855, but its formation dates to three decades earlier, when the city began to acquire the private estates above the dam at Fairmount to preclude the possibility that this land might one day be purchased for uses that would pollute the water supply. (The river did become polluted by industries farther upstream.) As the park commissioners explained in richly figurative language, "Trade and human industry had broken in upon the quiet of the rural scene [along the river], and had driven out the descendants of the ancient dwellers at the country seats, some years before the City authorities made the unwelcome discovery that their cup of water was in danger of becoming a poisoned chalice."[63] Philadelphia bought the first twenty-four acres in 1828, the same year that the state legislature also passed a law fixing penalties for anyone who put "any offal or any putrid, noxious or offensive matter, from any dye house, still house, brew house or tan yard or from any manufactory whatever" in the river near the works.[64]

Other cities followed Philadelphia in making the sites of their water-works into places where city dwellers could recreate amidst a reconstructed nature. A descriptive guide to the Cochituate system published the year it opened recommended a visit to the reservoir in Brookline, which its author described as an "artificial lake." "The prospect in the summer season is one of the most agreeable," the guide stated.[65] The thirty-acre body of water, tidily trimmed with a stone embankment and a walking path along its mile-long perimeter, was (and remains) a pleasant suburban setting. Like the members of Philadelphia's Watering Committee, Chicago officials wished to treat their city's waterworks as a civic space where nature was welcome. In 1857 Superintendent Benjamin F. Walker asked the commissioners for a "reasonable amount" to upgrade the site at Chicago Avenue. He proposed to devote this money to grading and fencing, as well as to placing shade trees in the sandy soil near the lake, where few such trees grew of their own natural accord.[66] A decade later the *Chicago Tribune* noted that the replacement of the 1850s system with the tunnel-fed Pumping Station and Water Tower would provide an opportunity to construct a lovely park that would surround the works and greatly improve the area. "Handsome trees and fountains will be planted and located in this park, which, it is earnestly to be hoped, will be 'a thing of beauty' and 'a joy forever,'" the paper said, quoting British Romantic poet John Keats's "Endymion."[67]

Residents and visitors responded positively to this integration of heroic technology and carefully constructed natural charm, especially in Philadelphia. Westcott's history recalled the appeal of Centre Square, with its engine house, landscaping, and water nymph. "The entire affair being unique," he wrote, "[it] was considered a great novelty, and one of the sights of the city which thousands flocked to see."[68] The Fairmount works became an even more popular "natural" attraction. Boston-based *Gleason's Pictorial Drawing-Room Companion* reported in 1851, "A ride to Fairmount, and a walk across the wire bridge [Charles Ellet's suspension bridge, completed in 1841], is the daily recreation of many citizens during the heats of summer, and is at once a convenient and economical pastime, inasmuch as some forty or fifty omnibuses are constantly plying to and fro."[69] Philadelphia educator, philanthropist, and local historian James Hosmer Penniman remembered firsthand the pleasure people took in the ensemble of buildings and landscaping along the river. "Philadelphians were more proud of the water works than of Independence Hall," he wrote. "They said one might as well visit London without viewing Westminster Abbey as come to Philadelphia and not see the water works."[70]

See the waterworks they did. Lafayette, during his much-heralded return to the United States in 1824–25, was one of the earliest of many illustrious visitors to view the improved Fairmount system.[71] Among them were several writers of current and future note who made their way to the spot and praised how effectively Philadelphians had turned their works into the focus of a marvelous attraction. In *Domestic Manners of the Americans* (1832), Frances Trollope's highly disparaging account of her recent travels in the United States, she recalled that she was unimpressed with Philadelphia's general appearance but declared that the Fairmount works deserved a fame equal to the celebrated arrangement of paddlewheels, pumps, aqueducts, and reservoirs that raised water from the Seine for the cascades and fountains at Louis XIV's grand chateaux at Marly and Versailles. "The vast yet simple machinery . . . is open to the public," she wrote, "who resort in such numbers to see it, that several evening stages run from Philadelphia to Fair Mount for their accommodation." Trollope singled out the setting ("one of the prettiest spots the eye can look upon"), the pumps ("enclosed in a simple but very handsome building of freestone"), and the reservoir ("magnificent" and "noble") for a rare effusion of praise.

Trollope commended how well the works combined the natural and the constructed. Behind the waterworks, she explained, was "a lofty wall of solid lime-stone rock," up which the pipes ascended to the reservoir. "From the crevices of this rock the catalpa was every where pushing forth, covered with its beautiful blossom. Beneath one of these trees an artificial opening in the rock gives passage to a stream of water, clear and bright as crystal, which is received in a stone basin of simple workmanship, having a cup for the service of the thirsty traveler." Trollope also took pleasure in Rush's *Water Nymph and Bittern*, which had been moved from Centre Square to Fairmount. She delighted in the way that

> a portion of the water, in its upward way to the reservoir, is permitted to spring forth in a perpetual *jet d'eau*, that returns in a silver shower upon the head of a marble *naiad* of snowy whiteness. The statue is not the work of Phidias [the famed fifth-century B.C. Greek sculptor of Athena and Zeus], but its dark, rocky background, the flowery catalpas which shadow it, and the bright shower through which it shows itself, altogether makes the scene one of singular beauty; add to which, the evening on which I saw it was very sultry, and the contrast of this cool spot to all besides certainly enhanced its attractions; it was impossible not to envy the nymph her eternal shower-bath.[72]

In *American Notes*, originally published in 1842, Charles Dickens was struck, as was his countryman Frederick Marryat, by Philadelphians' bountiful use of water, "which is showered and jerked about and turned on, and poured off everywhere." Of the waterworks themselves, he commented that they were "no less ornamental than useful, being tastefully laid out as a public garden, and kept in the best and neatest order."[73]

Seventeen-year-old Samuel Clemens, who passed through Philadelphia in 1853—a decade before his transformation into Mark Twain—wrote to his brothers Orion and Henry about his visit to Fairmount, which he reached from downtown Philadelphia in a six-penny omnibus. He admired the "fine dam" that "forms quite a nice water-fall," judging the park adjoining the works as "one of the nicest little places about. Fat marble Cupids, in big marble vases, squirted water upward incessantly." Like Trollope, he was charmed by Rush's "pretty white marble [as noted, it was actually wood painted white] Naiad," which he called "the prettiest fountain I have seen lately." Clemens continued, "A nice half-inch jet of water is thrown straight up ten or twelve feet, and descends in a shower, all over the fair water spirit. Fountains also gush out of the rock at her feet, in every direction." He also noted the memorial bust the city had recently placed nearby to honor the longtime water superintendent Frederick Graff (Clemens mistakenly called him Peter Graff), who had died in 1847. Seeing that a door to the engine house was open, the inquisitive Clemens entered and inspected the waterwheels before hiking up to the reservoir, which he called "a respectable-sized lake," from which he enjoyed the "magnificent view of the city."[74]

Figure 4.3. J. T. Bowen's 1838 view of the waterworks and the Schuylkill beyond. The presence of the well-dressed visitors illustrates how the Fairmount works became an attraction for both Philadelphians and tourists, a place where they could commune with a nature extensively reworked by humans. While not very distinct, the two Rush statues are visible on top of the mill-house entrances. Courtesy of the Library Company of Philadelphia.

One of the dozens of visual depictions of Fairmount is Philadelphia lithographer J. T. Bowen's prospect of the Schuylkill from the hill rising behind the works. A few well-dressed Philadelphians tour the buildings that housed the waterwheels and the office of the Watering Committee, while others mount a stairway up the bosky hill to an airy gazebo that provides a resting place and lookout. Those who have climbed farther turn to reflect upon the glorious scene. Rowing shells and sailboats glide along the placid water behind the dam, and lining the river are the riparian estates that became part of Fairmount Park. The afternoon sun, emerging from behind one of several dramatic clouds, radiates beams of light. A poem by Philadelphia journalist Andrew McMakin that accompanied a very similar view of the Fairmount works stated that in this place "art and nature with each other vie," creating a splendid harmony.[75]

Chicago's waterworks on Chicago Avenue was not surrounded by nearly as large or as gracefully fashioned a park as the Fairmount grounds, but it still was nicely landscaped.[76] It was the lake tunnel, however, that proved to be "an object of great fascination to the people of Chicago and to visitors from other cities." A voyage by tug to the crib at the tunnel's eastern end was described as "an indispensability," comparable to "the study of music among young ladies, or a journey to Mecca among the Moslems." Visitors "whose good opinions are worth cultivating" (i.e., journalists, business leaders, and dignitaries from other cities whom Chicagoans were always so eager to impress) were "of course expected to see the lion, or the leviathan of the deep."

General Ulysses S. Grant and his staff, fresh from victory in the Civil War, toured the tunnel in July 1865, when the "lion" was still under construction. So did "some of the prominent men of Boston," who were on a "tour of investigation from the 'hub,' to what was but a few years ago, on the rim of the universe of civilization." The Bostonians were accompanied by a "goodly company of Aldermen and other officials of our own city," wrote a reporter, "and a few men prominent in other than the governing walks of life," keen to show off this achievement that attested to the rise of Chicago to its self-proclaimed status as America's great inland metropolis. "When one is enabled," as these distinguished tourists were, "to witness the gigantic preparations, the mighty forces employed, the huge machinery brought to bear, for the effecting of the desired ends" of the collective enterprise of Chicago, "a due appreciation of its importance is felt, more than in any other way could be gained."[77]

Waterworks were not the only place in the cityscape that featured artificial constructions intended to retain or restore the presence of nature in urban life. So-called public gardens, which were in fact privately owned and charged admission, were sophisticated urban oases where one could take refreshment amidst arrangements of plantings.[78] Related initiatives, like the incorporation of the Boston Public Garden in 1838 by a group of wealthy benefactors, combined qualities of both the public gardens (including paid admission) and of the horticultural societies being organized in American cities.[79] Located on a muddy stretch of land between the Common and the yet-unfilled Back Bay, the Public Garden was the site in its early years of a building that had been turned into a conservatory housing tropical plants and exotic birds, a very different nature from the one native to the region.[80]

The creation of large and truly public parks in the middle decades of the nineteenth century, established by the community and open to all, was explicitly intended to protect against urbanization's further destruction of nature and to compensate for the damage done already. In this same period, so-called garden cemeteries not only provided places to bury the dead but also promised a consolatory escape, or at least the illusion of escape, from the quotidian world, which was literally walled out. These burial grounds, sometimes called rural cemeteries, were located beyond city limits, as local health ordinances increasingly prescribed. The first was Mount Auburn on the western edge of Cambridge, Massachusetts. It was established in 1831, to be followed five years later by Philadelphia's Laurel Hill, where Frederick Graff is buried. In 1860, when Graceland Cemetery officially opened two miles beyond Chicago's northern border, then at Fullerton Avenue, residents filled the special Chicago and Milwaukee Railway excursion trains that conveyed people to the event.[81] In these serene sanctuaries, city dwellers could (and still can) meditate on nature, including the great natural fact of mortality, which no artifice can defeat.[82]

All these "natural" settings were sure to include water as an essential design component, in the form of pools, ponds, and rivulets. Their planners resembled the artist that Melville's Ishmael describes, who desired to paint "the dreamiest, shadiest, quietest, most enchanting bit of romantic landscape in all the valley of the Saco," only to realize that no matter what the charm of the other elements he might include, "all were vain," unless the eye of the human figure in the painting "were fixed upon the magic stream before him."[83] The Frog Pond, the focus of the 1848 water celebration, was originally a naturally occurring body of water, but the first Mayor Quincy's administration placed a decidedly artificial stone curb around it as part of the continuing process of turning the Common, once mainly used as a shared grazing area, into an elegant urban park.[84] One of the things that the Chicago Board of Public Works did in the early years of Lincoln Park, which dates to the mid-1860s and consists substantially of filled land, was to fashion the lake that is now called the South Pond.[85]

The Fountain and the City

Of all the artificial aquatic features that cities put in place, the one with perhaps the strongest imaginative appeal was the ornamental fountain. Such fountains were, of course, nothing new, nor were they limited to cities or public sites. They were showplaces for kings and popes, as well as highlights on the grounds of the British landed estates that set a standard for American taste. William Wrighte's *Grotesque Architecture; or, Rural Amusement*—which was first published in London in 1767 and went through many reprintings over the next several decades—offered "plans, elevations, and sections" that explained how a person with sufficient means could construct "cascades" and "baths," as well as greenhouses, summer and winter "hermitages," "mosques," and "Moresque" pavilions, and both "grotesque" and "rustic" settings on his property. Wrighte specified about one of his "rural grottoes" that it "should be built of large rough stones rudely put together, so that the building may as near as possible imitate the beautiful appearance of nature." This, however, should not discourage one from including a statue of "Neptune on a rock, pouring out water, which descends under the pavement through

an arch, and forms a running stream," flanked "by side niches . . . ornamented with satyrs and other grotesque figures." To make this fantasia
on nature come to life, one element was, of course, indispensable. "The
situation," Wrighte pointed out, needed to be "near some water."[86]

The fountains that Wrighte described were intended for rural seats,
but they inspired smaller and simpler models favored by some prosperous city dwellers, especially once running water was available. Indeed,
backers of urban waterworks repeatedly asserted that these systems
would popularize fountains by making them far simpler and less costly
to install. Benjamin Latrobe envisioned Philadelphia's pipes as being
equipped with special plugs or cocks that, when opened, "will throw up
fountains playing to a hight [sic] proportionate to the elevation of the
reservoir." He predicted that, if built, his works would be able to drive as
many as twelve small fountains on a single street, and that their evaporating mist would clear and cool the air on stifling summer days.[87] In
1839 the Boston City Council's Joint Standing Committee on Water
pointed out that a new waterworks would make it much easier to bring
fountains to the city. The committee commented that "the convenience,
comfort and ornament of fountains have, in all ages and countries[,]
been felt and acknowledged, and that some amount of expense would be
justifiable, in a wealthy community, for these objects alone."[88]

As early as 1802, Philadelphian John Dunlap put in a fountain by
his home at Twelfth and High Streets, while wealthy merchant and
real-estate developer William Sansom did likewise at his residence on
Mulberry Street.[89] The Boston *Daily Evening Transcript* of May 26, 1849,
devoted a space two columns wide and over eight inches high to an engraving of what it called the "First Fountain that has been erected in this
city" following the inauguration of the Cochituate system. Its owner was
engineer R. H. Eddy, who in 1836 had prepared the third in Boston's long
series of water reports (Eddy had recommended that the city draw its
supply from Spot Pond). Eddy's was actually the earliest *private* fountain
fed by Cochituate water, since the first was the great jet in Boston's Frog
Pond. As the paper explained, Eddy intended his fountain, like the one
on the Common, to be enjoyed by all Bostonians. Situated in a 1,500-

square-foot lot of its own that was "tastefully laid out and planted with trees and shrubbery," it was "devoted to its use for the gratification of the public."

The Eddy Fountain, as it was called, was created abroad from the finest Italian marble at a cost the *Daily Evening Transcript* reported as over $1,000. It followed a classic design: a square column that rose from a basin supported three successively smaller *tazze*—wide and shallow bowls—the lowest and largest of which was five feet in diameter. On each of the four sides of the column below this *tazza* was a lion's head in high relief; in the space between it and the middle *tazza* were three dolphins, which, like the lions, spewed water from their mouths. Crowning the column, above the highest and smallest *tazza*, was a pineapple that emitted a five-foot spray. As the water descended, it successively filled and overflowed the rims of each of the three *tazze*, and then cascaded into the large basin in which the column was rooted. The *Transcript* recommended viewing the Eddy Fountain between noon and two in the afternoon, since that was when the sunlight sparkled most pleasingly on the moving water.[90]

American manufacturers took advantage of the market for urban fountains that arrived with city water. Mid-century New York hydraulic engineer G. B. Farnam published a series of editions of his *Description of Farnam's Patent Hydraulic Apparatus, for Raising Water*, which included all sorts of "practical information" on "supplying factories, towns, cities, &c. with water." He displayed an illustration of an ornate cast-iron creation that had won a gold medal at the Fair of the American Institute, a popular exposition held in New York beginning in the late 1820s.[91] Farnam designed his prizewinning fountain for a rural estate, but he furnished advice on how a city dweller might construct a more modest and far less expensive fixture from a vase, a stand, and hydraulic cement. Those who were willing to pay a little more might order one of several models that came fully assembled. For twelve dollars, the customer could purchase a fountain bearing relief portraits of Washington or Franklin. Twenty dollars would bring home a thirty-nine-inch tall figure of Hebe, goddess of youth and Ganymede's predecessor as cupbearer to the gods, from whose head emerged an adjustable spray. Those who wanted a fountain that alluded more directly to the natural world could choose ones with spouting swans and dripping lilies.

No matter what form they took, fountains used a lot of water, raising the question both of how much their owners should be charged and whether their hours should be restricted. In Philadelphia, both Dunlap and Sansom paid a considerable premium for their fountains—the annual water charge for each was twenty-four dollars, almost five times that paid by most home owners for domestic service.[92] By 1860, when private fountains were more numerous, the fee depended on the diameter of the jet employed.[93] Within a year after the Cochituate system opened, the water board received several applications for water from fountain owners. Since these were meant for "the gratification of the taste" and were not "objects of necessity," the waste-wary commissioners wanted to make sure that they did not divert water needed for "indispensable uses."

But if there was a surplus, the commissioners said, they were in favor of supplying private fountains, "as a means of promoting the public health and enjoyment, as well as of embellishing the public and private grounds of the City." With that in mind, they set a favorable wholesale rate.[94] R. H. Eddy asked that the water rent for his fountain be waived entirely, as did the proprietors of Beacon Hill's Louisburg Square for a fountain they placed in that posh private park. Both argued that their fountains served the public good since they benefited the entire neighborhood and passersby. The commissioners rejected these requests.[95] In 1856 the Chicago water commissioners charged between ten and sixty dollars a year for the eight private fountains on their water-assessment rolls.[96] Water administrators in all three cities agreed that there would be no water bill for fountains like the one in the Frog Pond, which were "exclusively of a public character, for the gratification of citizens generally, and for the embellishment of places of public resort," though their hours of operation could be "subjected to fixed limitation."[97]

It is possible to read these civic fountains, starting with William Rush's water nymph, as assertions of mastery over water and nature by city people, but to do so is to ignore the varied way in which such complex creations spoke to the relationship between the natural and the human-made in the urban built environment. As was the case in Centre Square, while these fountains celebrated human intelligence, skill, and power, they did not express the community's wish to dominate nature as much as its desire to ensure the flow of natural resources essential to the public good. Rush's maiden represented all that was pure, good, nurturing, and generous about the spirit of Philadelphia. This is not to say that the idea of controlling nature was absent in Centre Square. Ruling the scene, after all, was the implacable (if unreliable) steam engine that pumped Schuylkill water through the bittern's beak and into the air. But Latrobe's water temple that hid the steam engine from sight indicated that the city's goal was less to conquer nature than to civilize it, to Philadelphia's enduring advantage.

When trying to convince Bostonians to approve a public water system, Walter Channing described how the public fountains that this system would help make possible would transform the city into a more humane community. "Construct fountains in public places," Channing urged. "Give to the public eye these beautiful objects, and you give tone and purity to the public heart. You increase the amount of safe pleasure; and he or they who do this are benefactors to men." Channing denied that such ideas were fanciful, declaring, "They have their foundation in the moral nature; and that nature, in these days of morbid excitement, of excessive interest 'in those things which do not profit,' that nature asks for food by which it may grow into that excellency, beauty, and honor, for which it was made."[98] Channing claimed that just witnessing the play of water in a fountain could revitalize the souls of city people, whose character was at constant risk of being cheapened by day-to-day matters, and elevate the tone of city life.

Mayor Josiah Quincy Jr. shared Channing's regard for fountains as an essential part of a program of civic enhancement by reintroducing nature to the cityscape. Quincy told the City Council in January 1847 that one of the many benefits of the waterworks then under construction was that it "will give us an opportunity to embellish our squares and public grounds with fountains, those most beautiful emblems of health and purity." Once water was flowing through the new system, it would be "in our power, at a very moderate expense, to enable our fellow-citizens to come from the crowded and dusty streets in summer, to gardens beautified with flowers, shaded with trees, and sparkling with fountains."[99] This vision appeared to become fact during the water celebration of October 25, 1848, when the first jet of Cochituate water rose high above the Frog Pond, which was located in the heart of Boston's most historic and revered patch of nature, the Common.

As had happened with Rush's statue in Philadelphia's Centre Square and the works at Fairmount, and would happen with Chicago's Water Tower, the fountain in Boston Common became an arresting landmark and a symbol of its city as a flourishing urban community. "The fountain while playing is distinctly visible, as a striking object, from the Boston and Worcester Rail Road, as far off as Brookline, and the easterly part of Brighton." the *Daily Advertiser* reported in the heady days following the introduction of Cochituate water.[100] Three years later *Gleason's Pictorial Drawing-Room Companion* featured a large drawing of several dozen well-dressed citizens by the Frog Pond. Some engage in polite conversation, while children roll hoops on the grass nearby. One gentleman points at the fountain with his cane as he speaks with his female companion. Others simply gaze at it in admiration.[101] A similar image appeared on the cover of the sheet music for George Schnapp's *Cochituate Grand Quick Step*.[102]

According to *Gleason's*, "This superb ornament to the city, and the Common especially, is a rare beauty when in full play," making the park's walkways "unequalled, as a public promenade," by any other place in the country. The fountain jet formed "a pure and sparkling column, unrivalled by anything of the kind we have ever chanced to meet. It is seldom that its twilight performance is not witnessed by a large and delighted audience of citizens, rich and poor, gentle and lowly, old and young."[103] Swedish social reformer and author Fredrika Bremer was one of many visitors to the city who also found it delightful. She particularly admired the view of it from the steps of the State House. "Below lies the extensive green called 'Boston Common,'" she explained, "in the middle of which is also a beautiful fountain, which throws up its water to a great height."[104]

FOUNTAIN ON BOSTON COMMON.

Figure 4.4. This illustration appeared in the July 26, 1851, issue of the Boston-based *Gleason's Pictorial Drawing-Room Companion*. It is typical of depictions of urban water-works installations as genteel retreats in the growing city. Courtesy of the Northwestern University Library.

The Frog Pond fountain, like Rush's two Fairmount statues of 1825, expressed mutually enclosing aspirations. The powerful jet thrust into the air and then cascaded earthward, at once defying and obeying the law of gravity. It proclaimed both the power and glory of natural forces and the desire to shape those forces into a form determined by human intention and skill, to give structure to the fluidity of life. This fountain also conveyed the hope that city water and city life would never be entirely denatured and desanctified of their transcendent qualities. As much as city people wanted to turn nature's water into their servant, they also wished to exalt its untamable spirit and irreproducible purity. The fountain stood for the waterworks that made it possible, which in turn represented the city as a center of authority, but also as an urban community where not just the individual and the collective but also the natural and the built environment were in benevolent balance. "We rejoice with unmixed joy that we have made such wise use of intellectual power as to have accomplished so noble a work," the *Daily Advertiser* said of the Frog Pond fountain. "We rejoice that the moral nature has declared itself in so munificent a work, which can only be for good." This work was a constant source of civic pride and imaginative delight that enraptured those who witnessed it: "And there it is on the Common, springing as on fairy foot into the free air, as free itself," the *Advertiser* marveled, "—yes, dancing in the sunbeam as if twin born with the light. How sweet! How good!"[105]

The Boston water celebration commemorated the exhilarating reconciliation of the apparently opposing forces of art and nature. The water poised in the pipes below the city was a full hundred feet lower than at Lake Cochituate and eager to display its natural potential energy by shooting out of the fountain and into the air. "What a power was at that moment by our side," the paper gushed, "laboring to declare itself,—Yes, to burst away from the narrow limits in which man had compressed it for our daily easy, and safe use, and to spread confusion, dismay, danger all around." But all the undisciplined willfulness inherent in water was now to serve and delight city people without compromising its own power and freedom. The jet "sprang forth, and on the lofty level of its own far off height declared its beauty and its power, a chartered libertine, in the beautiful autumn evening sky."[106]

The resolution of the opposing forces of people and nature did not end there. Less than two weeks later, God and nature again seemed to sanctify the work of humans, as the dawn sunlight on the fountain spray created "as beautiful a rainbow as was ever seen in the heavens." The column of water was "as white and pure as the driven snow."[107] The city, like the fountain, could embody in healthy harmony the best aspects of dominion and deference, stability and freedom, utility and beauty, stasis and flow, physicality and spirituality, rationality and wonder.[108]

Twenty years after the water celebration, Boston waterworks official and historian Nathaniel J. Bradlee listed the thirteen different kinds of sprays that the Frog Pond fountain could create, from a six-inch column that reached almost a hundred feet in height to one that took the form of a small vase, depending on which fitting was employed. He also listed the twenty-one public fountains that were currently using Cochituate water. Perhaps the most intriguing of these was also the first work of statuary to be placed in Boston's Public Garden, which in 1859 became a free and open public park whose re-landscaping included the graceful lagoon that is still there today.

The thirty-foot Ether Monument, unveiled in 1868 (and rededicated in 2006 after a complete restoration), was a collaboration between the architectural firm of Ware and Van Brunt and sculptor John Quincy Adams Ward.[109] Ward's statue of St. Luke's Good Samaritan comforting a stricken man sits atop four marble columns, which rise from an ornate granite base that is enclosed by a square basin filled by water flowing from the mouths of lions' heads on each of the four sides of the base. As Bradlee explained, the monument was "erected as an expression of gratitude for the relief of human suffering by the discovery of the anæsthetic properties of ether," whose earliest practical use in a medical procedure occurred in 1846 at nearby Massachusetts General Hospital.[110] One of the several inscriptions on the monument is from Revelations 21:4, "Neither shall there be any more pain."

Dr. Walter Channing had been one of the first physicians to employ ether in an operating room. The week before the water ceremonies on the Common in 1848, the Boston *Daily Evening Transcript* carried a favorable notice of his 400-page *A Treatise on Etherization in Childbirth, Illustrated by Five Hundred and Eighty-One Cases.*[111] The play of natural water in the Ether Monument symbolized the triumph over the suffering that God had meted out to womankind ("in sorrow thou shalt bring forth children," Genesis 3:15) for Adam and Eve's disobedience. The control of pain, like the control of water, asserted that humans had reversed a major consequence of their eviction from Eden, here in the same city whose founders so steadfastly believed in original sin.

5

THE URBAN BODY
AND THE BODY OF THE CITY

*The Sanitary Movement, the Temperance Crusade,
and the Water Cure*

"We read of the fabled ages of gold, silver and iron. Ours is the age of *water*," Boston physician W. M. Cornell announced in the *Boston Medical and Surgical Journal* of May 1845. "Cold, pure, soft water," Cornell continued, "the universal solvent, the best drink for man and beast, the great promoter of cleanliness, health and happiness."[1] He was here referring to three immensely popular contemporary cultural phenomena, to each of which water was central: the sanitary movement, the temperance crusade, and the water cure. Sanitary reformers claimed that water was the best deterrent against disease-breeding filth, temperance leaders hailed it as the salutary alternative to demon alcohol, and water-cure practitioners declared that it could remedy almost any ailment.

All of these groups viewed urban life as prone to ills and evils that water would effectively combat; cities were, they argued, dirty, intemperate, and enervating places full of dirty, intemperate, and enervated people. Whether they believed that city living brought down otherwise clean, sober, and hearty residents or that the inherently flawed individuals who gathered in cities threatened the physical and moral well-being of the rest of the population, sanitarians, temperance champions, and water-cure advocates shared the conviction that the physical well-being of city life as a whole and that of the separate lives within it were interdependent. Their discussion of this connection merged the relationship between the individual and the collective and of the natural and the human-made to create a third pair of categories, the (individual natural) human body and the (collective human-made) body of the city.

The City Embodied, the Body Citified

The comparison of a city to the human body is a venerable idea. Both cities and bodies are readily understood as complex organisms consisting of analogous parts—parks, for example, are commonly called the city's lungs, roads its arteries.[2] A city's (or a society's) people are often understood as constituting its body.[3] We are also fond of attributing personality traits to cities—for example, staid Philadelphia, proper Boston, and "stormy, husky, brawling" Chicago—and assigning them gender.[4] The city-as-body metaphor gained additional power during the unprecedented urbanization of the nineteenth century, often in connection with the construction of new water systems. Philadelphia journalist and poet Andrew McMakin wrote of the works that his city built along the Schuylkill, "There matchless FAIR MOUNT rolls its waters down / Through iron veins, beneath the thirsting town," personifying both the waterworks and the city while likening water pipes to blood vessels.[5]

The most salient discussions of the relationship between the actual bodies of urban residents and the metaphorical body of the city went well beyond such relatively simple analogies, however, in describing the two kinds of bodies as mutually and continuously creating one another. As individuals crowded into growing cities and transformed a natural setting into a built one, they were themselves changed by the circumstances of the urban life they were fashioning. They worked at city occupations, interacted through city institutions, and conducted themselves according to city customs and practices—including drinking city water. In the process, they became city people.[6]

The interconnection between the collective urban social body and the condition of those who constituted it underlay an uneasy sense among some reflective observers that both might be changing for the worse. In a well-known passage from *Notes on the State of Virginia*, Thomas Jefferson employed a body metaphor to argue that urbanization would undermine the United States, which in his opinion owed its integrity and stability to the fact that it was an agrarian nation. Jefferson warned that "the mobs of great cities" would have the same malignant effect on the political and social well-being of democratic America "as sores do to the strength of the human body."[7]

Even some who otherwise looked favorably on nineteenth-century city life were alarmed by what they saw as its pathologies. They worried that cities drew undesirable individuals in disproportionate numbers and that urban existence transformed some people of naturally sound constitutions and upright spirit into sick people and bad citizens. The concern went beyond moments of crisis, as when yellow fever or cholera struck their city with special virulence, bringing death and spreading terror, or when a riot broke out in the streets, destroying property, inflicting injury, and deepening class and ethnic divisions.[8] They wondered whether even the normal circumstances of city life were both the cause and result of precarious health conditions and anti-social behavior.

Sanitarians, temperance leaders, and water-cure advocates often expressed such ideas in terms of water, whether literally, as when they identified it as a bearer of pestilence and pollution that permeated everywhere in the urban body, or, more abstractly, when they used figurative language to convey their sense of how overwhelming, unpredictable, and disturbing were what they saw as objectionable social developments, notably immigration. At the same time, however, they looked to clean water as the salvation of the imperiled individual and collective urban body. In modes that ranged from reasonable and scientific to frantic and wishful, they placed enormous faith in the ability of water to mitigate the hazards of city life by exerting control over the bodies of both the city and all of its gathered inhabitants. And if that proved impossible, water offered ways in which they could defend their own bodies from the threats they saw around them.

Bathing the Urban Body

The rise of the sanitary movement in Europe and the United States during the early decades of the nineteenth century was directly tied to large-scale urbanization. The sanitarians were physicians and social reformers who believed that removing filth was the best strategy for fighting disease, and they focused much of their attention on urban life because they believed that the body of the city was especially prone to illness.[9] They promoted the new idea of public health, with its emphasis on fighting disease and improving the quality of life across an entire community through organized prevention. Filth was the fell enemy, and "cleanliness" was the much-repeated battle cry.

In 1834 Dr. Charles Caldwell won Harvard's Boylston Medical Committee prize for his essay on plague and yellow fever, in which he argued that the best health policy lay "in CLEANLINESS."[10] By mid-century this was common wisdom, if not yet common practice. A *Chicago Tribune* reporter on an inspection tour of the city in 1855 was shocked to find alleys "entirely blocked up by mountains of manure, filth and garbage," gutters "full to overflowing of stagnant and green water," and outhouses "running over with filth of the most pestiferous description." He wrote his prescription in italics rather than Caldwell's capitals. "*Let us cleanse the city!* that is our first duty. Let us do it with all our might."[11]

The experts agreed that a clean and healthy urban body above all required lots of good water.[12] The backers of waterworks repeated the logical point that a plentiful supply that reached everywhere would wash away unsightly and dangerous urban grime. On Christmas Eve, 1804, with the late summer and early autumn yellow-fever season now safely past, the Philadelphia Board of Health spoke proudly of "the unusual exemption which [Philadelphia] enjoyed, from those diseases that are produced by air contaminated with exhalations from putrid substances," in spite of the fact that during the preceding warmer months "such diseases were uncommonly violent, prevalent and mortal, in different parts of the adjacent country." While the board believed that this "exemption" was due partly to the timely application of quicklime on "multifarious nuisances in every quarter," it also credited the new "Schuylkill pipe water" with being "indispensable for the effectual cleansing of the streets and gutters."[13]

Shortly after it was organized in 1847, the American Medical Association appointed a Committee on Public Hygiene, which issued its findings two years later, declaring emphatically that "much disease incident to poverty may be relieved by copious and never-failing supplies of pure water." The improved health conditions enjoyed by Philadelphia and New York after they constructed their waterworks convinced committee members that "the introduction of an abundant supply of water is so intimately connected with the health of a city, that the municipal authorities should rank this among the most important of their public duties."[14] As cholera threatened Boston in 1849, the city's Committee on Internal Health later reported, local authorities instructed residents "to cleanse their house drains with Cochituate water." In addition, "Many yards, lanes, and by-places, in different places, were also daily drenched" from the same source.[15] The following year, the 1850 *Report of the Sanitary Commission of Massachusetts*—one of the foundational documents in the history of public health in the United States—also endorsed washing the city, using vivid imagery to impress upon readers the dangers of dirt. It described cholera as a predatory beast that "bounds over habitation after habitation where cleanliness abides . . . whilst it alights near some congenial abode of filth or impurity."[16] This study, whose primary author was Lemuel Shattuck, was modeled on Edwin Chadwick's pathbreaking *Report on the Sanitary Condition of the Labouring Population of Great Britain*, published eight years earlier.

Caldwell, Shattuck, and the many others who sounded the call for sanitation by clean water believed in the miasmist theory of disease (also known as the climatist, filth, atmospheric, or anti-contagionist theory), which located the source of illness in the vaporous "miasmas" or "exhalations" emitted by damp and decaying organic matter and, some thought, living bodies.[17] While miasmas might occur naturally, as in marshlands and swamps, most public health leaders considered cities prolific generators of deadly emissions because they contained polluted "bodies" of water (e.g., Philadelphia's infamous Dock Creek, Boston's Back Bay before it was filled, and the blighted Chicago River), piles of garbage, and accumulations of human and animal excrement, all in close proximity to a concentrated population.[18]

While there certainly is a connection between filth and disease, the miasmist theory was arguably as much or more a moral and aesthetic concept as a scientific one. "In cities as well as among individuals, cleanliness has reference to morals as well as to comfort," Mayor Josiah Quincy Sr. stated in arguing for washing Boston's streets. "Sense of dignity and self-respect are essentially connected with purity; physical and moral," he added, personifying the urban built environment as a living being by stating, "A city is as much elevated as an individual by self-respect." Quincy again personified the city as a moral being when, upon leaving office in 1829, he advised, "In the whole sphere of municipal duties, there are none more important than those, which are related to the removal of those substances, whose exhalations injuriously affect the air. A pure atmosphere is to a city, what a good conscience is to an individual—a perpetual source of comfort, tranquillity, and self-respect."[19] In his opinion, building a water supply that cleansed the urban body of the sources of miasma would at the same time uplift its character.

Since the well-being of the urban body and of the individual body were so intertwined, it was just as important to wash the city dweller as the city. Cleansing the individual body would keep it from becoming vulnerable to disease and hazardous, not to mention offensive, to the rest of the social body. America was not a bathing culture, however. Well into the nineteenth century, regular washing of more than one's face and hands was uncommon, even among many urban residents who considered themselves refined, and there were few places where one could take a full body bath or shower.[20] In the "postscript" to his original 1798 Philadelphia waterworks proposal, Benjamin Latrobe puzzled over the fact that governments in the United States, unlike those in "despotic countries," did not provide "all ranks of men" with the "convenience" and "wholesome pleasantness" of bathing, and that "the American people do not indulge themselves, in the smallest gratification, as salubrious as it is innocent, of this kind." Latrobe commended Americans' industriousness and "our attention to more serious pursuits," but he found resistance to bathing "blameable," especially in climates like that of Philadelphia, whose steamy summers made the practice "almost an absolutely neces-sary means of health." He also thought that publicly owned bathhouses might be "a source of a large revenue" to the municipal budget, as well as a civic amenity that would draw people to the city.[21]

Hoping to shake the national indifference to personal hygiene, many of the antebellum period's myriad popular advisers on health and man-ners preached the gospel of bathing. Writing in the mid-1830s in his monthly magazine, *Moral Reformer and Teacher on the Human Constitu-tion*, Dr. William Alcott (who was related to reformer Amos Bronson Al-cott, father of author Louisa May Alcott) berated the nation's residents for ignoring the importance of washing their bodies regularly in clean water. "And as to bringing our bodies in contact with this cleansing liq-uid in any considerable masses," he wrote, "we, in general, only touch it once or twice a day with the tips of our fingers, and perhaps apply it grudgingly to the ends of our noses, (especially if warmed,) and a few square inches of surface in their immediate neighborhood." Alcott re-jected out of hand "the strange [though hardly rare] belief" that " 'dirt is healthy' " and that "an unclean skin" possesses a "salutary tendency." Like Josiah Quincy Sr., he blurred the line between sacred and secular purity with his admonition that, in refusing to wash fully and regularly, his fellow citizens were "not only ungrateful to Heaven, but traitorous to ourselves," even "suicidal."[22]

Books and magazines written by and aimed at middle-class women joined the cause. In the 1840 *Lady's Annual Register, and Housewife's Al-manac* (which, like Alcott's *Moral Reformer*, was published in Boston), Caroline Gilman, who wrote popular poetry and fiction as well as house-hold advice, dismissed the notion that bathing was simply too much trouble.[23] Among the illustrations in the August 1849 issue of *Godey's Lady's Book* was an engraving titled "The Bath of Beauty." It portrayed a voluptuous young woman of privilege from an unspecified Islamic cul-ture—"some favored child, or still more favorite sultana"—attended at her bath by "her dusky maid in waiting." While this image reflected the period's fascination with the Orient (and echoed Latrobe's association of bathing with "despotic countries"), *Godey's* used it to comment unfavor-ably on American personal habits. The magazine observed that "the time is coming when the daily or semi-daily bath will be considered, even in this country, conducing, as it does, not only to personal neatness, but to absolute health."[24]

The spread of bathing was both a cause and an outcome of the construction of waterworks. In 1839 the Boston City Council's Joint Standing Committee on Water predicted that a comprehensive water-supply system "would favor the use of the bath, and in this way would promote health, and the virtue of personal cleanliness, more than will readily be conceived by those unaccustomed to this simple, yet inestimable luxury."[25] The committee was correct. Once waterworks were built, bathtubs became more common, if only in a portion of the small minority of households that could afford them. Inventors and manufacturers produced and marketed a broader selection of sinks and tubs, as well as urinals and water closets.[26] Water company records indicate that better hotels deemed it important to their business to provide such "modern conveniences" for their guests.[27] These same records reveal the growing if still limited presence of privately owned bathhouses open for a fee to members of the public.

One public bathhouse that began to do business in 1801 near the center of settled Philadelphia, at Third Street just north of Arch, was greeted as "a valuable improvement to the city." Its appointments included hot and cold showers, a hot air and vapor "apparatus," and "special accommodations for ladies." When its owner moved to larger and more elegant quarters in 1806, he advertised such features as individual window vents and white marble tubs imported from Italy. He assured customers of the healthiness of the premises by inserting the line: "No sick persons admitted."[28] Swaim's Philadelphia Baths, a three-story structure with separate entrances for men and women, opened at Seventh and Sansom in 1828. It housed forty-four baths and showers, plus a swimming pool for children. By mid-century there were five bathhouses within the city limits.[29] Even before Bostonians built a central waterworks, they also had their choice of several baths that served the paying public.

To sanitarians, this kind of public bath did nothing to cleanse the most problematically dirty urban bodies. They were much more concerned about the working poor, who lacked the means to go to fashionable

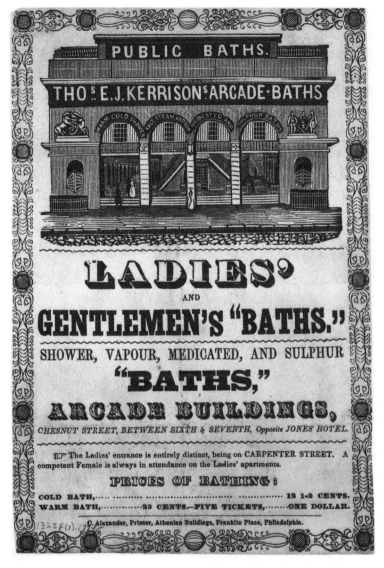

Figure 5.1. Thomas E. J. Kerrison's bathhouse occupied this Philadelphia building from 1845 to 1849. The building was originally constructed in the mid-1820s as a shop gallery. Located at 615–619 Chestnut Street, it was very near Independence Hall. Like other "public" baths at the time, Kerrison's Arcade Baths served only "ladies" and "gentlemen" who could afford it. As indicated, a cold bath cost 12½ cents, a warm one twice as much. Note the variety of baths available. The fine print advises that the ladies' entrance is "entirely distinct" from the men's, and that "a competent Female is always in attendance on the Ladies' apartments." Courtesy of the Library Company of Philadelphia.

establishments like Swaim's. Since a large number of the poor were immigrants, reformers often lumped these two categories together, frequently adding "Catholic" to the mix. While most health reformers agreed that this portion of the urban population needed cleansing, the reasons they gave reflected a range of political and social attitudes. Some took the point of view that the poor were dirty by personal habit and choice, which made them an especially intractable threat to the urban body's health and morality. The growing discussion of the importance of cleansing the foreign-born paralleled the spectacular mid-century rise in immigration from abroad.[30] The common charge against immigrants was that they were the kind of people who were likely to become, in Jefferson's language, sores on the urban body politic. A full decade before the potato blight brought Irish immigrants to Boston in large numbers, Boston mayor Theodore Lyman stated that the "course of life" of too many foreign-born residents was "the fatal and teeming source of epidemic or malignant diseases" that would undermine the body and spirit of the rest of the town.[31] Lemuel Shattuck's and Jesse Chickering's statistical surveys indicated that by the 1840s Irish immigrants were overrepresented among Boston's sick and insane, who, since many of them were cared for at public expense, imposed a hefty financial burden as well as a serious health hazard on the community.[32]

Anti-immigrant commentators used a water-related figure of speech that became commonplace also in xenophobic discourse of the late nineteenth and early twentieth century. The term "flood of immigrants" (and similar language) took on a life of its own, revealing and reinforcing a view of a major social development as a natural catastrophe that threatened to overwhelm the urban body. The most zealous nativists charged that foreign governments were taking advantage of America's receptivity and generosity to newcomers by inundating the nation's cities with their own undesirables as a matter of policy. In an 1844 speech on pauperism, R. C. Waterston, the minister who would bless the groundbreaking at Lake Cochituate two years later, condemned Europe for "pouring in among us the poorest of her population."[33] Lydia Sigourney, one of the most successful American authors of the 1830s and 1840s, began her story "The Harwoods" (which appeared in an 1848 collection titled

Water-Drops) by stating, "The flood of emigration which beats against the shores of the United States, seems to have no ebb-tide."[34] Shattuck joined the many who employed this kind of imagery. "Some poor-houses have been emptied, and their inmates have been transported to America,—to Massachusetts!" he exclaimed in the 1850 Sanitary Commission report. This "stream of immigration" (Shattuck's term) was becoming fuller each year, so that "Massachusetts seems to have resolved itself into a vast public charitable association."[35]

The poor and foreign-born, especially impoverished Irish Catholics, became objects of fear and anger in Boston because they seemed to threaten a supposedly stable and settled order. On October 9, 1848, a little more than two weeks before the Cochituate works opened, the *Daily Evening Transcript* quoted a recent article from another publication that chided the city for a recent increase in crime. "Here is old Boston," it read, "the renowned Puritan city—once so orderly and respectable, with its venerable board of selectmen, overseers of the poor, fire-ward, &c; O, how fallen! How degraded!" Using a synonym for "flood," the *Transcript* attributed the unfortunate situation to "the influx of foreign crime, ignorance and destitution," citing records that showed "conclusively" that almost 90 percent of "ruffianly outrage, intoxication and riot occur among those, who received their training far away out of the sphere of New England influences."[36] These foreigners corrupted the proud character of the civic body of Boston, which Shattuck identified with the essential fluid that circulated through the bodies of high-principled New Englanders and that throbbed with alarm today through the mortal being of their descendants. "Every man in whose veins courses any puritan blood," he wrote, "as he looks back upon the events of the past, or forward to the hopes of the future, is appalled and astounded."[37]

Water would help alleviate the threat. Shattuck championed bathing the poor as a way to scour their slatternly characters as well as their dirty bodies, pointing out that this could only happen if clean water was made available to them. Running water, laundry facilities, and bathhouses would enable these people to wash their homes, their clothes, and themselves, and would thus "promote among them habits of cleanliness and better modes of living."[38] Shattuck here implied that water would also

improve their character. Few likened washing the body of the individual
and the body of the city to moral and spiritual cleansing so effusively as
the Presbyterian clergyman Walter Mears. "A river of bright, pure water, poured in thousands of branches along every street and by every
door, is a truly delightful persuasive to something very near akin to virtue," Mears wrote in 1866. Water was an evangelical force of prodigious
positive influence. "People yield to the persuasive; they luxuriate in the
cheaply got cleanliness of their persons and homes," he explained. "The
filthier ones around them are shamed and stimulated. They, too, must
have their share. And so the demand increases, and civilization and virtue are promoted in the action and re-action of increasing supply and
increasing demand; it being difficult to apportion the praise accurately
between the supply and the demand."[39]

Civic leaders who were more sympathetic to the poor than Shattuck
or Mears were just as zealous about the benefits of bathing them. Walter
Channing maintained that Boston's unfortunates were victims rather
than perpetrators of the defilement of the body of the city, and that they
were in fact more at risk than the privileged native-born. He and others
who worked with the poor advanced an environmentalist argument that
is usually associated with turn-of-the-century Progressivism.[40] If some
Bostonians were debased and degraded, Channing contended, the dirty
city made them that way. "*A condition can never be a cause*," he angrily declared in an 1843 talk on pauperism, "and a voluntary, a moral, an intellectual being, can hardly be the sole agent in the production of his own
deepest misery."[41] This was also the conclusion of the city's Committee
on the Expediency of Providing Better Tenements for the Poor, on which
Channing served, along with Dr. Henry Ingersoll Bowditch, who worked
as a dispensary physician in South Boston's blighted Fort Hill immigrant
neighborhood. The committee's 1846 report contended that just as the
"physical nature" of people who lived in overcrowded and filthy dwelling became "blunted, and hardened to the impurities around them, . . .
so their moral nature gradually accustoms itself to the sight of evil, and
ceases, at last, to be offended, at what was originally shocking to it."[42]

In the 1830s and 1840s, participants in the public discussions of
city water frequently cited the waterless plight of Boston's disadvan-

taged and the consequences for the body of the city as a whole. Testifying before the legislature in 1839, one Boston resident reported that indifferent landlords told Irish families "to get water where they can."[43] Dr. Bowditch recalled at the same hearings that he had witnessed Irish women fighting for the slender supply furnished by a local pump. His experience as a practicing physician among the poor had "forced upon me daily occasion of observing, that there was the want of some place where they could free themselves easily from the impurities upon them, by a copious supply of soft water."[44]

"We all know how scarce pure water always is, and how often it becomes most difficult to get that, which is barely tolerable for use," Walter Channing declared in his speech on pauperism. He then asked, "How can the poor acquire or preserve habits of cleanliness under such absolute destitution of its means?"[45] His 1844 *A Plea for Pure Water* asked privileged Bostonians not to recoil from the dirty underclass but to give them the means to wash themselves. He advised his readers that the position they took on furnishing an abundant supply of clean water to the poor revealed the state of *their* character and *their* Boston. "I see this project to supply this city with pure water in its moral relations," Channing remarked. "It is in these that its real importance lies." If one looked at the cost of building the Cochituate system in these terms, the financial objection that opponents raised "ceases to exist." No person could live, let alone be clean, without water, and so it was profoundly wrong to tolerate a situation where those with little or no money could not get good water. Likewise, no well-to-do Bostonian could know "and no body can learn what deprivation means, who does not see the actual workings of a system which denies to the people the use of water."[46]

At the height of the 1845 water debates, other prominent Bostonians equated supplying clean water to all residents with purifying the spirit of the civic body. In a speech at one of the Faneuil Hall meetings on water, merchant, manufacturer, and philanthropist Abbott Lawrence asserted that furnishing pure water at no cost through public pumps to those of limited income was a high social obligation. "In a city like this, of ample means," he stated, "it is a matter of duty and conscience, of humanity and patriotism, which a devotion to the public welfare demands of every citizen, of all men of all kinds to make this provision for the wants of the whole community." According to Lawrence, who sounded like some early twenty-first-century advocates of universal health care, wealthy Bostonians "have no excuse to withhold this provision from those less favored than ourselves." To do so would be no different, and no more morally acceptable, than withholding food from the famished.[47]

Both sanitary reformers who blamed the dirty poor for their condition and those who saw them as victims of the indifferent city agreed that delivering water to everyone would, among its other benefits, protect the economic health of the community. They warned that a single "sweeping epidemic may injure the property of the City, to a greater amount than the entire cost of an aqueduct to supply the City with pure water."[48] Channing appealed to the self-interest as well as the consciences of propertied taxpayers. Since "*an abundant supply of pure and fresh water directly promotes health and longevity, and as surely tends to diminish, or prevent, pauperism*," he told them, it also reduced the drag on city budgets caused by the ill and indigent. The more prosperous might not directly experience or even witness the deprivation caused by lack of water, but they would certainly feel its consequences when "the tax-gatherer comes for their portion of what supports the State Prison,—the House of Correction,—the City Watch,—the Fire Department,—the House of Industry, and the City Jail, which their personal neglect of great interests directly produces."[49]

By the time Lemuel Shattuck prepared the 1850 state Sanitary Commission report, the harrowing mortality and morbidity rates in poor immigrant neighborhoods had sensitized him to the intense suffering of the people who lived there, and not just to the danger he believed they posed for others. He even took some reformers to task for what he charged was their hypocrisy, criticizing those who opposed capital punishment while they ignored the "social murders and suicides" caused every day because of unsanitary conditions. If these do-gooders expended "the same zeal, labor and money" on "diffusing correct sanitary information among the people, in removing the causes of disease which prey on them, in propagating sound sentiments relating to life and health, and in elevating the physical, social and moral condition of man," he asked, "how many more lives might be saved?"[50] Shattuck repeated Channing's point that it was a far better economic as well as ethical bargain to spend public funds now on the water that would cleanse dirty people rather than face the more financially and morally costly consequences of not doing so. Getting water to the poor would reduce the costs of dispensaries, workhouses, and orphanages while ameliorating the character and reputation of Boston as a community.

To expect and encourage the immigrant poor to cleanse their bodies and clothing with any thoroughness required more than just building a water-supply system, Shattuck explained. Promoting the public good of widespread personal cleanliness depended on the availability of reasonably priced bathing facilities that, far from making the kind of profits that Latrobe predicted, might well have to run at a loss. This meant that they would have to be publicly owned and subsidized. Shattuck pointed out that the twelve private bathing establishments he counted in Boston only served people of means, given where they were located and that they charged between twelve and a half to twenty-five cents admission. He recommended that much more affordable *"public bathing-houses and wash-houses be established in all cities and villages,"* specifically for use by working people.[51] The modest deficits these facilities might incur because of the necessarily low prices they charged should not discourage their continued operation and expansion, since the individual and social benefits would be so considerable.[52]

Promoters of bathing the poor claimed that the "dirty" wanted to be clean and that bathing was an important step in their becoming good citizens. The *Boston Medical and Surgical Journal* cited a story in the *London Times* of an experiment at an English asylum for the homeless that "most satisfactorily proved that the very lowest of the poor will gladly avail themselves of facilities for personal cleanliness." The inmates were hesitant at first, "but as soon as they had felt the refreshment of the warm water and clean clothes, they eagerly availed themselves of it." The positive results were dramatically apparent: "Many a countenance became pleasing which before was disgusting, and no doubt was left that soap and warm water could work wonders with the poor."[53] During the 1845 water debates, the Boston *Daily Advertiser* published a submission from "W. C."—possibly Walter Channing—that included an instructive anecdote. An acquaintance had told "W. C." of a group of illiterate boys ranging in age from ten to sixteen who had not been able to go to school because their families needed the income they could earn. The opportunity had now arisen for them to get an education, but they did not enroll because they did not have the water they needed to make themselves presentable. *"They could not be made or kept clean."* If they went to school as they were, "they would be shunned by all other children. They would be driven away." Another person told him of an evening charity school where the instructors "absolutely have caught infectious diseases from the boys" whose bodies were afflicted by "extreme, unmitigated filth."[54] Without water to bathe in, these children would remain a health danger to themselves and others and would never have a chance to be educated into becoming better Bostonians.

Stories of this sort were potentially persuasive to immigrant-wary social observers like the Reverend Waterston, since he saw education as an effective way to "wash" the foreign-born. He hoped that they might "be persuaded to send their children to our public schools, and . . . be there trained to habits of honesty and usefulness." They would break the bad habits they learned at home and become good Americans, that is, people of whom Waterston approved: "In this way the children of emigrants will blend with our own population, they will grow up to love and honor our institutions, and will soon be numbered among the most worthy citizens of our land."[55] According to "W. C.," this could only happen if these young people first had a chance to bathe, and the construction of a waterworks would provide them this chance. Three days after the Cochituate system opened, the *Daily Evening Transcript* ran a short article titled "Cheap Baths, &c." that enthusiastically predicted, "Now that we have the whole of Lake Cochituate ready to flow into the city, there is no good reason why there should not be aplenty of cheap bathing establishments, laundries and washing houses, established in convenient locations." It added, "Public baths might be established, much to the advantage of the laboring classes."[56]

This "advantage" might be a matter of opinion. Even the presumably indisputable value of cleanliness, like so much of the cultural discussion about city water, became part of the debate on urban democracy. The person who called himself "Rabble," and who exulted over the defeat of the Long Pond proposal in the May 19, 1845, referendum on the grounds that it would give "despotic power" to the proposed water commissioners, also suspected the motives of those who were so eager to cleanse the many. "Rabble" condemned city water as a scheme by a devious upper class interested only in maintaining its own status and authority by keeping the rest of the population under its thumb. He contended that the real purpose of the "oligarchy" and "aristocrats of Beacon and Park streets, the conspirators of Temple Place and Collonade [sic] Row" who were behind the Water Act was "to pour out their wealth like water, if not to grind, at least to wash the faces of the poor." In his pun-soaked prose, "Rabble" accused Boston's elite of wishing to supplant "the landmarks of our constitutional liberties" with "water-marks, hydrants, hose and pipes—pipes, which like their designing authors, wriggle their tortuous way in darkness underground, loosening the very foundations of our city."

"Rabble" himself poured it on: "Our civic faces are to be scrubbed by our nursing civic fathers, unless we squall lustily and vindicate the 'freedom of the soil,' which is our American birthright." He pointed out that the masses were commonly known as the "Great Unwashed," and that these people were not ready to "resign the very title-deed of our democratic principles, without a struggle against this deluge of oligarchy." As for the Long Pond plan's advocates, whom he mockingly dubbed "magnates 'of the first water,'" he accused them of bribing public officials to get their way, adding that their expressions of concern for the health of the people rang hollow.[57] "Rabble" celebrated the rejection of the Long Pond waterworks as a defiant defense of the rights of the people to think, speak, and live as they wished.

The Cold Water Army

The temperance movement was immensely popular in America, whether measured by the number and variety of organizations that tried to prohibit or discourage the consumption of intoxicating beverages, or by the volume of essays, sermons, pamphlets, periodicals, tracts, broadsides, fiction, poetry, songs, and illustrations that warned of the dangers of alcohol. By the second and third decades of the nineteenth century, temperance groups were everywhere. Even tiny and rough-edged Chicago could boast of one, founded in 1832, a year before the settlement's first town government was instituted. It was started by twenty-seven-year-old newcomer Philo Carpenter, who traveled the last leg of his journey from upstate New York in a canoe that was reportedly towed across Lake Michigan by Indians, arriving in a town beset by cholera. An advertisement in the *Chicago Democrat* late the next year announced that copies of a temperance almanac were on sale in the store that Carpenter had opened in a rented log cabin on Lake Street.[58]

While temperance advocates focused on the eradication of alcohol rather than filth, they resembled sanitary reformers in claiming that they were trying to protect both the individual and the social body by cleansing them of a physical and moral evil, since Americans' staggering amount of drinking inflicted enormous damage on private and civic life. Shattuck's Sanitary Commission report stated that it was "universally acknowledged" that alcohol was filling "the cup of suffering."[59] The temperance movement overlapped with sanitary reform in several other ways. People who strongly supported sanitation measures were also highly likely to be active proponents of temperance. Health experts pointed out that in addition to the direct harm that drunkenness inflicted on the body, it also made people more susceptible to serious illnesses. The physicians who investigated the 1866 cholera outbreak in Boston, for example, placed "intemperance" ahead of "uncleanliness" on their list of "exciting causes" of the disease.[60] Temperance spokespersons were even more prone than sanitarians to view their efforts through the framework of evangelical Protestantism, and to claim that their aim was the spiritual redemption as well as physical betterment of the individual and the community.[61]

Urban temperance leaders, especially those also concerned about public health, believed that the problem of drinking, like that of miasmas, was more widespread and entrenched in the city than in rural and small-town America. Walter Channing said that alcohol found a less hospitable home outside the city, since "moderation and temperance are the peculiar blessings of a country life."[62] He here joined others in employing an idealized vision of "country life" as naturally clean and healthy, as opposed to urban centers, where, according to Channing, "artificial wants unite with those which are more strictly natural." The combination of heightened competition and economic uncertainty of city life "produces and constantly ministers to anxiety and disquiet," he explained, from which people sought relief in alcohol. They were easy prey for the opportunistic liquor dealer, who, because the local population was such a mix of strangers, could too easily refuse to recognize his complicity in the misery he caused. A person who profited from the sale of beer, wine, and whiskey was "not apt to stop and ask what may be the effects of gills or pints sold hourly to men, of whom he may never hear, and never has known."[63] Nativists used the temperance cause as a way to voice their conviction that immigrants and the poor were drunkards by nature, not by force of circumstances, but they agreed with Channing that intemperance was a bigger problem in urban areas than in the country, and that city life exposed even supposedly "better" people to dissipation.[64]

The temperance movement celebrated clean water as the proper beverage for a healthy person, regardless of class or background, and, by extension, for a sound society. Members of some groups referred to themselves as soldiers in "the cold water army" or "the cold water phalanx." Even more fiercely than the sanitarians, these water warriors viewed the cityscape as a virtual battlefield, especially on holidays like the Fourth of July, which, as Krimmel observed in his second painting of Centre Square, had become times of very heavy public drinking.[65] On occasion the campaign for temperance turned the city in to a real battlefield. In 1855 Chicago mayor Levi D. Boone, who had been the candidate of the anti-immigrant Know-Nothing Party, decided to limit alcohol consumption by enforcing existing laws, while the like-minded Common Council raised license fees. This was an attack not just on alcohol, but also on the city's large foreign-born community, in whose lives social drinking played an important part. It led to the Lager Beer Riot of April 21, 1855, in which an attempt by the police to repress a protest by German and Irish Chicagoans resulted in deadly violence.

Proponents of urban waterworks, whether militant supporters of temperance or not, contended that the objectionable flavor and suspect quality of the water currently available in wells and other local sources were important factors in prompting people either to enhance its taste with alcohol or to reject it entirely in favor of beer, wine, and spirits. The Boston City Council's Joint Standing Committee on Water stated in 1839 that if residents could drink the pure country water that a new system would deliver, this "would diminish the temptation to mix other liquors with it, and would consequently aid the cause of temperance, and therefore the public health."[66] In his appearance the same year at the state legislative hearings on Boston water, Dr. James Jackson estimated that "one half the young men who mix something with water, begin on account of the water not being good." In his opinion, "Intemperance often begins with mixing something with the water." Another Boston resident attested to this, recalling, "When I was an apprentice, I used to mix spirit with water, when it was so bad I could not drink it without."[67] A supply that was—as always promised—sweet-tasting would prove as effective against intemperance as it would against dirt, reducing drinking and, as a result, the social and financial costs of drunkenness.[68]

In the preface to the second edition of his 1834 report on the prospects of building a waterworks for Boston, engineer Loammi Baldwin Jr. noted that improved water-supply systems would assist the work of temperance societies, which, he felt, should in turn gratefully support their construction. "While striving to expunge from the community a loathsome indulgence in vitiating liquors," Baldwin wrote, "their benevolent labours would essentially tend to more practical benefits, by promoting a free and copious introduction of pure, soft water as a substitute." Baldwin could not resist a humorous wink at the moral righteousness of temperance reformers: "In fact, so gratifying is this simple luxury to the taste of those unaccustomed to its enjoyment, that on trying it, many would agree with the lady who exclaimed, in her enthusiastic devotion to the cause of temperance, 'what pity 't is not a sin to drink it.'"[69]

Water ceremonies became occasions to hail temperance along with other civic virtues that city waterworks affirmed. In Boston, only Cochituate water and lemonade were served at the aqueduct groundbreaking in 1846, and in his address to the crowd on the Common in 1848, Mayor Josiah Quincy Jr. boasted that the immense task of constructing the system had been completed "without the stimulus of intoxicating liquor," referring to the common practice of providing alcohol to laborers on such projects, including the Philadelphia waterworks decades earlier.[70] A number of people in Quincy's audience wore ribbons imprinted with the words, "Cochituate Water, Introduced Oct. 25, 1848," along with the quotation, "O That's the Drink for Me." The source was the temperance hymn, "The Drink for Me," which appeared in a songbook published by the Massachusetts Temperance Union.[71] Later in the evening, after the official events were completed, there was a great temperance meeting in Faneuil Hall, which was scheduled to follow Daniel Webster's speech to Whig loyalists. Two decades later, the *Chicago Republican* began its coverage of the opening of the lake tunnel in 1867 with a paean to the crystalline water that Chicagoans would soon be enjoying, which it called "a greater advocate of temperance than laws that legislatures frame, or societies that men form."[72] The *Tribune* predicted that the new water would "abolish one extensive excuse for tippling, whereat the temperance people will be exceedingly glad," though it also noted that the site of the waterworks on Chicago Avenue was "not exactly the one which an advocate of temperance would have regarded as *apropos*, since it was hard by two breweries."[73]

Not everybody was convinced that constructing a large new public waterworks would end drunkenness. Prior to the approval of the Cochituate system, some of its critics ridiculed the attempt to identify the project with temperance. "A Selfish Taxpayer" complained, "Not satisfied with all ordinary help, the movers in this matter have thought proper to hitch on the Temperance cause, at the head of their team." Claiming that he supported the fight against alcohol, "A Selfish Taxpayer" asserted that the quality of local water had little to do with alcoholism. To say otherwise offered drinkers a facile excuse by lending credence to the false notion "that they are justifiable in their attempts to improve bad water by the addition of rank poison."[74] Intemperance was a sign of weakness of character, not of the quality of the water supply.

The association of intemperance with the individual drinker's integ-

Figure 5.2. Some of those attending the Boston water celebration of October 25, 1848, pinned ribbons like this one to their clothing, declaring their opposition to alcohol. The illustration likens Cochituate water to the pure goodness of a country well. The line "O that's the drink for me," comes from the temperance hymn "The Drink for Me." There were several temperance groups in the waterworks parade, one of which carried a scroll that was inscribed with the Pledge, the oath of total abstinence. Courtesy of the Massachusetts Historical Society.

rity, coupled with the repeated failure in numerous cities to pass or enforce laws that effectively regulated or prohibited alcohol, helps explain an important dimension of the temperance movement in which water also took a significant place. Without giving up on the campaign to stem the supply of intoxicating beverages, temperance groups concentrated on reducing the demand. They appealed to individuals to demonstrate

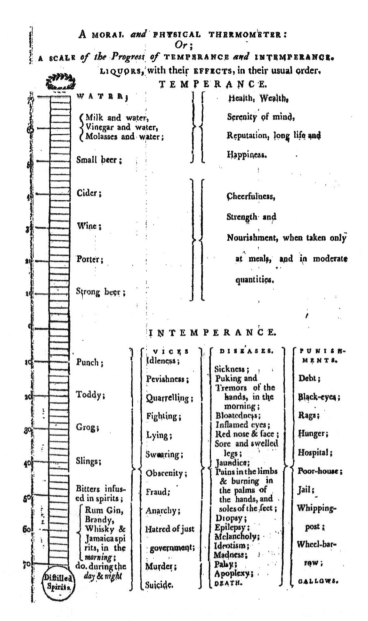

A MORAL _and_ PHYSICAL THERMOMETER:
_Or;_
A SCALE _of the Progress of_ TEMPERANCE _and_ INTEMPERANCE.
LIQUORS, with their EFFECTS, in their usual order.

TEMPERANCE.

WATER;		Health, Wealth,
{ Milk and water, Vinegar and water, Molasses and water;		Serenity of mind, Reputation, long life and
Small beer;		Happiness.
Cider;		Cheerfulness,
Wine;		Strength and Nourishment, when taken only
Porter;		at meals, and in moderate
Strong beer;		quantities.

INTEMPERANCE.

	VICES	DISEASES.	PUNISH-MENTS.
Punch;	Idleness;	Sickness;	Debt;
	Pevishness;	Puking and Tremors of the hands, in the morning;	
Toddy;	Quarrelling;		Black-eyes;
	Fighting;	Bloatedness; Inflamed eyes; Red nose & face; Sore and swelled legs; Jaundice;	Rags;
Grog;	Lying;		Hunger;
	Swearing;		Hospital;
Slings;	Obscenity;	Pains in the limbs & burning in the palms of the hands, and soles of the feet;	Poor-house;
Bitters infused in spirits;	Fraud;		Jail;
{ Rum Gin, Brandy, Whisky & Jamaica spirits, in the morning;	Anarchy;	Dropsy; Epilepsy; Melancholy; Idiotism; Madness;	Whipping-post;
	Hatred of just government;		Wheel-bar-row;
do. during the day & night	Murder;	Palsy; Apoplexy;	
Distilled Spirits.	Suicide.	DEATH.	GALLOWS.

Figure 5.3. Famed Philadelphia physician Benjamin Rush's "Moral and Physical Thermometer" first appeared in the _Columbian Magazine_ in 1789 and was reprinted many times in the years that followed. "Water" is at the top of the scale, and "Distilled Spirits" is at the bottom, with other drinks at different grades in between. Water is the drink of "Health, Wealth, Serenity of mind, Reputation, long life and Happiness," while alcohol is the choice of those whose lot includes "Vices," "Diseases," and "Punishments." Courtesy of the Northwestern University Library.

their moral backbone by renouncing alcohol completely and uncondi-
tionally. The total abstinence movement was impressively broad. Speakers at countless small meetings and large rallies urged attendees to take
the Pledge (it was usually capitalized), a brief oath in which they swore
to refrain absolutely from drinking alcohol and from selling or otherwise
furnishing intoxicating beverages to others.[75]

While many city dwellers, including sophisticated people who saw no
harm in a convivial glass of fine wine, thought that total abstinence was
unnecessarily extreme, it attracted a large and varied following. In 1833
the leadership of the Massachusetts Temperance Society set complete
renunciation of alcohol as a condition for membership. Meanwhile, significant numbers of immigrants and workers joined existing total abstinence organizations or formed their own groups. Josiah Quincy Jr.
made abstinence a hallmark of his administration. After he learned that
City Council members planned to serve "wine or other intoxicating liquors" at a public dinner, he reversed his earlier decision to attend and
preside. "It is the known & settled policy of the Commonwealth to dis-
countenance the use of these articles in all forms, except as medicines,"
Quincy lectured the event's organizers, "and the disuse of them on Public occasions has become almost universal."[76]

The total abstinence wing of the temperance movement thus emphasized self-discipline rather than the control of others. The insistence that
one's personal rejection of alcohol must be unconditional and without
exception reflected a fixation on purity similar to the one underlying
sanitary reformers' emphasis on cleanliness. This obsession with pure
bodies also manifested itself in the contemporary social ethos that declared that a single act of extramarital intercourse—even if it was the
result of seduction or rape—irreversibly "ruined" an otherwise virtuous
woman forever, and in the racial code that defined as black a person in
whose veins coursed a single drop of Negro blood, with all the legal and
social consequences that implied. Unlike the fallen (sometimes called
"polluted") woman or the person of mixed blood, however, an intemperate individual had the power to reclaim his or her sullied body by im-
bibing nothing but cool refreshing water. This raised water from a mere

thirst quencher to a magical elixir that could restore the besotted self to a pristine state.

What helped make water seem magical were its transparent simplicity and its association with unspoiled nature. Some of the claims for the sources delivered by waterworks were drenched with rhetorical excess, revealing how powerful was the appeal of introducing natural goodness to the city. The Boston *Daily Advertiser* hailed Lake Cochituate for being located far from the urban world of factories, smoke, and "noxious trade." The lake, it wrote, "spreads itself to the rain and sunshine remote from those occupations and that concourse of busy life, which could discolor its crystal stream." When Cochituate's best "leaves the bosom of the lake to visit the capital," the *Advertiser* claimed, "it goes like a darling child from the maternal mansion, with all the aids and appliances of the most tender solicitude, to preserve unsullied the purity of its birth." Once arrived, "it is entrusted to the protection and kindness of its new friends, and theirs be the blame if, in the intimacy of their dwellings, it find aught that defileth."[77]

To temperance crusaders, water—the "darling child" of a nurturing nature—could redeem both city life and city people. Total abstinence resonated among those who feared that mastery of experience was constantly under siege within the urban throng.[78] One could not direct the course of city life, but perhaps it was possible to resist being contaminated by it. To swear not to let a single drop of alcohol enter one's body and to permit only water to pass one's lips was to declare that one could successfully negotiate an impure and shifting world defined by contention and compromise. William Alcott spoke of how the individual must resist "the multiplied evils" not just of disease but also "the errors of the family, the school, the factory, the counting room,—indeed in all the varied employments and modes of life."[79] When Walter Channing described temperance reform as "the moral power of a community, or of a nation, distinctly and efficiently directed to a single specific object," he perhaps was more on the mark than he knew.[80] Alcohol served so well as this "single specific object" because it could be made to stand for everything that could go wrong in an inherently precarious milieu of "artificial wants" and "anxiety and disquiet."[81] And water, pure and sweet, was the single specific object that could make all things right.

Universal Panacea

While never as popular as the crusade against alcohol, the water cure attracted a large following, including a number of prominent people on both sides of the Atlantic, and especially women. Sharing the sanitary and temperance movements' belief in the value of cleanliness and of a bland diet that shunned alcohol (and even such "stimulants" as coffee and tea), the water cure exalted water even more than the others did by claiming that plain water, properly applied internally and externally, could rescue the body from virtually any physical or mental ailment. As opposed to sanitation and temperance, however, the water cure was not part of a program of social reform. Although both practitioners and patients often attacked—with some justification—conventional medical practices, they frequently did not want to change these as much as claim that the water cure provided a superior alternative or at least a helpful supplement to other forms of treatment. In addition to its many other reputed therapeutic benefits, the water cure had special appeal for city people as an antidote to the toll they thought that city life was exacting on their bodies.

The belief in the therapeutic qualities of water crosses all cultures and eras. Colonists in the future United States were familiar with European mineral springs whose water reputedly had healing power, so that William Penn was pleased to learn that there were similar sources close by his new city.[82] By the end of the eighteenth century, a number of American entrepreneurs had opened spas, several modeled after fashionable health resorts in Britain and on the Continent. A person might "take the water" by drinking it, bathing in it, or both. Philadelphian Benjamin Rush, perhaps the most renowned physician in late eighteenth- and early nineteenth-century America, and also a great believer in the moral as well as physical benefits of temperance, commended the salutary effects of mineral water in a pamphlet promoting the Harrogate spa, located about four miles from Philadelphia.[83] Rush listed the many maladies that would respond to this therapy, among them hysteria, palsy, epilepsy, gout, diarrhea, want of appetite, colic, obstructions of the liver and spleen, chronic rheumatism, piles, female obstructions and weaknesses, worms, skin ailments, kidney and bladder problems, and gleets (morbid discharges, such as pus, from a wound, or as a result of chronic gonorrhea).[84] Commercial springs like Harrogate increased in number and size during the nineteenth century, sometimes bottling their water for retail sale to customers in cities. In 1831 John Bell, another eminent Philadelphia physician and health expert, published *On Baths and Mineral Waters*, which surveyed the chemical content and medicinal attributes of many different kinds of waters, including sea water.[85]

Both Rush and Bell also discussed the value of hydropathy, the more technical name for the water cure. The line between mineral-water treatments and the water cure was not always very distinct, but mineral-water spas and purveyors usually attributed the effectiveness of their water to its specific and unique chemical composition, while hydropaths credited the water itself, not anything that might be dissolved in it. Proponents of the water cure recommended simply putting the body in contact with clean water. Like mineral-water therapy, hydropathy's lineage goes back to ancient times, and it was well known to Americans. Rush said that bathing in plain water could ease many of the same conditions he stated were treatable with mineral water, to which he added heart palpitations, asthma, Saint Vitus' dance, rickets, and headaches.[86] The middle decades of the nineteenth century witnessed a major surge in the popularity of hydropathy, first in Europe, with the United States quick to follow. This began with the work of an Austrian farmer, Vincent Priessnitz, from the Silesian town of Gräfenberg, who developed his version of the water cure in the course of rehabilitating his broken body after a terrible accident. By the 1840s he was receiving over a thousand patients a year at Gräfenberg. Testimonies to Priessnitz's success in healing the sick and injured brought him an international reputation, along with plenty of apostles and imitators.

Although there were variations among practitioners, most types of water cure (the term describes both the method and the place where it was administered) shared certain features. The water used was preferably cool or no warmer than tepid. It was applied in a number of ways: wiping the body with a wet cloth or sponge; dripping or gently pouring water on it; spraying water directly at a specific tumor or locus of pain; or placing a part of the body in a basin (e.g., the sitz bath) or a small bathtub (known as a half bath). One could also undergo a pail-douche, during which attendants poured buckets of water on the patient in a room equipped with a slanted floor and a drain. Or these attendants might wrap a person's body, mummy-like, in a wet sheet and blankets, to induce sweating. The wrapping was often accompanied by energetic rubbing, followed by a naked full-body cold plunge. A patient might follow a course of treatment that combined several types of baths, along with drinking large amounts of water.

Publications devoted entirely to the water cure circulated widely, if not to the same extent as temperance literature. Some authors of the many available all-purpose household guides included instructions on

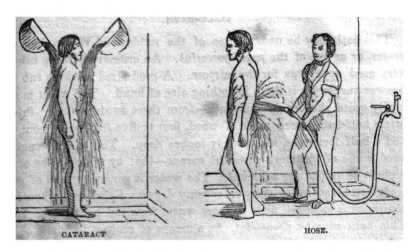

Figure 5.4. In the cataract bath, water was poured over the patient's entire body, while the hose bath enabled an attendant to aim at a particular spot. These and the other illustrations here are from Dr. Joel Shew's 816-page volume, *The Hydropathic Family Physician*, issued in 1854 by Fowlers and Wells, the most prolific publisher of books on alternative health treatments and subjects such as phrenology and mesmerism. Courtesy of the Northwestern University Library.

Figure 5.5. As its name suggests, in the half-bath only the lower body was immersed. At a water-cure establishment, one would take several kinds of baths in conjunction with each other. The baths were combined with exercise and a bland diet, including lots of water. Courtesy of the Northwestern University Library.

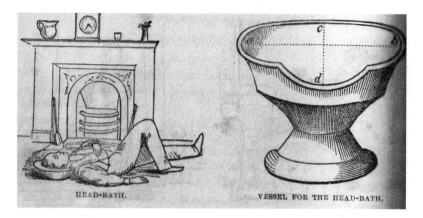

Figure 5.6. There were baths for different parts of the body (including an eye bath), with special equipment to facilitate the treatments. Shew specified that the water cure must be taken only where there is "a good supply of PURE SOFT WATER." He also prescribed that it be in a distinctly non-urban setting: "A mountain location of the proper kind is probably the best, all things considered, for the majority of patients; although lakes, rivers, and the sea-shore often afford pleasant and suitable places." Courtesy of the Northwestern University Library.

LEG-BATH.   SITTING-BATH.

Figure 5.7. As these illustrations indicate, with the proper tubs and containers, one could take the cure in the convenience and privacy of one's home, and without even having to remove most of one's clothing. Courtesy of the Northwestern University Library.

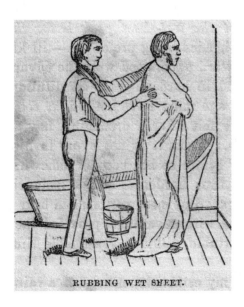

RUBBING WET SHEET.

Figure 5.8. Attendants might rub the patient's sheet-covered body, as pictured here, or wrap the patient in layers of wet sheets and blankets, to induce sweating before a bracing cold plunge. Courtesy of the Northwestern University Library.

administering the cure at home.[87] Orson S. Fowler—arguably the era's most prolific publisher of works on alternative health and medicine (Fowler is perhaps best known for his association with phrenology) at a time when the field was very crowded—advised in his much-reprinted *Physiology, Animal and Mental* that the hearty rubbing of the body with water, "properly applied," was "an almost sovereign panacea."[88] Joel Shew, one of several leading American hydropaths who had been converted by a visit to Gräfenberg, wrote several books on the subject and in 1844 became the first editor of the *Water-Cure Journal.*

Shew's inventory of the ills that water could vanquish or ameliorate was even longer than Rush's. You name it: heartburn, hiccups, bleeding from the stomach, acute gastritis, cholera, constipation, headache, toothache, earache, canker of the mouth, inflammation of the throat, clergyman's sore throat (from "excess of speaking"), erysipelas (a serious skin rash), heart palpitations, brain fever, apoplexy, palsy, pleurisy, pneumonia, whooping cough, venereal disease, gout, rheumatism, convulsions, colds, scrofula, cancer, consumption, female complaints, and dropsy (the collection of fluid in a body cavity or joint, now called edema). Shew even recommend water as a treatment for hydrophobia, and he claimed that it could both prevent and induce vomiting. He said it could also correct psychological and personality disorders, such as nervous excitability, insanity, melancholy, and obstinacy. Once wrapped in a wet sheet, a distressed person bent on suicide would choose to live, and an individual who was determined to avenge a personal injury "would soon begin to feel the promptings of charity, and conclude that, after all, he will be upon the side of forgiveness and mercy."[89] Water could reduce inflammations and accompanying pain, revive "debilitated, withered organs," clear the stomach, brace the system, and infuse "new life throughout all its borders." As a therapeutic remedy, Fowler proclaimed, "water excels all other agents combined."[90]

The water cure's monomaniacal emphasis on the value of treating any and all of the internal and external parts of the body with water in order to return a patient to health combined some key aspects of the sanitary and temperance movements in intensified form. Here, too, the active ingredient in water was its purity, which was crucial to a comprehensive program of personal physical and moral regeneration in which salvation depended not on faith but on effort, in the form of an extended physical regimen. Water could not work an instant miraculous healing of the kind attributed to some purportedly sacred sources, such as the one at Lourdes, France, revealed to Saint Bernadette in 1858.[91] Hydropaths similarly distinguished not only between the water-cure and the mineral-water therapies, but also between their methods and the dozens of patent-medicine formulations guaranteed to restore health and vigor, such as Swaim's Panacea, which was manufactured in the same Philadelphia establishment that housed Swaim's Baths and marketed around the country.[92]

Indeed, part of the attraction of undergoing hydropathy, like that of taking the Pledge, lay in submitting to its ritualized pseudoscientific discipline. "We do not see the cleanly, sober, industrious, temperate and healthy man, committing evil deeds," Shew asserted in support of the water cure. "The quarrelsome, the licentious, and the wicked, we shall find, all have their improper dietetic and physical habits, and that these have preceded the moral evils such persons practise."[93] The reward for good habits went beyond simple cleansing of the body. William Alcott stated that the "great rule" to observe in taking a cold bath was "to use it as to secure a glow on the surface immediately after its use," an outward physical aura that could be read as a sign of inward virtue.[94] Of course, water did not cure cancer or the sundry other diseases promoters and patients said that it did. But to be bound inside dampened sheets and blankets from which one emerged bare and sweaty and then immediately endure a cold plunge was, like total abstinence, to wrap oneself in the idea that one had performed an efficacious act of self-redemption. While urban sanitary improvements, including piped-in water, counteracted the health dangers of living in a dirty place and among the unwashed many, the water cure healed the ill effects of what even some sophisticated observers worried was the overly refined life characteristic of cities. The spartan regimen seemed to reverse the damage that modern habits visited upon the body and spirit.

Even respected conventional doctors agreed that the therapies promoted by hydropathy could be beneficial in this way. Dr. John Collins

Warren stated in his medical advice manual, *Physical Education and the Preservation of Health*, that many highly regarded activities enjoyed by the urban well-to-do were unnatural and thus inimical to good health. "Nature has destined that the physical and intellectual education of man should be conducted in very different modes," Warren wrote. The "culture of the mind" entailed the imposition of "an artificial system" on humans, while the nurture of "the physical faculties is fully effected by the power of nature." As a result, "our state of civilization," in the course of improving the mind, "raised obstacles to the development of the physical powers."

Warren wondered if industrious working people were in fundamental ways better off than the city's educated elite, who were too sedentary. "Observe the sinewy arm of the mechanic," Warren directed. "The muscles are large and distinct; and when put in motion, they become as hard as wood, and as strong as iron."[95] Warren's wording evoked the physically and morally "mighty," if socially humble, figure idealized in Henry Wadsworth Longfellow's recently published and immensely popular poem, "The Village Blacksmith." Though Longfellow was an urbane man of letters and distinguished Harvard professor (appointed by university president Josiah Quincy Sr.), he expressed great if also patronizing admiration and envy for the humble blacksmith's "large and sinewy hands" and "brawny arms" (as "strong as iron bands"), which resemble the burly branches of the spreading chestnut tree under which he plies his strenuous trade.[96]

Dr. Warren's ironic use of the term "civilization" criticized urban culture at its vaunted best. He condemned the "artificial state" that "continually thwarts the course of the native dispositions of the animal economy." Warren feared that this would lead to "a degenerate and sinking race," like the higher classes in France before the Revolution "and such as now deforms a large part of the noblest families in Spain."[97] City life was subjecting the American social body, as Jefferson had feared, to the ills of the Old World, though the main problem was not the urban mob but the most cultured citizens. Whether he fully recognized it or not, this learned and citified grandson of a Massachusetts farmer was delivering a kind of sermon against urbanization that was a secular variation on the Puritan jeremiad. As a corrective, Warren recommended some of the same therapies conducted by hydropaths, among them the eye bath. Warren said that this would undo the detrimental effects of the "prolonged application" of the eyes to reading matter, especially "the badly printed newspapers and cheap publications of this country." Such reading, typical of city life, was "a violence, which cannot be practiced without doing mischief sooner or later." The best remedy, besides rest, was to expose ailing eyes to fresh air and plenty of cold water, the latter by means of a "very fine continuous stream of cold water driven upon the naked eye-ball."[98]

Unlike the publications of the sanitary and the temperance movements, water-cure literature did not suggest that restoring one's individual body was a way of improving urban life, or even that urban life could be improved at all, whether by the introduction of pure water or any other means. The visitor to the water cure was instead seeking a refuge from the city in which to recuperate from its unhealthy effects without doing anything to amend their causes. This did not mean that one necessarily had to leave the city to take the water cure. What is thought to be the first one in the nation was established in New York by Shew and two associates, including his wife, Mary Louise Shew, who prepared hydropathy guides for women.[99] More soon appeared in other cities. Christian Charles Schieferdecker's forty-room "hydriatic institution" in Philadelphia was situated where Chestnut Street meets the Schuylkill, the same spot as the water intake of Latrobe's original 1801 works.[100] In 1856 Dr. James E. Gross, who had practiced at a like institution in Madison, Wisconsin, started his Lake View Water Cure, located by the Lake Michigan shore less than five miles north of downtown Chicago. His advertisements in the *Chicago Tribune* and the *Water-Cure Journal* informed readers that they could conveniently reach his establishment by either omnibus or private carriage, thanks to a plank road that furnished "one of the finest drives" in the area. Designed "in the best style," the Lake View Water Cure could accommodate up to eight guests. "Board and Treatment" cost ten dollars a week. By the early 1860s, Dr. Gross had in-town competitors, including the Hygieo-Therapeutic Retreat and the Water-Cure and Homeopathic Infirmary, both on State Street.[101]

Most serious devotees of hydropathy, however, viewed the treatment as a way to remove themselves from metropolitan life. The leading water cures were in the countryside, surrounded by nature and supplied by local water sources. Given that some patients took up residence for weeks or even months, the cure was limited to those who could not only pay the fees but also get away from their regular routines for a prolonged period. At the conclusion of *The Water-Cure Manual*, Joel Shew strategically placed an advertisement for his "Water-Cure Institution," which he opened on the North Shore of Long Island at Oyster Bay, a two-and-a-half-hour steamboat ride and a world away from the turmoil of Manhattan. "For copiousness and purity," Shew told the reader, "it is believed the springs of this place are unequalled in any institution yet formed either in this or the old country," including Priessnitz's Gräfenberg.

The setting of Shew's water cure was almost as important as its water. "The beautiful roads, the shady walks, the fine sand beach, winding in various directions about the shores, the pure fresh air, coming from the Long Island Sound, and, above all, the great number of springs here to be enjoyed, seem to render the place a most desirable one." For those who could not make this trip, in-town treatment might still be effective, but only if the water came from a pure source. Shew judged rainwater acceptable, but he specifically warned against places that used an underground supply that was subject to the "drainings of the filth of cities." "All manner of filthy and disgusting substances find their way into the wells," he explained, "and troublesome diseases have been known to have directly resulted from the use of the water."[102]

Harvard awarded the 1846 Boylston Essay Prize to physician and naturalist Samuel Kneeland Jr. for his "On the Medical Uses of Water." He praised the "therapeutic agency of simple cold water," even if Kneeland specified that treatment must be administered in a "comprehensive and well-regulated institution" supervised by a conventional physician whose "moral and intellectual qualifications are acknowledged to be of the first order," not, he added in an unveiled reference to water cures and other alternative methods, in "a catch-penny concern, under the care of a self-dubbed Doctor."[103] Kneeland stated that the kind of "comprehensive and well-regulated institution" he recommended should be located "apart from the smoke and din of cities, among the beautiful hills of New England." While the urban patient in a country water cure would benefit from "the therapeutic action of water," he explained, "the complete change of hygienic conditions at such an institution could not but be extremely beneficial in the numerous diseases of modern refinement." The negative connotations of "refinement" here indicate that Kneeland agreed with John Collins Warren that the habits of wealthy urban residents were inimical to their well-being. Kneeland continued:

> The votary of fashion, turning night into day, and exciting his worn-out powers by stimulating meat and drinks; the fine lady, living in a hot-house, and breathing the external air perhaps once a month, with her loss of appetite, her vapors, and her capricious temper; the merchant, whose ears are greeted from morning to night only by the clink of dollars and cents, whose nose is refreshed by the odor of musty bank-bills, and whose only care is about the rise and fall of stocks; all these (and their name is Legion,) will find themselves at such an institution in a pure air, obligated to keep regular hours and lead a regular life as to food and exercise—financial, social, and domestic troubles will there leave them in peace, and enable them to devote the proper time and attention to the restoration of health.

The city's supposedly civilized customs and rewards simultaneously overstimulated and deadened the senses. Life in urban society disrupted the "regular" (i.e., natural) rhythms of day and night, of breathing in and out, as well as the flow of pure liquids through the body. Kneeland joined the alternative practitioners he otherwise criticized in noting that "change of habits and scene is the secret, in the immense majority of cases, of the success of our mineral waters," along with "the therapeutic agency of simple cold water." In his professional opinion, it did the individual body a world of good to flee the city any time it could for a long immersion into the purity of nature and nature's water.

Some popular spas in the countryside drew criticism for not being sufficiently different from city life. This was more likely to be the case with the fancier mineral springs than water cures, though the distinc-

tions between the two varied from one spa or cure to another. Kneeland specifically advised against visiting "Saratoga and other fashionable
places."[104] Even *Godey's*, which followed the latest trends, commented
unfavorably on the more stylish retreats that were too similar to city
life to be effectively restorative. "We rarely have allowed ourselves to
take exception to prevailing fashions," the July 1850 issue observed in
an article titled "Chit-Chat upon Watering-Place Fashions," but there
was one current custom that the magazine felt "duty bound" to criticize.
"This is the prevalent style of great display in dress at the *table d'hote*
of a crowded watering-place, where half the people are strangers, and
openly remark upon the taste which one displays." Even worse were the
impertinent "penny-a-liners," the gossipy society journalists "who will,
perhaps, relieve the tameness of their next letter by a description of the
toilets to the minutest detail."[105] Far from surrendering themselves to
country purity, city people, with encouragement from entrepreneurial
proprietors, had colonized such establishments, transforming them into
outposts of the worst kind of urban social spectacle. A year later an issue of *Gleason's Pictorial Drawing-Room Companion* included an article
that condemned the crowded scene at the cure in Lebanon Springs, New
York, midway between Albany and Pittsfield. The author advised readers to stay at home.[106] Instead of being returned to their better natural
selves, they would find themselves subjected to the same hazards that
they were trying to escape.

Withdrawal into Water

For city people, washing one's body, abjuring alcohol for nature's nectar,
and taking the water cure all involved a removal from city life into the
salutary benevolence of water. To bathe was to cleanse from the surface
of the skin the dirt and sweat of everyday experience. To follow a vow of
total abstinence was to prevent worldly weakness and evil from invading
the self and subduing the best qualities of one's character. To take the
water cure was to insulate the body from daily life entirely by immersing
and wrapping it in cool and bracing natural purity.

   In a very real sense, the advent of comprehensive waterworks also offered an unprecedented opportunity to withdraw from the city. But the way to do so was, paradoxically, by connecting to the urban grid, since people obtained the water they used for drinking, cooking, and bathing through the new network of water pipes, which linked their bodies to the body of the city. This paradox was most striking in the modern wonder of domestic plumbing, which made the individual experience of tending to bodily needs more private and isolated. To have a home with access to running water that supplied the latest models of fixtures meant that one could live far more completely and conveniently within one's residence than when it was necessary to rely on a well and a privy. At the August 1846 groundbreaking ceremonies at Long Pond, Water Commissioner T. B. Curtis wittily alluded to an additional possibility when he enumerated the benefits of the new waterworks. He stated that it was the commissioners' aim "to furnish every body in Boston with a means of the 'water-cure,'" without having to travel across the ocean to Gräfenberg or just the 125 miles to the noted one in Brattleboro, Vermont. Or, he might have added, even out the back door. Curtis quipped that the commissioners hoped "that every man in the city . . . would soon be a knight of the *bath*."[107]

For those who were able to pay for them, the bathtub and the toilet became the most intimate spaces in the home, removed from the buzz and hum of the world outside, although it was a connection to city water that made this separation from urban life possible.[108] Indoor plumbing was a sign of status and success, a way not just to distinguish but also to distance oneself from other city people, especially the immigrant working poor, who still had to get their water from the promiscuous public hydrant and relieve themselves in rank and overflowing outhouses. Ellis S. Chesbrough observed, "The favor which [indoor plumbing fixtures] meet in the better class of dwellings . . . is such, that nothing probably could induce their occupants to go back to the old fashioned privy, with its open cess-pool."[109]

The lead editorial in an issue of the *Chicago Tribune* from the summer of 1868, titled "What Chicago Needs," complained of the terrible odors emanating from the river, the city's streets, and all the other places "where filth stagnates," which made the warm months especially unpleasant and unhealthy. The author regretted the city's general lack of attractive public facilities, "first, and above all, public baths, drinking fountains and the convenience necessary to cleanliness and a healthful life," and concluded that sinks, tubs, and toilets in private homes only contributed to the problem by cultivating self-absorption among the more fortunate. "Chicago's wealthy men, being provided with bathing-rooms in their houses and with means to purchase refreshing beverages," the editorial complained, "take no thought of the poor who are suffering for the lack of these things, or of the consequences to the sanitary condition of the city of its crying deficiencies."[110]

Be that as it was, it is hard to fault those who took to "these things" as quickly as possible. Indoor plumbing enabled individuals to believe that they could mount some defense against the contamination of their bodies by city life and by real and imagined filth of all kinds, including the dirt their own bodies accumulated and produced in the course of being alive in this whirling world. The miracle of the bathtub and the water closet was not just that by taking city water one could remove oneself from city life, but that they also made one's natural body odor and waste—reminders that nature itself is "dirty" and that existence is inescapably corporeal—seem to disappear. This kind of contact with water thus involved an attempted escape both from the nature of which water was such an important part and symbol, and from the city whose waterworks made this water so available. The prosperous urban citizen's private plumbing positioned him or her as a self-exiled ruler over the urban and human bodies' filthy processes. Whether washing oneself in the bath, sipping a glass of water from the tap, soaking an ailing limb, or sitting atop one's "throne," the individual body thus extracted itself from the urban body and even from nature while immersed within nature's liquid or obeying its call.[111]

6

~~~~~~

THE FLOW OF TIME

City Water as Cultural Anticipation

~~~~~~

Nineteenth-century city life marched to a new sense of time. Its participants became accustomed to living by the dictates of the clock and the
artificial rhythms of an increasingly scheduled workday. The continuous
expansion and alteration of the built environment and the changes in
the size and nature of the population made even the recent past seem
irrelevant. This was especially true in Chicago, whose urban history was
both so brief and yet so intense, as it grew exponentially on waves of
newcomers. Since most residents were born and, in many cases, raised
elsewhere, they had had little local heritage or memory. Taking office in
March 1847, Mayor James Curtiss, who himself was in his thirties when
he came to Chicago from Maine, stated, "We are not delegated to act for
an old and established community—for a community whose interests
and whose character have been firmly settled by a gradual and almost
imperceptible process." On the contrary, he explained, "we are to furnish
precedents and speak for, to bind and control, the affairs of a city which
but yesterday had no existence."[1] Bostonians and Philadelphians with lo-

cal roots believed that they were rich in settled traditions, but they were a shrinking minority by the time Curtiss was sworn in. The only urban history that appeared to matter was the future that cities were making with each new day. On a deeper level, however, urbanization instilled a new understanding of the relationship between the past, present, and future, one that was clearly revealed in the ways in which they dealt with the exigencies of city water.

Cultural Anticipation

It was only natural that people living in rapidly expanding cities developed a distinctively forward-looking state of mind, a way of seeing that can be called *cultural anticipation*.[2] As opposed to cultural memory, cultural anticipation consists not of broadly held understandings of the nature and meaning of the past, but of shared notions about the future—not recollections of what happened, but expectations of what will be. Cultural memory influences cultural anticipation, since the attitudes and beliefs that past experience inculcates always inflect how individuals conceptualize the future. Likewise, the way people think about the future often affects how they remember the past.

As the growing populations of urban centers came to include more and more people who arrived from many disparate places, the cultural memory of cities like Philadelphia and Boston, like so much else about urban life, became more fragmented, and the desire to honor and preserve the past less powerful, even in long-settled families. The heirs of merchant and patriot John Hancock were not above selling tons of earth from his large holdings on Beacon Hill to be used as fill in order to create new and valuable real estate. The excavation undermined the foundation of architect Charles Bulfinch's 1789 monument to the Revolution, a sixty-foot column on which an American eagle defiantly perched, so that it had to be removed in 1811 because it no longer had reliable ground to stand on.[3] Some Philadelphians wished to demolish parts of Independence Hall in order to build new offices. After Benjamin Franklin's death, his daughter and son-in-law first converted his home into a boardinghouse and then tore it down in order to subdivide the property.[4] Shared cultural memory in young Chicago was very slight because there was little common past to remember, except insofar as that past involved the decision to leave somewhere else.

Both cultural memory and cultural anticipation have a literary quality; the first consists of a version of what happened, the latter a prediction of how things will unfold. The growth of Philadelphia, Boston, and Chicago fed on the hopes of tens of thousands of men and women from other places, who wrote themselves into the future of these cities. Their actual experience upon arrival qualified and reconfigured their outlook, which in turn influenced future experience and the expectations it conjured. As a result, cultural anticipation, like virtually every aspect of nineteenth-century urban life, was conflicted and changing. Public expression that tried to move the city in a certain direction by enlisting others to share certain goals focused on articulating a credible and appealing idea of how taking a particular step would lead to a better future. This is what proponents of different city water proposals did again and again. They regarded waterworks not only as a key component in a physical infrastructure but also as statements about the urban future.

In this period of constant alterations, most social observers and leaders envisioned urban experience as moving in a straight line of either advance or decline. With things happening so quickly in so many places throughout the nation, however, even impressive local expansion might not be enough to keep cultural anticipation rosy if progress seemed less robust than in other cities. Time had its own watery qualities in that it was, as has often been said, like a river, and its shifting currents at any historical moment seemed to favor some communities over others. In 1805 the *Aurora* described Philadelphia as "the chief city of the United States, in point of size and splendour," but conceded that "it now fills but the second rank in respect to commercial importance." In recent years "the trade of America" had "flowed more freely into the open channels of the bay of New-York."[5] In spite of Boston's record pace of growth in the 1830s and 1840s, local historians admitted that New York had seized control of the future, leaving them to reconcile themselves to the fact that their city was a regional rather than a national hub.[6]

In contrast, Chicagoans believed that time would soon fill their city's cup of eminence as full as Lake Michigan, and that at some not very distant day Chicago would become the country's leading city. Mayor Walter Gurnee proclaimed in 1852 that the recently opened Illinois and Michigan Canal, combined with "the certain completion of the great lines of railroad which are converging to this point," promised that, "with proper attention to our municipal affairs, [Chicago's] progress will in [the] future be even more rapid than heretofore, and will realise the hopes of the most sanguine."[7] Fifteen years later, a journalist writing from New York reported that businessmen there looked to Chicago "as the source whence are to flow streams of trade the coming autumn, and in fact for a long future." He noted that it had not been long since "a New York merchant would hardly recognize a Chicago merchant."[8]

In spite of the vagaries of time, civic leaders had confidence that an enterprising city could do a great deal to control its fate if, as Gurnee said, it paid proper attention to its affairs. Like its natural setting and resources, time's value lay in how well it was used. In this sense, it was less a current than a kind of currency that, like it or not, one was constantly spending. The task at hand was to spend time wisely. Among the best ways a city might do so was by paying careful and effective attention to water.

Booster Visions

Proponents of water proposals in all three cities often identified the community's historical progress through time as correlating with the condition of its water. Aware of interurban competition, they pointed to the projects they favored as an effective means of intervening for the better in the flow of their city's history. Amid the transformations and instability of accelerating urbanization, they claimed that the waterworks they advocated would ensure the local population's role as actors in time by meeting their city's present needs and its ambitions of continuing greatness. They claimed that the waterworks they supported would not just provide water but also serve the vital purpose of enabling a city to fulfill its highest possibilities.

This also meant that to fail to construct the system they recommended was to betray these possibilities. Speaking in support of internal improvements in 1829, Boston mayor Harrison Gray Otis used the history-as-flow metaphor to warn that if Massachusetts and Boston hesitated to undertake such major projects soon, "the streams of our prosperity will seek new Channels."[9] Five years later, Mayor Theodore Lyman Jr. told the Boston City Council that committing resources to city water was one of those "items of expense that belong to a civilized and enlightened condition of society."[10] A city that neglected to recognize this, Lyman and others argued, would condemn itself to stunted growth and civic embarrassment.

Similarly, a growing city must not just provide its citizens with excellent water; it must also do everything it could to make sure that local and national public opinion of its water was highly positive. Current reputation and future prosperity depended on providing water of solid reputation. According to Lyman, every self-respecting and ambitious commercial city was "obliged to bid" for population and resources from elsewhere, and "pure and good water cannot be placed low in the list of inducements or temptations to any class of persons to settle in a place."[11] Engineer James Baldwin emphasized this point when he dissented from the 1837 report in which two colleagues recommended that the city take its water from Spot and Mystic Ponds rather than from Long Pond. Baldwin not only felt that the quality of the other two sources was inferior, but also that a significant number of others shared this view, and that their attitudes must be taken into account. Just being thought to have suspect water, whether or not this was the case, Baldwin advised, "is a stigma which the guardians of the public welfare should be anxious to remove."[12] Two years later he made a similar point when he stated that Mystic Pond was badly polluted by all the tanners and curriers who dumped their waste products into it. Boston would jeopardize not only the health of residents but also the city's future good name if it introduced water that was impure or even "not popular." People would identify Boston with its water source and spurn both, so that a new system, "instead of becoming the pride and boast of our city, will be a sad failure and a lasting disappointment."[13]

In 1862 the editors of the *Chicago Tribune* maintained that "the Water Question in this city is one of pressing importance, and that it must be correctly solved, or be left open at the risk of making our town uninhabitable."[14] Even if its water was in fact plentiful and good but was believed to be of low quality, that would lower the city's appeal to the eastern investors that Chicago wished to impress. Two years later, the paper reprinted a recent article by a waggish *New York Evening Post* correspondent who expressed his distaste for mixed drinks made with Chicago's "swamp water," wishing that the city had a Croton Reservoir or Lake Cochituate (so much for the idea that good water would remove a major reason for the consumption of alcohol).[15] The 1867 report of the city's Board of Public Works, which announced the recent completion of the lake tunnel, confided that the board had always believed "that to fail, from any cause, would have been disastrous to the city and themselves."[16]

But if cities feared what would happen if they did not take effective steps to improve their water supply, they found courage and motivation in the assurance of what would result if they seized the day. Good water would draw new people, which would benefit the city, since, as Lyman explained, "every accession of population adds to the value of land and every description of property."[17] Josiah Quincy Jr. countered the assertion by enemies of the Cochituate proposal that it would cost far more than estimated (a contention that proved correct) when he stated, "My own opinion is, that the increase of population and the value that will be given, by the supply of water to these lands, will more than provide for the difference, if any should occur." Quincy also argued that the ramifications for the city's standing would be severely negative if it chose the wrong source. "Boston 'is a City set upon a hill that cannot be hid,'" he explained, echoing not only John Winthrop's most famous turn of phrase (taken from Matthew 5:14) but also Winthrop's reminder that, given the settlement's high and holy mission, "the eyes of all people are upon us."[18] Once voters endorsed the Cochituate system, however, Quincy expressed confidence that its construction would make up for time lost to other cities that had already taken action and would prove that Boston remained true to its lofty sense of purpose.

The common argument of water boosters was that the more boldly a city met the challenge of providing a plentiful supply, the greater the rewards would be over time. The more water the better. Advocates of waterworks likened "watering" the city to tending an insatiably thirsty plant so that it would grow as large as possible. There was no way to overdo it, since every drop would be generative. They were convinced that building an optimistically capacious waterworks was a self-fulfilling prophecy of growth. A city that truly believed in its future would dare to construct a system commensurate with its grandest conception of itself, and the resulting abundance of water would transform expectation into actuality. This in turn would convince people of the rightness of the original ambitions, leading to an even more glorious future in which conception and realization would keep pushing each other ever onward and upward.[19] To build a waterworks was to master time by entering into an insuperable alliance with the future.

A main reason the water debates dragged on for so long in Boston and became so contested was the opposition mounted by those who offered more conservative projections of future needs. Proponents of building big had no patience with what they considered to be tightfisted and shortsighted caution and prudence.[20] City attorney Richard Fletcher declared that it was unacceptable to tell Bostonians "that they could possibly *get along* as they were, and that they would not actually perish without a further supply." The small-minded and defensive outlook implied by "get along" and "not actually perish" would render city life narrow and mean. The future of Boston as the leading metropolis of New England tasked the city's citizens with adopting a strategy that supplied them with water "freely and liberally, and without stint." Fletcher described a properly supplied Boston as overflowing with water: "[Water] should be seen all about us—in fountains before the State House, on the people's Common, in the public squares, it should be like the air we breathe, everywhere, in doors and out of doors, ministering to the comfort, cleanliness and health of every living thing." As things currently stood, the city's residents were sadly accustomed to "a scanty supply of impure water," which made some (i.e., opponents of a new system) skeptical of the need to construct a works, but once everyone enjoyed "the blessedness

of an abundant supply of pure water,—of the little band who now oppose this measure, there could not a man be found, who would undertake to utter a single word against it."[21] City water *was* city life.

The members of the Faneuil Hall Committee strongly disagreed with
anyone who suggested that Boston did not need as much water as it
could possibly get. They belittled those who would "parsimoniously cut
and carve, and calculate upon a provision only for our present wants"
when it was "a moral certainty . . . that but little time can probably elapse
before new enterprizes [sic] must be resorted to, to meet our wants."
Opponents of the Long Pond plan—"who in their real or affected regard to *prudence* and *economy* would have us measure and estimate by the
gill the amount of water that may possibly be made to answer"—were
small-minded and wrongheaded. The city's future depended on a "liberal
policy." At this crucial moment, "half-way measures in this important
matter" were "adverse to the true interests of Boston." The committee
praised New York for building a works that could meet the demands of a
population five times its current number, telling voters that with a generous water supply Boston's population would double or triple in size.[22]

Chicagoans also maintained that it was impossible to build a waterworks that was too big. In 1858 Water Superintendent Benjamin F.
Walker pointed out how the city's short history was one of continuously
exceeding expectations. "The unexampled prosperity and increase of our
city in wealth and population has of course set at defiance all calculations based upon what Chicago was six years ago," Walker declared, "and
any prudent calculation for the future might fall as far short of the reality as those already made. I think, therefore, there is very little danger of
our being likely to extend the works faster than the wants of the city will
require."[23] As noted earlier, in 1870 the Chicago Board of Public Works
explained that while its policy in laying new pipes was to give precedence
to settled areas that guaranteed immediate revenue from users, it was
also a good idea to place them in sparsely inhabited neighborhoods, since
new businesses and people would go wherever water was supplied. This
would in turn lead to new construction of buildings that would take the
water and increase waterworks revenue, which would spur further expansion.[24] Four years later, when a second lake tunnel seemed to prom-

ise enough water to support a population of a million and a half (at this time about 400,000 people lived in Chicago), the board maintained, "It cannot be denied that this invaluable source of general health and comfort is a significant factor in the increase in population."[25]

Borrowed Time

A large public waterworks demanded a capital investment of far more funds than were in the municipal coffers. If a city decided to construct its own system, it would need to borrow money—a lot of money—and in so doing enlarge its budget to a heretofore unimaginable degree. The decision to issue bonds on this scale for a public improvement was in its own way as momentous an event in American urban history as that of building the actual waterworks, since such borrowing would underwrite not only water but also other major initiatives. In addition, going purposely into significant debt as a matter of policy, like the construction of a centralized system, caused city people to focus their attention on the nature of both urban time and urban community.

A city (or any entity) that issues a bond is dealing in time as well as money, since it is borrowing funds that it promises to repay by a fixed date in the future. The interest on the bond is what it costs to have use of the principal for an agreed-upon amount of time. Cities have commonly covered smaller expenses with short-term loans of under a year. This is sometimes termed, metaphorically, floating debt. But for a project like a waterworks, they take on what is known as funded debt, usually much larger sums that they borrow for a period of years.[26] Before the 1790s, when yellow fever convinced Philadelphians that they must raise money for a waterworks, there had been only limited need for municipal borrowing. Infrastructure—mainly roads, bridges, and public buildings—was financed with current taxes, donations, subscriptions, special lotteries, and the sale of city-owned real estate.

As urbanization accelerated, so did funded debt for large undertakings, with waterworks leading the way. Bond issues expanded in earnest in the 1820s and 1830s, so that by 1840 the total of all municipal debt in the United States had risen to approximately $25 million.[27] Bonds would soon be used not only to pay for major infrastructural improvements but also public schools. Individual projects were often separately incorporated by the state legislature—this was true of the publicly owned Cochituate system and Chicago City Hydraulic Company, though not Philadelphia's first works—which would authorize borrowing up to a specified limit. An important reason to establish these semi-independent corporations with their own funding authority was to enable cities to get around the borrowing caps set by their charters.

By 1801 the anticipated cost for dealing with all of Philadelphia's city debts and expenditures was $64,136. Of this, $8,000 was due to installment payments on Nicholas Roosevelt's engines and other construction costs, with another $4,470 to cover interest on the water loans. Other significant estimated costs were for lighting and watching the city ($16,000), cleaning it ($9,500), and paying its officers ($7,200).[28] In 1806, when the city spent just over $96,000 from its annual tax fund, almost half of that went to the waterworks, including $4,279 in water loan interest, $10,000 to pay back a loan taken out in order to compensate Roosevelt, another $10,000 for distributing Schuylkill water, $16,682 to maintain the engines and for employees' salaries, and some $3,500 in repairs and improvements on the Centre Square engine house. The waterworks' disproportionate strain on the municipal budget would continue for over two decades.[29]

The town of Boston resisted borrowing money, but from the incorporation of Boston as a city in 1822 to the end of the first Josiah Quincy's activist mayoralty six years later, net municipal indebtedness multiplied almost ten times, to $670,000. This figure peaked at over $1.5 million in the early 1840s, followed by determined belt-tightening during the next few years that cut this sum by two-thirds. Then it rose to new heights, thanks to the Cochituate waterworks. Boston hoped that by 1853 the Cochituate system would be paying for itself, or even running at a small profit that could be used to reduce the principal on the loans. Instead, while that year water receipts were almost $214,000, expenditures were close to $345,000. The works turned a profit in 1861—a very slender $2,000 (on receipts of over $380,000)—but the budget then went back into the red, forcing Boston to borrow more money and hike taxes. The project raised Boston's net funded debt to nearly $5 million in 1849, with almost $3.8 million (or over 75 percent) of it being water debt. This figure approached $7 million four years later, and about $6 million of this was related to water. For the next few years, the net funded debt of the city continued to increase, but the proportion of it due to the waterworks declined.[30]

The new city of Chicago, which was incorporated in 1837, was averse to going into debt.[31] The city charter authorized it to borrow up to $100,000 annually, but there was little expectation that it would. Between 1849 and 1868, however, as the population grew more than twelvefold (from about 20,000 to over 252,000), annual government spending accelerated more than ten times faster, from about $45,000 to more than $6 million. The city's debt climbed at an even more stunning pace, from $20,338 in 1848 to over $14.1 million in 1871, almost 7,000 percent. As of the end of March, 1869, the year the Water Tower was completed and two years after the lake tunnel opened, the city had spent $3,146,383 on the construction of the water system, of which bonds paying 6 and 7 percent interest provided $2,677,362. Servicing this loan was by far the largest single annual expense incurred by the works, far outstripping the cost of salaries for water administrators and employees, office rent, and repairs. In 1871 Chicago was paying almost $290,000 annually on what was by then $4.82 million in water bonds.[32]

The borrowing demanded by waterworks inspired bold and ingenious invocations of history and time that characterized debt as a bequest or endowment for which future residents should be grateful, not as a burden they should resent. The 1799 report of the Philadelphia Watering Committee explained that repaying the loan over time would render the current and future annual expense minimal "by dividing it between the present and many succeeding years." True, this placed a financial obligation on Philadelphians who were yet not members of the community that took out the loan, but this was only fair, because they would share in the advantages provided by the waterworks for which the money was borrowed. The Watering Committee pointed out that people living in the city in years to come would be "justly made to pay, in some proportion, for the benefit they would receive."[33] A water loan should be viewed as a gift to them, not an encumbrance, since it did no less than make a greater Philadelphia possible. This logic expanded the urban community to include people not yet arrived, who would live in the more expansive metropolis that would come to be as a direct result of yet-unbuilt water-works. This splendid waterworks and the money foresightedly invested in it would draw these people to the city and keep them there by offering them membership in a community that, because of the waterworks that made it so large and prosperous, would easily and gratefully redeem the debt assumed for its future benefit. Such thinking rationalized a city's borrowing money it did not have by simultaneously projecting a con-stituency that did not yet exist.

In Boston as in Philadelphia, supporters of borrowing to build a pub-lic waterworks described the water bonds as a legacy that would benefit future residents. Josiah Quincy Sr. unhesitatingly recommended mak-ing use of the recently incorporated city's legal authority to go into debt when this capability was applied to purposes with clear long-term ben-efits.[34] In reply to those who believed that this was an inherently bad policy, Quincy compared borrowing to a sword, stating that the key to each lay in how it was used. While both issuing a bond and brandishing a sword were dangerous when done by incompetent fools and irrespon-sible madmen, they were likewise "very safe, innocent and useful in the hands of the wise and prudent." Switching comparisons, he maintained that a shrewd loan "is no more a just object of dread, to a city, than a debt created for seed wheat, is to a farmer; or than a debt for any object of certain return, is to a merchant."[35] For his part, Mayor Josiah Quincy Jr. stated that the borrowing he advocated to erect a publicly owned water system was both prudent and economical, given that it magnified the promise of Boston's future.

In *A Plea for Pure Water*, Walter Channing attributed a special mean-ing to a debt that was "the price, the purchase money of a great and *permanent* interest" like the Cochituate works. This was because the water-supply system was not to be constructed "for ourselves, for this city as it is, or for this age." No, it was "to be made also for the coming time, the long future, a blessing beforehand, and in the beginning and completion of which the future has no concern." He scoffed at skeptics who used scare tactics "to prevent one of the most important measures ever attempted by any city." In answer to the question that doubters raised about who would pay the debt and how, Channing declared, "I say it is the debt of the city in all the days of its being, till it is paid." One had to think of all those who constituted the Boston of tomorrow as well as of today, and their relation to each other. "We are working, in all we do," he proclaimed, "for our successors, and paying for them, too, and we cannot help it." Current Bostonians were the "great parent" of the future. Channing implied that the gift of debt tied the lives of present-day Bostonians to those of future residents, and so it extended the former's existence. They would live on in the future's positive appreciation of their wisdom.

"Whose debt then is it?" he asked. "It is the debt of all who are, or who may be, till it is paid." As such, water bonds were a trust fund of sorts that made possible the aqueduct by means of which generations to follow could draw Lake Cochituate's water. Rather than resent the water loans, "most grateful will the long future be that we created such a debt, which, in its product, will be to that future so great a blessing."[36] The following year, shortly after the public waterworks plan was temporarily defeated, the *Daily Evening Transcript* reprinted a comment in the Philadelphia *Gazette* that chastised Boston voters for withholding support. The comment compared them unfavorably to the citizens of New York, who had a decade earlier endorsed the Croton works. "We hope the Bostonians will soon learn to view the matter correctly and to look beyond themselves," it read, "to remember that they should legislate a little for those who are to come after them, and to endure a burden, that a good may be secured to the city and its denizens, when they shall have departed."[37]

As elegant a solution as long-term urban debt is in theory, it can be a lot messier in practice. In deciding whether or not to take on debt to pay for water that they deemed necessary for their cities to thrive, Philadelphians, Bostonians, and Chicagoans had to take into account not just the known expenses of building a system, but also the hypothetical costs of fires unextinguished, epidemics unchecked, and advantages lost to other cities if they did not construct a waterworks, or if the smaller one they did build instead was unsatisfactory or inadequate. Like all loans, selling municipal water bonds entailed risks and, as a result, plenty of tough choices, not to mention circumstances beyond the borrower's control, starting with the state of the economy.[38] Even if a city could calculate with accuracy how much water it likely needed—and none did—it was difficult to know in advance how much borrowing might cost since this depended on the unpredictable financial marketplace.

A city seeking funding had to convince potential bondholders that its credit was good. In planning new works and issuing bonds to fund them, Philadelphia, Boston, and Chicago expressed their belief in how large they would grow if they had the system for which they were borrowing money to build. Purchasers of the bonds in turn confirmed this belief—albeit for a price that varied with the degree of perceived risk and the alternatives that potential lenders had for their capital. When Chicago took out what A. T. Andreas called the city's "first great loan," the $250,000 in twenty-year 6 percent bonds it negotiated in 1852 to construct its first publicly owned water system, the momentous transaction was front-page news. As Andreas put it, "The Chicago of that day commenced to draw confidently upon the Chicago of the future."[39]

To gain investor confidence and lower the cost of the money that they needed, cities established so-called sinking funds (another financial term, like "float," that links money and water). These were sums from annual revenue sources that were set aside to help pay off long-term debt and, in some cases, cover additional expenses that might arise. The establishment of a sinking fund was another way to try to manage the future. In 1807 Philadelphia became one of the first of many cities to employ this device. In order for sinking funds to work as hoped, two upbeat assumptions about the urban future had to come true: that the city's revenues would exceed its continuing expenses, and that the government would, as Philadelphia's did, follow through on its commitment to contribute to them and to use the funds as specified.[40] Waterworks planners anticipated that within a reasonably short time the new works would turn a profit and would help pay off the long-term debt that the city had assumed. But even in Chicago, where (unlike Philadelphia and Boston) this proved to be the case, the water system continued to take a leading place in the city's rising debt obligations.[41]

In their early efforts to borrow, both Philadelphia and Boston discovered it was sometimes hard to find lenders at all, which was why Philadelphia proffered three years of free service for residents who purchased its first water bonds. Boston had originally planned to enlist buyers in Europe, and it sent two different agents abroad for this purpose, hoping to secure a 4 percent loan. But competition from borrowing by railroad companies (which also had enormous upfront costs), recent defaults by some states, and unsteady financial conditions in Europe thwarted this plan. Although Mayor Josiah Quincy Jr. was able to report that Boston was paying less than 6 percent on $3 million in bonds, which was lower than prevailing rates, he admitted that this was still "higher . . . than would be required in ordinary times on the securities of the City of Boston." He also claimed that his administration was doing the best it could to time the market in securing an additional authorized loan of $1.7 million. The plan was to sell five-year bonds in the expectation that Boston would be able to refinance on more favorable terms when these came due.[42]

Critics of building a large new public waterworks had to choose their words carefully, since they did not wish to appear pessimistic about the prospects of their city. They sometimes attacked a proposal indirectly by questioning income estimates as overly sanguine. The Long Pond opponent (and Spot Pond supporter) who called himself "Temperance" observed sarcastically in 1845 that the city's population would never reach the figure of 250,000 (which it did by 1870) cited by a backer of the Water Act, "unless the seventy-five acres of land in the Common is appropriated for dwellings, and the spirit of emigration [out of Boston] is vetoed."[43] Lemuel Shattuck predicted that the population would not increase much above its current size because many local people would find the emerging suburbs a more attractive place for their homes than the city. They might well conduct their businesses in Boston, he reasoned, but "they will be compelled to have their dwellings in the adjoining towns, and in places easily accessible by our railroads," because they wanted to live where pure water and air "can be freely enjoyed."[44] (As it turned out, both Boston *and* its suburbs grew rapidly.)

More convincing and worrisome were the warnings that heavy debt would foreclose rather than enable a shining urban future. These warn-

ings were loudest and most numerous in Boston, which was the only
one of the three cities where a privately funded alternative to a public
works appeared to be a real option. During the waterworks discussions
of 1838, 164 taxpayers petitioned the mayor, aldermen, and Common
Council, citing their alarm "at the prospect of having the debt of the city
increased in a two or three-fold ratio, for the purpose of supplying the
city with water." They held that the Panic of 1837 had made the present "a time of great commercial distress." It was far preferable to entrust Boston's water needs to privately funded companies rather than
increase the city's debt or its residents' taxes, especially since it was not
clear how many people would switch from their current water sources.[45]
Opponents of public systems mounted their most effective argument
with voters when they pointed out the size of the debt that building a
waterworks would incur. In Boston they repeatedly complained that re-
deeming this debt would be very difficult because the proposed works
was so much bigger than what was needed, which meant that too few
people would have to pay for too much water. This would put a heavy
burden on the Boston of the present that would hinder its growth.

Boston mayor Jonathan Chapman noted that the public had long been
mindful of the need for a waterworks, and that there was no doubt that
"something must in time be done in reference to this important matter." Now was not the right moment, however, since this project "must
involve a very considerable outlay, and it cannot but be admitted that
some doubts may reasonably be entertained as to its pecuniary results,
for at least a considerable amount of time." Chapman cited as implicitly
backing his position none other than the elder Josiah Quincy, who as
mayor held a liberal view of investing in new projects but whose recently
published history of Harvard observed that while "those who limit and
economise are never so acceptable to mankind, as those who enlarge and
expend," there was "no higher obligation [that] rests upon history, than
to do justice to men on whom these unpleasant and unpopular duties
devolve." Chapman stated in 1840 that while there was little doubt of
the importance of building a water supply, the high cost and uncertain
prospects of this venture were such that "no prudent government would
enter into it" without a strong majority of representatives and the citi-

zens who elected them behind it. At present, he argued, "the public mind is not yet ready to sanction the undertaking by the City Government." A year later he said that although it would be "pleasant and exciting" if the city were able "to embark in large and striking enterprises," in his judgment "the homelier and less captivating duty awaits us, of husbanding resources and superintending details."[46]

In 1845—by which time financial conditions had improved considerably, and it was clear to all that the city had little choice but to do something on a major scale about its water supply—opponents of the Long Pond proposal nonetheless continued to warn that debt endangered the city's future. An attorney for opponents of the Long Pond aqueduct told the state legislature that a public system "will involve a needlessly large expenditure of money, bring upon the city a heavy debt, and subject them and coming generations to an onerous taxation to pay the annual interest on the debt, which will not for many years, if ever, be met by the income from water rents."[47]

Several early Chicago mayors maintained that the city's brief history magnified the implications of every action, and so financial discipline was the best policy. In 1848 Mayor James Woodworth described the current debt, which was less than $10,000, as "extremely embarrassing and mortifying," not to mention "highly prejudicial" to the "pecuniary interest of the city."[48] Speaking two years later, Mayor James Curtiss was upset that the municipal debt had increased to over $50,000, leaving Chicago "to hobble along and take care of itself as it best may." The city's credit, the measure of the world's faith in its future, was "practically worthless" unless it paid its bills.[49] In 1851 future mayor John Wentworth led the opposition to the proposed publicly owned Chicago City Hydraulic Company on the grounds that it would cause Chicago to borrow heavily.

The many advocates of a public water system responded vigorously to such hesitations about borrowing. In 1838 Boston mayor Samuel A. Eliot stated that members of the Joint Standing Committee on Water believed that "the financial embarrassments, which will occur to every mind as a plausible objection to an immediate beginning [of constructing a public works], are nothing more than temporary evils." He also forwarded the argument, made decades earlier in Philadelphia, that the

project would benefit the economy by providing jobs. There was "no better means of alleviating the distresses of those who depend upon labor
for support, than by furnishing them with employment."[50]

In Chicago, as elsewhere, leaders who favored building a public system distinguished between annual operating shortfalls, which were to
be strenuously avoided, and taking on large-scale debt to fund worth-
while long-term capital-intensive improvements. A community's being
ready, willing, and able to borrow money in order to pursue such plans,
they asserted, was a sign that it took its responsibilities to its future seriously. In 1852, the same year as Chicago's first major water loan, Mayor
Walter Gurnee said he assumed that citizens did not wish to forgo those
improvements—he also listed schoolhouses, markets, prisons, sewers,
streetlights, and the police and fire departments—that "mark the progress of the times, and give character to the city."[51] Five years later, John
Wentworth, now mayor, stated that while he was determined to put an
end to ill-advised short-term borrowing to cover high city expenses due
to inefficiency and corruption, the Chicago of years to follow would benefit from long-term borrowing for major infrastructural projects.[52]

A few days after the 1848 celebration of the introduction of water
into Boston, the water commissioners sent to the offices of the *Daily
Evening Transcript* "a pitcher of the genuine Cochituate," which the paper
approvingly found to be "soft, inodorous, and untasteable." According to
the *Transcript*, "The value of such a blessing, freely dispensed throughout
our city, is not to be calculated in dollars and cents; for it has relations
inestimable with the moral and physical welfare of generations present
and to come."[53] In late November 1849, by which time he was no longer
mayor, Josiah Quincy Jr. used similar phrasing to evoke the future in a
short speech at a ceremonial inspection of the continuing work on Lake
Cochituate and the aqueduct. He noted the great difference between the
kinds of projects for which taxpayers of Boston and Massachusetts had
agreed to go into debt and the bloody and destructive wars for which
foreign nations borrowed money. Yes, the waterworks was expensive,
but the voters of Boston would not now part with the Cochituate system
for double its cost. He then offered a toast: "*The Debts of Boston and of*

Massachusetts—Expended in procuring blessings for this age, and blessings for all coming time."[54]

Manifesting Destiny

"Procuring blessings" is a revealing phrase. A "blessing" is usually a positive benefit that may be requested or desired by the recipient but in any
case is bestowed by another. "Procuring," on the other hand, involves the
exertion of active effort to get something on one's own. Quincy's toast
restated the view that successful communities were the ones that made
timely use of the opportunities afforded them, while settlements of
lesser imagination and energy let opportunity—and time—slip away.

Some conveyed similar ideas by referring repeatedly to their city's
"destiny," which, they believed, city people were currently fulfilling by
recognizing and realizing the possibilities offered by all their blessings.
The language of destiny at once provided a reassuring explanation for
how history was unfolding and an endorsement for projects like the Cochituate system. Josiah Quincy Sr. emphasized how much greatness was
there for the taking by using the plural when he explained what rewards
the future held for his city if his fellow citizens made certain steps. "The
destinies of the City of Boston, are of a nature too plain to be denied, or
misconceived," he declared in 1824. These "destinies" were inscribed, as
others also contended, in the abundance of natural resources on which
the city could draw (including Long Pond) and, "above all," in "the character of its inhabitants." This argument elevated the claim that urban
growth was in accord with nature to viewing this growth as a primordial
force in itself that was driving history. People and capital were streaming
to Boston not by some accident but "by a certain and irresistible power
of attraction."

Having flattered his fellow Bostonians by praising their "singular degree of enterprize [*sic*], and intelligence," Quincy reminded them of their
duty. They must be "willing to meet wise expenditures and temporary
sacrifices, and thus to cooperate with nature and providence in their
apparent tendencies to promote their greatness and prosperity." Great

public works projects would at once accomplish the worthy goals of "improving the general condition of the city, elevating its character, multiplying its accommodations and strengthening the predilections, which exist already in its favor," while also serving the practical purpose of "patronizing and finding employment for its laborers and mechanics."[55]

When the members of the Faneuil Hall Committee predicted that a public water system's successful completion would double or triple the population, they added that this was "destined" to happen "at no very distant day."[56] Josiah Quincy the younger also spoke of urban destiny multiple times. He exhorted Bostonians to remember "that we are the fathers of the generations that will succeed, and that we have not the apology of being ignorant of the probable destiny of our City." After presenting a detailed case in support of the construction of the Cochituate system, which he called one of "the duties that await us in the year which we have entered," Quincy reflected, "But we cannot be faithful to the present, without casting our eyes toward the future."[57] This view turned the mundane act of building waterworks into a solemn obligation, the alternative to which was tantamount to spurning Providence.

The same kind of rhetoric that described building water systems as the heroic fulfillment of destiny provided the basis for equally hyperbolic assessments of their historic significance. After admitting that appropriations for 1801 would be $6,000 more than the previous year, the Philadelphia City Council's Committee on City Debts and Expenditures informed other members of the council that this increase was due to the expense of completing Latrobe's plan, which was nothing less than "one of the greatest works ever undertaken by this, or any other city on the continent."[58] The Boston waterworks parade of 1848 passed under a series of fourteen tablets suspended on ropes over the streets of the parade route. Garnished with evergreens and flanked on both sides with American flags, each tablet briefly described a step in the long march to building the new water system. The parade was thus symbolically reenacting Boston's long journey to the accomplishment of its exalted historical purpose.[59] As the Lake Michigan tunnel was nearing completion, the *Chicago Tribune* hailed it in a headline as "The Eighth Wonder of the World."[60] Ten days later, it called the tunnel "the glory of the nineteenth century," as well as "the pride of America," and, once again, "the wonder of the world."[61] Shortly before a second lake tunnel began to deliver water in 1874, the Chicago Board of Public Works claimed that the two tunnels would soon be able to provide a population of a million and a half people with "a drinking water as pure as that enjoyed by any city in the world, while the facilities for its supply are unsurpassed."[62]

Sometimes the allusions to destiny referred to the example of other leading cities in order to promote a new waterworks proposal. According to the Faneuil Hall Committee, the point of building a larger system than Boston needed at the moment was to transform it into one of the greatest metropolises in the world, contending that Boston could equal or surpass other well-known places near and far if it only had the will to do so. "Why should not Boston attain to as great a population as Liverpool, as St. Petersburg, Vienna, or Paris?" the committee asked.[63] A year earlier Henry A. S. Dearborn—who had been collector of the Boston Custom House and served in both the Massachusetts legislature and the United States House of Representatives—told those who might be hesitant about the scale of the Long Pond plan that Marseille was currently constructing a publicly owned aqueduct more than twice as long.[64]

Civic leaders also compared what their city was accomplishing with what had been done by famed cities of the past. They explained that their purpose was to demonstrate that the goal was to meet the historical benchmarks of great urban achievement and to take a place of honor in the memory of future generations. Waterworks served such claims so well because of their close association with the most admired cities and nations throughout time. When a new system for Boston was still in doubt, Walter Channing wrote, "I honor those ancient States which made such noble provision for the public want in regard to Pure Water." He asked, "What has the latest civilization, and which is so vaunted, done, which approaches to that ancient provision for the hourly recurring wants of man?"[65] At the reception following the 1846 groundbreaking for Boston's new system, John Quincy Adams turned classical history to the purposes of cultural anticipation in his toast to the waters of Lake Cochituate: "May they prove to the after ages of the city of Boston as inspiring as ever did the waters of Helicon to ancient Greece."[66] The obvious precedent, of course, was Rome, whose grandeur was so closely identified with its magnificent aqueducts. American waterworks advocates agreed with Goethe's reaction to the Roman example, "What a noble ambition it showed, to raise a tremendous construction for the sake of supplying water to a people."[67]

Channing praised New York for the nobility of its ambition in building the Croton reservoir. He complained that while many in Boston were overly fixated on the cost overruns New York experienced, "not a word has been said of the real character of that splendid monument to the civilization, and to the philanthropy, of our sister city." All should appreciate that "*It is a national work*. It is a work which will carry into the time long to come, the enlarged intellect which planned, and the noble spirit which completed it. Who does not thank that munificent city that it has made such noble use of its honorable wealth? [A]nd what community would shrink from the office of following so worthy an example?"[68] Boston should be ashamed if it did not meet the challenge of its historical moment by building a magnificent aqueduct.

On January 2, 1846, by which time the construction of the Cochituate system finally seemed certain, Boston Common Council president Peleg W. Chandler had no doubts about its historical significance. "The introduction of pure water into our city is not to be regarded with a narrowness of view, that looks only to the present," he stated, but as "a great moral spectacle." He counseled his colleagues that they could best perceive their obligation to Boston's future by looking back to Rome's crowning achievements in hydraulic engineering, which served its people in so many ways every day of their lives. These "appeal to the highest feelings of our nature, whilst we gaze on the Colliseum [*sic*] with awestruck amazement alone."[69]

Enduring Monuments

The designers and supporters of nineteenth-century American water systems often spoke of them as monuments. The *Daily Advertiser* pronounced early in 1846 that the Cochituate works would bring everlasting honor and glory to the city of Boston. Two weeks after the introduction of the water in 1848, the paper predicted that "not Boston only but Massachusetts and New England will be proud of this structure as one of the noblest monuments of civil architecture existing in the country."[70] To describe waterworks this way was to reinforce the idea that they were statements from the present to the future, physical reminders of how much the latter owed to those forebears who demonstrated their sense of history by thinking ahead. "Like the ornaments of the desert," the *Daily Advertiser* stated, referring to the pyramids of Egypt, the Cochituate aqueduct "will be the admiration of future generations."[71] When work began on the system, Josiah Quincy Jr. admitted to "the pardonable expectation that our successors, as they enjoy the benefits of our work, may remember that on the 20th of August, 1846, this great undertaking was commenced."[72]

At the 1879 dedication of the twenty-foot statuary monument to Quincy's father (who died in 1864 at the age of ninety-two), Unitarian minister Samuel K. Lothrop stated his hope that the work would "abide for long years and successive generations," so that it may "ever and always speak to us, and to the successive generations as they pass, of his worth, and virtues, and usefulness." Quincy drew praise for going beyond "the necessities of the present" in his fidelity to the idea "that Boston was destined to be a metropolis."[73] Sculptor Thomas Ball had dressed his figure of Quincy in a coat over which was draped a cloak suggestive of a Roman toga.

Waterworks ceremonies also emphasized the meaning of these projects as messages from the present to the future. At the Cochituate reservoir groundbreaking, the shovel wielded by the participants bore the inscription, "With this spade the first earth was removed, in constructing the LONG POND AQUEDUCT, By Hon. Josiah Quincy, Jr.," followed by the names of the water commissioners and of the members of the City Council Water Committee.[74] As part of the Beacon Hill Reservoir dedication, a copper box twelve inches square and six inches deep was sealed in the cornerstone. Among the contents of this time capsule were plans of Boston and adjoining towns, charts of Long and Spot Ponds, and waterworks documents dating back to the initial water report of 1825, as well as various souvenirs of the times, such as money, newspapers, current statistics, and official documents.[75] In addition, the box contained two silver plates on which were engraved the names of city officials and participants in the construction of the works.[76] Just before Chicago mayor John B. Rice set the final block in place to complete the Lake Michigan tunnel, he and other members of his official party pushed a few coins into the damp cement. So that the future would know just when the great deed had been finished, on the surface of the block were carved the words "Closed Up Dec. 5, 1866."[77]

The builders of water systems wished to attest to future generations that they were masters and not victims of time. Like the Roman aqueducts, their waterworks would impress and inspire people many years hence. An emphasis on permanence ran through the repeated insistence on how long waterworks structures would last and the limitlessness of the quantity of water sources they would provide. "These people built for eternity," Goethe marveled upon visiting Rome, and the promoters of public water projects in Philadelphia, Boston, and Chicago claimed that they had acted in the same spirit.[78] We see this in Benjamin's Latrobe assertion that the Schuylkill supply would be "inexhaustible" and "perpetual," in Josiah Quincy Jr.'s insistence that the Cochituate aqueduct would have "a great and permanent effect," and in the Chicago Board of Public Works' declaration that the Water Tower would be "an enduring monument" that "may last for ages."

Similar assurances about the life span of a new system recurred throughout discussions of urban waterworks in all three cities. In 1812 the Philadelphia Watering Committee claimed that Fairmount "presents a very favourable surface for constructing reservoirs of such capacity as may be required or thought advisable, which at all times can furnish as great a supply of water to the city as can possibly be wanted."[79] When the 1838 report of the Boston City Council's Joint Standing Committee on Water argued for public ownership, it stated that this was "too important a business to be suffered to be affected by the calculations of private interest, which it is certainly possible might be injurious to the permanent character of a work which ought to be begun with reference to the future wants of a great and growing city."[80] The use of the phrase "permanent character" and the reference to Boston's enduring greatness indicate that the committee believed that the city was building a waterworks that was, as Henry Williams enthusiastically predicted, "fitted and calculated for Boston *as she will be centuries to come.*"[81]

Central to the importance of the Roman aqueducts and a key quality to emulate, Josiah Quincy Jr. observed at the dedication of the Beacon Hill Reservoir, was their longevity. "Why may not this edifice be of equal perpetuity?" He pointed out that the block of granite the city was laying "has existed from the creation, unchanged by fire, unmoved by earthquake," adding, in an allusion to Revelation, that it would "exist till time shall be no more."[82] In his speech at the celebration on the Common the following October, Quincy cited biblical and classical precedents when he declared, "There are no works of man so permanent as those connected with the supply of water." He predicted that "like the generations of men, a constant succession in this stream will make it permanent." Bostonians, he said, "will look back with gratitude to the men of this age, whose foresight and energy secured an unfailing wellspring, for themselves and their descendants."[83] A description of the gate house of the Brookline reservoir stated, "Every thing is of granite or iron, as if it were intended to last till the end of the world."[84]

Such open-ended predictions continued. Quincy's successor, John Bigelow, said that the waterworks "promises to be a permanent memorial of the public spirit and judgment of its authors, and of the skill and energy of those, under whose auspices it approaches completion."[85] A few months later, the *Daily Evening Transcript*'s description of the Beacon Hill reservoir concluded, "As an architectural ornament, and characteristic 'lion' of Boston[,] it will take the lead, and last as long as time does."[86] A *Chicago Tribune* reporter's glowing account of an 1854 visit to the city's first public waterworks called the installation "a credit to the city" that would "ever remain a monument to the genius, skill, and foresight of its projectors and executors," while the words of all the "croakers" who doubted the success of the project would soon be forgotten, as would be the croakers themselves.[87] Fifteen years later the Board of Public Works would describe the new Water Tower as being "of a massive and permanent character."[88] These expressions of certainty that a particular water source or waterworks would forever meet the needs of a healthy, prosperous, and growing city represented more than the enthusiasm of the moment. They also revealed a powerful form of wishful thinking that, while admitting that the course of history was a never-ending and always-changing stream, believed that one's city could nevertheless make an everlasting mark in it. In this view, a city that had secured the flow of water had taken command of experience, determining the course of the future rather than allowing the future to decide for itself.

In their talk of the permanence of new water systems, nineteenth-century American urban leaders seemed to assume that their cities' greatness would now be without end. This assumption drew heavily on a faith—Jefferson and other critics of urbanization notwithstanding—that the nation's cities were not like those of Europe but would be a confirmation of the country's exceptionalist character. To hold such an opinion was to claim that—unlike the most honored urban civilizations of the past—American cities would never decline and fade away but, like their new water supplies, live in perpetuity, which was another way of saying that they would outlast time.

It was therefore as important to distance one's city from the precedent of Rome as it was to cite it. While Rome presented the most honored example of urban triumph, and the one most closely tied to waterworks, it also was the most well-known instance of a great city's decline and fall. The lesson of Rome occupied the minds of such cultural observers as painter Thomas Cole, whose five-part *Course of Empire* (1833–36, New-York Historical Society) could be read as a warning about America's future.[89] The triumphant parade in the third canvas, *Consummation*, which in many respects resembles contemporary American urban civic celebrations, is an unwitting prelude to the catastrophic barbarian invasion that follows in the fourth painting in the series, *Destruction*, and the erasure of the imperial city depicted in the final painting, *Desolation*. The goal for rising American cities was to find a way to emulate the civic probity of the Roman Republic without succumbing to the inner decay that doomed the Roman Empire. Shortly after Mayor Josiah Quincy Sr. told

Bostonians that Boston's "future greatness" was "written on the face of nature" and that it was their duty "to be true to their own destinies" by building waterworks and other improvements, he also instructed them that they needed to avoid "the corrupt and superannuated forms of ancient despotisms."[90]

In the long letter to the *Daily Advertiser* in which he interpreted the defeat of the Long Pond plan in the May 1845 referendum as a triumph over Boston's elite and their desire to dominate the "dirty" masses, "Rabble" concluded that the very fact that the city was trying to build an enormous waterworks meant that it was succumbing to "the corrupt and superannuated forms" to which Quincy alluded.[91] The fall of Rome remained on Boston's mind after the vote was reversed in the referendum of the following year and construction finally began. In his Thanksgiving Day 1846 sermon, Nehemiah Adams sounded like his Puritan predecessors when he told the members of his congregation that they must not allow the new supply to become a source of earthly pleasure and pride and, in so doing, forget all they owed to the divine Giver of the water. History demonstrated that "the cities of the old world are full of the ruins of aqueducts and baths and marble palaces, to show us that God is angry with those who abuse his blessing to the purposes of mere self-indulgence."[92] At the dedication of the reservoir on Telegraph Hill in South Boston, Mayor Bigelow described a mighty aqueduct like the one that carried Cochituate water as "a monument erected upon the highway of time. Successive generations sweep past it. On it may be read the story of a city's greatness, or the record of its decline."[93]

Urban Americans wanted to believe that with enough water at hand the future would be one of ever-rising glory. When they hailed the benefits of water, they sometimes seemed to think that they could escape history and time altogether, at least insofar as these included an inevitable declension. Like Ponce de León searching for the Fountain of Youth, they hoped that the water systems they built might provide an antidote not just to dirt and disease but also to mortality itself.[94] Mixed into the mortar that held together the bricks and stone of which engine houses, aqueducts, reservoirs, water towers, and fountains were made was a wish that the preeminence of one's city and the memory of those who built it might never fade.[95]

There is an irony, of course, in the fact that the planners and promoters of waterworks wanted to inscribe their mastery of history in, of all things, water, which, like time, resists all attempts to hold it in place. In 1821, as he lay on his deathbed in—all too aptly—Rome, twenty-five-year-old John Keats expressed his belief in the futility of human striving when he dictated his epitaph, "Here lies one whose name was writ in water." When Boston local historian James Spear Loring proposed that Keats's epitaph might be applied to Josiah Quincy Jr., however, he intended that it would carry exactly the opposite meaning. Loring wished to proclaim how permanent would be Quincy's legacy precisely because of his role in building the Cochituate system. What made this accomplishment all the more remarkable was the elusive medium in which he worked it. "It has been felicitously said of the younger Quincy," Loring stated, "that he has written his name in water, yet it shall last forever. The imaginative vision of posterity shall see it written in letters of light, in the rainbows of the fountains."[96] Convinced, like so many others, that mastery of water meant mastery of time itself, Loring overlooked the fact that rainbows are an optical illusion whose magic and mystery are inseparable from their ephemerality.

7

EPILOGUE

It took until the 1870s for city water to begin to catch up with city life.[1] In 1830, about half of the nation's ninety communities of 2,500 or more (sixty-seven of which had populations of under 10,000) were served by waterworks. As the number and size of cities rose over the next four decades, this percentage dropped briefly but then jumped to 64 percent by 1880, reflecting the surge in urban infrastructure projects over the preceding ten years. By the last two decades of the century, the list of new works was growing even faster than the urban population.[2] This did not mean, however, that the water supply stayed ahead of needs in large cities. The builders of the early systems may have imagined that they were providing for the long future, but they seriously underestimated the demands of even the short one. And the same issues they encountered earlier remained central to discussions of city water.

Three Cities Revisited

Philadelphia, Boston, and Chicago quickly outgrew their original successful works and found themselves in a continuous race with demand. Philadelphia installed a fourth waterwheel at Fairmount in 1827, a fifth in 1832, a sixth two years after that, and two more in 1843. It laid a second main from the works to the city in 1829. During the 1850s it began to convert its waterwheels to a more efficient turbine design. With the 1854 consolidation of all of Philadelphia County into the city of Philadelphia—which more than tripled the population of the municipality to over 400,000 people and vastly expanded its footprint from 2.5 to 140 square miles—the city assumed centralized control of the waterworks that had served Spring Garden, Northern Liberties, and Kensington, and it built a pumping station in West Philadelphia.[3] Numerous other changes would follow, including the construction of four additional pumping stations between 1870 and 1907. By the early twenty-first century, Philadelphians obtained their water from three of these, all much improved since they opened: the Belmont and Queen Lane installations, located farther up the Schuylkill from Fairmount, and the Torresdale works, along the Delaware in the northeastern section of the city. The Fairmount works was retired in 1909, by which time both its condition and that of water it delivered had deteriorated badly.

The development of Boston's water system similarly tracked annexation and population growth. In 1862 the Cochituate Water Board urged the construction of a new reservoir, choosing a site in Chestnut Hill, about a mile northwest of the Brookline reservoir. This project, delayed by the Civil War, was completed on October 25, 1870, the twenty-second anniversary of the opening of the Cochituate works.[4] In the same year, the city took over the Mystic Lakes system when it annexed Charlestown. But Boston needed more good water than what was available nearby. It completed an aqueduct to the Sudbury River in 1878, and it added several other new reservoirs, including a much-altered Spot Pond. This was still not enough, however, and the water needs of the broader metropolitan Boston area soon led to a new kind of administrative solution. In 1895 Boston and twelve neighboring communities joined in the formation of the Metropolitan Water Board, which was chartered by the state legislature. The board built the Wachusett Reservoir by damming the Nashua River near Clinton, Massachusetts, northeast of Worcester. This was completed by 1905. The Wachusett project was tiny compared to the creation of it successor, the 400 billion-gallon Quabbin Reservoir, situated eighty miles west of Boston, which serves around 2.5 million people and over five thousand large industrial users in over sixty communities. Construction of the two dams at Quabbin began in 1936, and the filling of the reservoir took from 1939 to 1946. The original Cochituate Reservoir and the aqueduct between it and Boston ceased operation in 1951.[5]

Chicago officials scarcely had time to celebrate the completion of the 1867 lake tunnel before they realized that the city required a second tunnel, which was in place by 1874. Chicago would add more tunnels and cribs and acquire other pieces of waterworks through annexation of neighboring communities. The largest single jump in the size of the city and its water system came with the annexations of 1889, which more than doubled the area of Chicago to 185 square miles and added 225,000 people, making the city the nation's physically biggest metropolis (it was already the second largest in population).[6] Today some twelve pumping stations throughout the city push about a billion gallons of Lake Michigan water into the mains daily.[7] This water passes first through the Jardine Water Purification Plant, just north of Navy Pier. One of the pumping stations is the 1860s castellated Gothic building on Chicago Avenue, from whose catwalk visitors can still view machinery in operation.[8]

During the latter decades of the nineteenth century, the theory and practice of waterworks construction became much more technically sophisticated, capital-intensive, and standardized than it had been when Philadelphia, Boston, and Chicago built their earlier systems. The period witnessed the emergence of the first generation of formally trained hydraulic engineers, specialists who drew on each other's knowledge and experience in building works to more consistent specifications, though each system was necessarily adapted to local circumstances. In 1881 the American Water Works Association was established "for the exchange of information pertaining to the management of water-works for the mu-

tual advancement of consumers and water companies, and the purpose
of securing economy and uniformity in the operation of water-works."[9]

Urban water managers continued to speak proudly of their ever-
broader responsibilities. In its 1876 report, the Chicago Department of
Public Works bragged of a new engine that could pump 100 million gallons a day (30 million more than its predecessor and almost 60 million
more than average daily consumption at the time). The greater capacity
of the works would enhance "protection and safety from widespread and
devastating fires," which was a concern very much on Chicago's collective
mind five years after the great conflagration of 1871 had destroyed so
much property. The mighty engine would also promote "the vast commercial and mechanical interest of our city" while bringing "to nearly
every household an abundant and inexhaustible supply of this healthful,
cleansing and purifying agent." The department congratulated itself for
the "completeness" and "perfection" of Chicago's water system, which
it attributed to the "energy, wisdom, and skill of those in charge of this
great work."[10]

In the midst of such triumphs, Chicago and other cities continued
to struggle with a multitude of challenges posed by faulty or outmoded
equipment, unpredictable financial conditions, and professional and political conflicts, not to mention the difficulties always inherent in moving millions upon millions of gallons of water everywhere it is needed
and wanted in an ever-changing large city. Writing in 1905, Chicago city
engineer John Ericson listed all the problems caused by "the feverish
haste with which additions had to be planned and executed in order to
meet the constantly growing demands, the defects in organization of the
municipal departments, the uncertainty of the future trend of increases
in population and manufacturing centers," all of which "made it nigh
impossible and beyond human intelligence to plan an ideal system for a
metropolis of many millions of people."[11]

The Shifting Collective

Public ownership of urban waterworks, in which Philadelphia, Boston,
and Chicago had taken a leading role, came to be the rule, if not without

exceptions. By 1880 almost half of American communities with a population of 2,500 or more were served by public systems, and by 1924 the
fraction was up to nearly three-fourths. In a continuing trend, the larger
the city the more likely it was to have a public system. By 1890, 43 percent of American cities owned their own works, but these delivered water to two-thirds of the aggregate urban population. Seven years later,
forty-one of the fifty largest urban centers consumed public water.[12] Like
Philadelphia, Boston, and Chicago, other cities were motivated to build
their own works by doubts—now amply verified by experience—about
the desire and ability of private corporations to provide so many people
with water of sufficient quantity and quality at an acceptable price. The
rise in the number of large public systems also reflected the greater willingness of residents of late nineteenth-century cities to accept a bigger
role and operating budget for city government and public agencies.[13]

The repeated expansion of existing waterworks may not have inspired
the kind of citywide excitement with which Boston had greeted the Cochituate system in 1848 or Chicago had with the inauguration of the lake
tunnel in 1867, but significant additions were still officially heralded as
affirmations of civic pride and as triumphs of the collective good. After
the completion of the second lake tunnel, the Chicago Board of Public
Works declared that the vast and ingenious components of the city's water system not only revealed the skill and dedication of their designers
and builders, but also "reflect[ed] luster no less on the public spirit of
Chicago, which demanded so novel and so stupendous enterprises."[14]

Such rhetoric of civic unity notwithstanding, the urban collective
defined by those who shared a system's water became much larger and
more diverse.[15] At the same time, the bureaucracy running the works
expanded and became more complex. Even in places where the directors
of waterworks were elected, citizens had little actual control. The reor-
ganization of Chicago city government in the mid-1870s replaced the
Board of Public Works with the Department of Public Works, which was
now part of the executive branch of the city government rather than a
semi-autonomous agency. Its titular head was the mayor, though it was
still closely supervised by members of the City Council, who were especially eager to assert their authority and influence over this department

because of its large revenue stream, operating budget, and number of appointed employees.[16]

Perhaps because of all the debate that preceded the approval of the Cochituate system, Boston tried to put city water outside of politics. This began with the creation of the Cochituate Water Board, whose commissioners were supposed to be above the fray.[17] This did not bring them close to the people, however. The distance between Boston citizens and the administrators of the water supply only increased with the creation in 1895 of the Metropolitan Water Board, whose members were selected by the governor. This change reflected two reforms that were hallmarks of the Progressive era: the entrusting of large public responsibilities to experts rather than politicians, and a regional approach to urban services like water and sewerage. The first chief engineer of the Metropolitan Water Board was the highly regarded Frederick P. Stearns, who had once assisted Ellis S. Chesbrough and would soon work on the most titanic hydraulic project of its time, the Panama Canal.[18]

Individuals continued to consume water as if they had little sense of connection to the collective or to anyone's needs other than their own. Through the late nineteenth century and into the first decades of the twentieth, per capita water consumption kept rising. Chicago remained one of the thirstiest cities on the planet—its per person daily water use went from 112 gallons in 1880 to 264 gallons forty years later (in Philadelphia during the same period this figure jumped from 68 to 171 gallons, in Boston from 87 gallons to 126 gallons).[19] Although water officials made numerous attempts to discourage waste, their primary strategy for avoiding periodic shortages (sometimes called "water famines") was to keep expanding capacity. Better water meters appeared, but until the twentieth century their cost continued to make them impractical for most customers.[20] As recently as 2010, the residences of about a third of Chicago home owners were still charged by a long-standing flat-rate formula based on lot and building size and the number of plumbing fixtures.[21] And, in an action that recalled Boston's rejection of the Children's Friend Society's appeal for free water and Massachusetts General Hospital's appeal for reduced rates, Chicago mayor-elect Rahm Emanuel announced in 2011 that he would fulfill his campaign promise to end the practice of providing water gratis or at a discount to charitable and religious groups.[22]

Chesbrough's upbeat 1869 report on Chicago's new water system described the lake tunnel, the Pumping Station, and the Water Tower as affording "a feeling of security with regard to the supply of water to the city that has not been enjoyed for years before." When other pending improvements were in place, he predicted, "the city will be in a better condition with regard to the water supply than ever before." By this time, however, he knew all too well how quickly his fellow citizens could deplete supposedly inexhaustible reserves, so he avoided the open-ended promises of eternal sufficiency that had been common in statements by waterworks officials in earlier years. "The constantly increasing demand for water has been such, heretofore," he explained, "as to give no rest in these matters. What proves to be sufficient and satisfactory one year ceases to be so the next; and the only safety appears to be in planning for several years ahead."[23]

Reconfiguring the Natural World

Nature stood little chance against the advance of urbanization. While cities budgeted significant resources for parks and, especially by the late twentieth century, became more focused on environmental concerns, they kept claiming as long as they could that what they were doing in altering the natural world somehow enjoyed nature's and God's blessing. Even some city people who should have known better voiced this position. In 1874 Philadelphia attorney Eli K. Price, who two decades earlier had become the first chairman of the Fairmount Park Commission, contended that the chemical composition of the Schuylkill fortuitously neutralized the pollutants from anthracite coal mines upriver, forming a coating inside lead water pipes that shielded residents from lead poisoning. Price likewise maintained that the abundance of Schuylkill water, the richness of the region's coal deposits, and this chemical reaction were all signs of Providence's "munificence and kind protection," for which Philadelphians were duly appreciative. According to Price, "The God of nature has done this beneficently and wisely for human welfare and happiness, and all who can appreciate these blessings will with grateful hearts give Him thanks."

Price advised that these blessings made it incumbent on local residents "to do much more and better to meet the increase of population, to supply their growing wants and achieve the demands of a rising civilization." The "more" and "better" he prescribed included the further development of Fairmount Park and the construction of intercepting sewers that would help keep pollution out of the Schuylkill above the waterworks. "Failing in this," Price warned, "we fail in duty to the Giver of boundless benefits, fail in an intelligent self-protection, nay, become criminal in the fulfillment of our obligations to our families and society, in not saving the thousands who will perish that might be saved by a provident foresight, an intelligent skill, and a liberal comprehension and administration of municipal affairs."[24]

The most remarkable alteration of the natural world by a nineteenth-century American city took place in Chicago, which had put so much faith in the ability of Lake Michigan's vastness to dilute all human-made filth into harmlessness.[25] To rescue residents from the terrible and periodically deadly consequences of their disregard for all the damage they had inflicted on nature by their continued contamination of the Chicago River, officials decided that the proper course of action was to reverse the current of the river away from the lake. In other words, as a result of the damage that humans had inflicted upon nature, they decided to alter nature rather than their own damaging habits. The city's first major effort to accomplish the reversal, one that met with very limited success, was the so-called Deep Cut. This project, completed in 1871, deepened the Illinois and Michigan Canal and installed pumps in order to draw water from the South Branch into the canal.[26] Engineers finally succeeded in redirecting the river away from the Lake Michigan water supply with the construction of the Chicago Sanitary and Ship Canal, a mammoth project that took over seven years. The canal opened in 1900, sending the Chicago River's befouled waters toward the Mississippi. The agency in charge was not the city but the Sanitary District of Chicago (which later became the Metropolitan Water Reclamation District of Greater Chicago), an autonomous body created by the Illinois legislature in

1889, the same year that the Massachusetts legislature established the Metropolitan Sewerage Commission.

The language of conquest that hailed the start of the Sanitary Canal's construction revealed that earlier ideas about man and nature remained alive and well. Excavation began on September 3, 1892, with another "shovel day." Speaking from a platform erected near Lockport, about thirty miles southwest of downtown Chicago, Sanitary District trustee and hydraulic engineer Lyman Cooley told the five hundred guests who had been conveyed to the spot by a special Santa Fe Railroad train that the canal did not so much alter nature as restore it to an earlier condition. He explained that "but yesterday" in geological time the conformation of the region tipped westward, draining local rivers from rather than toward the Great Lakes. The canal would correct the shift that had reversed the earlier natural flow. Cooley conceded that the assembled might well marvel at what he called "the strange mischance" that tilted the river to Lake Michigan, but he then declared, "We are here by right of primogeniture, to claim all that which we should possess and have the energy and purpose to acquire." After pausing for the applause that greeted these words, Cooley resumed his exposition on man and nature. It was completely appropriate for human "creative intelligence" to "remedy nature's caprice," which would be "a progressive step to all the needs of the future." Chicago possessed "a virile, hustling, adult population who should show a cleaner bill of health than any other." In cutting the canal, he explained to more cheers, "we sever the gordian knot" that nature had capriciously tied and "loosen possibilities of which statesmen have dreamed for a century." This heroic project "set the gauge which shall govern the waterway of a continent."[27]

Other undertakings directly connected to the construction of waterworks also involved remaking the existing landscape. The building of Chicago's Jardine Water Purification Plant reconfigured the adjoining lakefront's shoreline, parks, and beaches. Sometimes the alteration of nature took place far from the city being served. Perhaps the best example of creating an artificial natural retreat was the construction of the Quabbin Reservoir in central Massachusetts, which required the obliteration by submersion of several small Massachusetts towns. The reservoir is thus a kind of third nature, that is, a "natural" retreat created where once a built environment had displaced the original landscape. Eighteen miles long, with 39 square miles of surface area and 118 miles of shoreline, Quabbin is the heart of a state park where Bostonians in search of communion with nature can fish, hike, watch birds and other wildlife, or just admire the scenery.[28]

Public fountains, as well as the public parks in which many of them are located, continue to be a way in which city people attempt to create a harmonious balance of the natural and the man-made in city life. In several instances, fountains have figured in visions of the ideal city that might be. The future Boston to which narrator-protagonist Julian West awakens in Edward Bellamy's 1888 *Looking Backward*, by far the most popular of the dozens of utopian novels from the period, has "large open squares filled with trees, among which statues glistened and fountains flashed in the late afternoon sun," just as the Frog Pond fountain did at the climax of the opening of the Cochituate system on October 25, 1848. Nowadays the Frog Pond no longer hosts the great jet that so excited the multitudes that day, but it is still a popular gathering place. In the summer months it serves as a wading pool, at the center of which is a more modest fountain with multiple small sprays, while in the winter it becomes a public skating rink.[29]

The elegant buildings erected by Benjamin Latrobe and Frederick Graff to house the Centre Square and Fairmount pumps in Philadelphia found their apotheosis in the enormous white Beaux-Arts exhibition halls of the 1893 Chicago World's Columbian Exposition. With its lagoons and canals dotted with gondolas, the so-called White City was an urban water park on the lakefront. Variations on William Rush's *Water Nymph and Bittern* fountain appeared in countless nineteenth-century city squares, many in association with a waterworks project. Some statuary featured just a nymph or modest maiden; others, like Evanston, Illinois's 1876 Centennial Fountain, included only a waterbird.[30] A modern equivalent of sorts to the original Rush statue in Philadelphia is the Swann Memorial Fountain in that city's Logan Square. Completed in 1921, it honors Dr. Wilson Cary Swann, founder of the Philadelphia Fountain Society. The fountain is the work of architect Wilson Eyre and sculptor Alexander Stirling Calder, son and father of the two other famed Philadelphia sculptors who shared Calder's first and last name. It is a complex work that includes several sprays, frogs, fish, turtles, and swans, the last of which offer homage both to Swann and the tradition of fountain water birds. Its three major figures, however, are humans—a girl, a woman, and a man—who respectively symbolize Wissahickon Creek and the Schuylkill and Delaware Rivers.

One of the major tourist stops in Chicago since the late 1920s is the Clarence Buckingham Memorial Fountain in Grant Park. Designed by Edward H. Bennett—coauthor with Daniel H. Burnham of the 1909 *Plan of Chicago*—who worked with sculptor Marcel Loyau and engineer Jacques Lambert, it is modeled on the Latona Basin at Versailles. The fountain and the four great sea horses that cavort within its pool represent Lake Michigan and its surrounding four states. Its central jet, one of 134 in all, can propel a shaft of water 150 feet into the air.[31] Since 2004 the Buckingham Fountain has shared the downtown lakefront spotlight with the Crown Fountain in Millennium Park. Designed by Jaume Plensa in collaboration with the architectural firm of Krueck + Sexton, it celebrates humans and modern technology much more than nature. The fountain consists of two 50-foot-high rectangular glass towers that frame a broad open space. The inner surfaces of the two towers are enormous video screens that display the unity of the collective with an ever-changing sequence of faces of local residents of all ages and races. Positioned precisely at their lips are sprays that create a shallow wading pool between the towers that has become, like the Buckingham Fountain, an exceptionally popular summertime attraction.[32]

Bodies of Humans, Bodies of Water

The sanitary campaign to wash the dirty city and its people, begun in the early nineteenth century, became more organized and powerful in the decades following the Civil War. Two key developments of these years were the construction of many urban sewer systems and the establishment by states and municipalities of the first permanent public health departments and boards. The national sanitarian movement also became much stronger. Chicago's Dr. John Rauch helped found the American Public Health Association in 1872 and was elected its president four years later. Sanitation was an important component of the Progressive reform agenda, which successfully promoted the construction and expansion of waterworks and greater efficiency in their administration. Public health physicians still had a difficult time finding broad support except in times of crisis, however, and the installation of sewers lagged well behind the building of waterworks, as did the recognition of the need for effective removal of wastewater in order to prevent pollution of the water supply.[33]

Bathing the individual body became a far more common practice. In 1860 Boston appointed a special committee "to consider and report what measures, if any, can be adopted to provide such facilities for cheap bathing as will induce all persons to avail themselves of the means provided," but the Civil War interrupted this initiative. In 1866 the city constructed six free bathhouses, though these were open only during the summer months. Other cities, including Philadelphia, also built seasonal facilities. These were frequently located along the shores of rivers, lakes, or oceans, sometimes near fancier private swimming and bathing establishments. More substantial bathhouses that provided piped and heated water and were meant to serve the working poor did not appear until the 1890s. Chicago opened its first year-round public bath in 1894. Boston followed suit in 1898, under the leadership of Mayor Josiah Quincy III, the grandson and great-grandson of the two prior mayors Josiah Quincy.[34] By the end of the nineteenth century, the modern domestic bathroom—consisting of a sink, tub, and toilet—became a standard feature of middle-class residences, but many urban working people did not have private bathing and toilet facilities in their homes for several more decades.[35]

In spite of measures, including the establishment of Fairmount Park, that were meant to keep factories away from the shores of the Schuylkill close to town, industrialization and urban growth increasingly polluted sources near Philadelphia. In Chicago, when heavy rainstorms drove Chicago River water into Lake Michigan, the consequences to public health could be very serious, even after the Sanitary Canal was built. While a number of European cities and towns were filtering their water supply by the middle of the nineteenth century, and though Ellis S. Chesbrough offered filtering as one of the several options he presented to the Chicago water commissioners in 1862, American cities were slow to adopt this technology.[36] The situation was worsened by the continuing belief on the part of engineers and politicians in the miasmist theory of disease and the cleansing powers of dilution, even for some years after the general acceptance of the germ theory by public health experts beginning in the 1870s and 1880s.[37]

Their eyes, noses, and taste buds convinced many city people that city water was not the crystal-clear elixir officials had promised, creating a growing market for private water filters among wealthy individuals and industries. An advertising brochure distributed by the Boston Water Purifier Company in 1881 appealed directly to such buyers. After noting the importance of pure water to health and manufacturing, it observed, "Many of the sources of public water supply are so contaminated with vegetable and organic matter that a reliable filter and purifier is a necessity." The company promised that its filters would not only remove "solid matter held in mechanical suspension," but would also achieve "the perfect elimination of vegetable and organic matter [that] has not been hitherto successfully accomplished," including the "many impurities not susceptible to the taste, sight, or smell, but are none the less fruitful sources of disease."[38]

Early filters did nothing to combat the microbes that were the source of many waterborne maladies. After experiments with chlorination and better filters during the first decade of the twentieth century proved promising, most major cities were soon treating their water this way, which brought a marked improvement in urban public health. The mortality rate from typhoid, the great scourge of the time, dropped drastically in Philadelphia, Boston, Chicago, and other major cities.[39] Today urban water systems have large staffs of scientists and technicians who use multiple methods to remove or neutralize dangerous non-organic as well as organic contaminants.

In the face of continuing urbanization, temperance leaders continued their assault on cities as a particularly contested battleground in what was essentially a culture war that continued through Prohibition and beyond. Water remained a key weapon. During the late nineteenth century, public drinking fountains became more common, though their main purpose was to relieve thirst rather than battle demon alcohol.[40] But teetotal organizations and individual benefactors frequently endowed prominently placed fountains that were expressly dedicated to temperance. The Sons of Temperance donated one to the 1876 Philadelphia Centennial Exposition, and San Francisco dentist Henry Cogswell alone sponsored some fifty temperance fountains between 1878 and 1890, including one (since removed) on Boston Common.[41] When the Women's Christian Temperance Union was organized in 1874, its members pledged to erect water fountains in their hometowns as a way of counteracting the temptations of the saloon.

Unlike sanitation and temperance, hydropathy was always a personal health therapy rather than a reform program, though it did challenge conventional medicine. Its popularity as a panacea peaked around midcentury. Although few certified medical professionals claim that water can cure the long list of diseases against which hydropaths said it was effective, many still prescribe varieties of water therapy for certain bodily conditions, such as fever and muscular injuries, and spas in both country and city attract water worshippers. One can argue that the multibillion-dollar vogue of bottled water is a modern-day version of the desire by hydropathy patients to maintain the purity of their bodies and their closeness to nature by withdrawing from the urban collective, or at least its drinking water. Purveyors of bottled water, even ones whose source is the public tap, appeal to this desire by favoring names and labels that evoke idyllic springs and snowy mountain peaks far from human habitation. Apart from all the other issues raised by bottled water—including the environmental costs of its production and delivery, not to mention the disposal of all those plastic bottles—there is considerable historical irony here. When one recalls how much time, effort, and money city people have expended so that low-priced and publicly owned water might flow at their bidding, it is striking to see so many today pay a great deal in order to lug their own private supply home from the market or to have it delivered by contemporary watermen driving large trucks, not to mention how willing they are to carry water around in store-bought bottles or personal containers.[42]

Time and Again

In his speech to the Philadelphia Fountain Society in 1874, Eli Price stated that the city "need not . . . have any apprehension of the sufficiency of her water supply, except it be in the supineness of her citizens in providing the means to bring it into use."[43] Time would be the city's everlasting ally as long as its citizens acted together to take advantage of all the gifts with which God and nature had blessed them. At the 1892 groundbreaking for the Chicago Sanitary and Ship Canal, Lyman Cooley asserted that this project represented the triumph of the force of Chicago's cultural anticipation over time as well as nature. Because Chicago was brave and bold enough to imagine an undertaking on this scale and then carry it out, it would defeat time. "He who set his conception as the limit of human achievement writes in his designs the obituary of his enterprises," Cooley warned. "This City, this State, this Nation, are but in youth," he continued, "and we can only dream of what they may seek to do in manhood and maturity."[44]

These were brave words, but only partly true. By 1900, if not earlier, the leaders of urban communities that had grown very substantially over the preceding century found it very hard to maintain convincingly that their cities were exempt from the effects of time. An established city might in some instances appear to renew its prospects by finding a bigger and cleaner supply, building larger-capacity pumping stations, and even changing a river's course, but it also had to confront issues such as what to do with outmoded facilities of diminished or little value, like Fairmount and Lake Cochituate, and how to counteract all the damage its residents had inflicted on nature, including their water-supply source. Civic leaders could not describe their cities as triumphant travelers on the stream of time without straining the truth.

Not that they stopped trying, even in places whose growth had slowed.[45] Meanwhile, rising younger metropolises heralded the dedications of their waterworks with the same rhetoric that had been heard before on such occasions. Los Angeles devoted two days in early November 1913 to the opening of its famed—in some views, notorious—Owens Valley Aqueduct. The ceremonies on Wednesday, November 5, took place at the spot where the water was to cascade into the northern reaches of the San Fernando Valley, while the next day's festivities began with a three-mile-long "monster industrial parade" from downtown Los Angeles to the new Exposition Park (now a complex of museums and the Memorial Coliseum), where they could witness an "athletic carnival" of different sports and the dedication of a commemorative fountain.

Tens of thousands eagerly gathered at the water gate, where, in addition to cheering speeches by officials including former governor George Pardee, Chamber of Commerce president Arthur Kinsey, Congressman William Stephens (formerly mayor and soon to be governor), and the hero of the hour, Chief Engineer William Mulholland, they thrilled to the strains of "California, Hail the Waters," composed and sung by soprano Ellen Beach Yaw.[46] "Lark Ellen," as she was called, was introduced by Harrison Gray Otis, publisher of the *Los Angeles Times*, which was unstintingly hyperbolic in its praise of the project and contemptuous of any opposition. The *Times* told readers that the water would speed the annexation and growth that would make Los Angeles—which in 1910 was still only the seventeenth biggest urban center in the nation, just ahead of Minneapolis and Jersey City—"one of the greatest cities of the country."[47]

Like Ms. Yaw, the speakers that day hit all the right notes. Their themes were the individual and the collective, humans and nature, the healthy urban body, and, most of all, the conquest of time. Pardee called the aqueduct "California's great monument to the genius, the pertinacity, the patriotism of the men who have created it—a monument to the courage, the undauntable spirit, the intelligence of a new people in a new city builded great in a new land." Over 230 miles long and traversing some of the most forbidding terrain in North America, the aqueduct was in his opinion greater than anything Rome had achieved. And it was accomplished by a still-young nation that, unlike Rome, abjured slave labor and relied instead on "the free will offering of a free people," who were determined to make possible "that which will conserve human life, banish disease, and make continually more habitable a great city."

"Pierced are the mighty mountains and riven are the barriers of the forbidding sun-baked deserts," Kinsey proclaimed, stating that this project overshadowed both the ancient Egyptian pyramids and the modern Panama Canal. Los Angeles, this "audacious and optimistic city on the far western rim of our continent," had courageously "mortgaged its future" in floating almost $25 million in bonds for a project worth immeasurably more. The city, "assured of a water supply ample for all time," could "now prepare for the splendid era of industrial expansion that approaches apace, and adequately plan for her future commercial greatness." As builders of waterworks in Philadelphia, Boston, and Chicago had asserted, Los Angeles had now taken command of its fate: "Our dream of dreams becomes a reality and our city writes in her proud history another wondrous page." For his part, the more laconic Mulholland told the crowd that this great waterworks was dedicated to "you and your children and your children's children for all time." Kinsey presented both the chief engineer and his associate Joseph B. Lippincott with loving cups, as Philadelphia had gifted Frederick Graff a century before. Much to the pleasure of all the onlookers, Mulholland lifted his cup to his lips for a symbolic drink of Owens Valley water on behalf of the people.[48]

As in Boston in 1848, the celebration climaxed with the unloosing of the natural liquid bounty. Mulholland raised a signal flag, and the gate was opened to the sound of booming cannons and aerial explosives. As they watched for the water to make its way down the mountainside, people stood expectantly on tiptoe, as Bostonians had when they awaited the explosion from the fountain on the Common. And, also as in Boston, suddenly it was there, "a roaring, rushing, waterfall growing larger every minute." According to the *Times*, "The great crowd went wild with delight. The noisy demonstration continued for a quarter of an hour, and still the people gazed with fascination on the foaming cascade." One enthralled citizen, with thoughts of posterity on his mind, collected a bottleful of Owens Valley water so that his newborn grandson might be the first person baptized with it.[49]

By this time older and bigger cities had become accustomed to taking a retrospective as well as prospective view of their water histories. In 1851 the Fairmount Water Works Committee observed the golden anniversary of the original Centre Square system with an eight-course banquet at the United States Hotel.[50] Public works departments periodically would pause in their reports of recent accomplishments and pending challenges to chronicle what had been done over the years, or publish such an account separately.[51] At its centenary in 1969, the Chicago Water Tower was declared a water landmark by the American Water Works Association, the first building so designated by that organization.[52] Soon enough Los Angeles, too, would look backward as well as ahead. On March 30, 1940, Rose Mulholland turned the first spade at the groundbreaking for the memorial fountain in honor of her late father. The site, near the Los Angeles River and Griffith Park, was where William Mulholland had lived some sixty years earlier when he was a young Irish immigrant ditch-tender at the outset of his life's work on city water.[53]

Today discussions of cities and water do not talk of endless supplies and limitless futures but of competing claims on limited resources, the need for conservation, the constant hazards of contamination, and the importance of sustainability. Cities (and nations) contend with each other, as well as with farmers and manufacturers, for this essential resource that they each desperately need but that is in too short supply to satisfy them all. The public relations departments of urban water companies list the steps these agencies are taking to sustain their sources and keep them healthy as they also offer information on how individual customers can limit their use of fresh water and dispose of waste safely. The waterworks at Fairmount, which after it closed was first converted into an aquarium and then into a swimming pool, has been brilliantly restored as the Fairmount Water Works Interpretive Center, where "you and your family can learn about watersheds and have a ton of fun doing it." One can "fly a helicopter simulation from the Delaware Bay to the headwaters of the Schuylkill River. Or visit Pollutionopolis, America's most contaminated and disgusting town, to see how a city can really mess up its water supply."[54] The meditation on city water flows on.

NOTES

Chapter One

1. William James, "The Stream of Thought," in *Principles of Psychology* (New York: Henry Holt, 1890), 224–90. James uses the phrase "stream of consciousness" on p. 245.

2. William Shakespeare, *Julius Caesar*, IV.3; *Othello*, V.2.

3. "It . . . should be acknowledged," Simon Schama writes, "that once a certain idea of landscape, a myth, a vision, establishes itself in an actual place, it has a peculiar way of muddling categories, of making metaphors more real than their referents; of becoming, in fact, part of the scenery." Schama, *Landscape and Memory* (New York: Knopf, 1995), 61. I make a similar argument, substituting "cityscape" for "landscape" and focusing on city water. I likewise concur with Lawrence Buell's assertion that "we live our lives by metaphors that have come to seem deceptively transparent through long usage." Buell, *The Environmental Imagination: Thoreau, Nature Writing, and the Formation of American Culture* (Cambridge, MA: Belknap Press of Harvard University Press, 1995), 3.

4. Ralph Waldo Emerson, "The American Scholar," in *Essays and Lectures* (New York: Library of America, 1983), 62.

5. Michael R. Haines, Table Aa684–698, "Urban and Rural Territory—number of places, by size of place: 1790–1990," in *Historical Statistics of the United States*, ed. Susan B. Canter, Scott Sigmund Gartner, Michael R. Haines, Alan L. Olmstead, Richard Sutch, and Gavin Wright, Millennial Online Edition, 1–102, http://hsus.cambridge.org.turing.library. northwestern.edu/HSUSWeb/toc/tableToc.do?id=Aa684–698 (accessed March 31, 2011); "United States Urban and Rural," Table 4: Population 1790 to 1990, in United States Census

Bureau, "Selected Historical Decennial Population and Housing Counts," http://www
.census.gov/population/www/censusdata/hiscendata.html (accessed December 17, 2011);
Campbell Gibson, "Population of the 100 Largest Cities and Other Urban Places in the
United States: 1790–1990," United States Bureau of the Census Population Division Working Paper No. 27 (June 1998), http://www.census.gov/population/www/documentation/
twps0027/twps0027.html (accessed November 20, 2010).

6. City of Philadelphia, Department for Supplying the City with Water, *History of the
Works, and Annual Report of the Chief Engineer of the Water Department of the City of Philadelphia* (Philadelphia: C. E. Chichester, 1860), 5.

7. Joel A. Tarr states, "Waterworks were often the leading sector in the creation of
modern urban infrastructure, the initial example of a planned, integrated system." Tarr,
"Building the Urban Infrastructure in the Nineteenth Century: An Introduction," *Essays in
Public Works History* 14 (December 1985): 65.

Chapter Two

1. Nathaniel B. Shurtleff, *A Topographical and Historical Description of Boston* (Boston:
Boston City Council, 1871), 25.

2. Ibid., 401.

3. "Franklin's Last Will and Testament," in *The Writings of Benjamin Franklin*, ed. Albert
Henry Smyth (New York: Macmillan, 1907), 10:503–6.

4. *Aurora and General Advertiser* (Philadelphia), January 13, 1802.

5. James Mease, M.D., *The Picture of Philadelphia: Giving an Account of Its Origin, Increase
and Improvements* (Philadelphia: B. and T. Kite, 1811), 17.

6. "Address of the Mayor to the City Council of Boston, January 1, 1838," in City of Boston, *The Inaugural Addresses of the Mayors of Boston* (Boston: Rockwell & Churchill, 1894),
1:222.

7. *Chicago Tribune*, December 28, 1850.

8. As Joel A. Tarr points out in "Building the Urban Infrastructure in the Nineteenth
Century: An Introduction," *Essays in Public Works History* 14 (December 1985): 63–65. See
also Letty Donaldson Anderson, "The Diffusion of Technology in the Nineteenth-Century
American City: Municipal Water Supply Investments" (Ph.D. diss., Northwestern University, 1980).

9. City of Chicago, *Seventh Semi-Annual Report of the Board of Water Commissioners of
the City of Chicago to the Common Council of the City of Chicago, December 18, 1854* (Chicago:
Democratic Press Steam Print, 1855), 39.

10. Estimates of the number of deaths caused by the fever vary. These figures are from
the introduction by Kenneth R. Foster, Mary F. Jenkins, and Anna Coxe Toogood to the
updated edition of J. H. Powell's 1949 study, *Bring Out Your Dead: The Great Plague of Yellow
Fever in Philadelphia in 1793* (Philadelphia: University of Pennsylvania Press, 1993), ix–x.

11. City of Philadelphia, *Report to the Select and Common Councils, on the Progress and
State of the Water Works, on the 24th of November, 1799* (Philadelphia: Zachariah Poulson, Jr.,
1799), 41, 43. Philadelphia's corporate charter of 1796 vested legislative power in a Select
Council of twelve members serving staggered three-year terms and a Common Council
of twenty members, who were elected every year. There was no property requirement for
voting. The Select and Common Councils together chose the mayor. Beginning in 1839, he
was directly elected by the voters. In Pennsylvania, these consisted of white male citizens

(blacks were excluded the year before) who had paid state or county tax within the past two years. Alexander Keyssar, *The Right to Vote: The Contested History of Democracy in the United States* (New York: Basic Books, 2000), table A.9, 354.

12. On the power of council committees, see Edward P. Allinson and Boies Penrose, "The City Government of Philadelphia," in *Johns Hopkins University Studies in Historical and Political Science*, vol. 5: *Municipal Government and Politics* (Baltimore: N. Murray, 1887), 37–38.

13. City of Philadelphia, *Report of the Joint Committee of the Select and Common Councils on the Subject of Bringing Water to the City* (Philadelphia: Zachariah Poulson, Jr., 1798), 5.

14. Benjamin Latrobe (1764–1820), who left England for the United States in 1795, lived in Philadelphia from 1798 to 1802. While he is perhaps best known for his contributions to the designs of the Capitol and the White House, these were only two of his many major architectural and engineering projects. Among the latter was his waterworks for New Orleans, where he died of yellow fever, the disease that had prompted Philadelphia to build its system twenty years earlier. His son Henry had been fatally stricken by the disease four years previously while engaged on the same project.

15. Edward C. Carter II, ed., *The Virginia Journals of Benjamin Henry Latrobe, 1795–1798*, vol. 2, *1797–1798* (New Haven, CT: Yale University Press, 1977), 380.

16. Carroll W. Pursell Jr. states that Latrobe's design for Philadelphia's waterworks was "the most ambitious use of steam yet attempted in this country." Pursell, *Early Stationary Steam Engines in America: A Study in the Migration of a Technology* (Washington, DC: Smithsonian Institution Press, 1969), 31–32. See also Darwin H. Stapleton, "Benjamin Henry Latrobe and the Transfer of Technology," in *Technology in America: A History of Individuals and Ideas*, 2nd ed., ed. Carroll W. Pursell Jr. (Cambridge, MA: MIT Press, 1990), 34–61.

17. At this time the city of Philadelphia was about two miles wide from east to west and a mile north to south. Centre Square was renamed Penn Square in 1829. It has long been the site of Philadelphia's massive Second Empire–style City Hall, on which construction began in 1871 and lasted over thirty years.

18. Latrobe estimated that a supplementary system, which would bring water from a nearby spring at a later date, would cost another $275,000. B. Henry Latrobe, *View of the Practicability and Means of Supplying the City of Philadelphia with Wholesome Water, in a Letter to John Miller, Esquire, from B. Henry Latrobe, Engineer, December 29, 1798* (Philadelphia: Zachariah Poulson, Jr., 1799), 6, 12–16. This document and several others relating to the Philadelphia waterworks are included in *The Correspondence and Miscellaneous Papers of Benjamin Henry Latrobe*, vol. 1, ed. John C. Van Horne and Lee W. Formwalt (New Haven, CT: Yale University Press, 1984).

19. City of Philadelphia, *Report on the Progress and State of the Water Works* (1799), 7.

20. *Address of the Committee of the Delaware and Schuylkill Canal Company, to the Committees of the House of Representatives, on the Memorial of Said Company* (Philadelphia: John Ormrod, [1799]).

21. *Remarks on a Second Publication of B. Henry Latrobe, Engineer, Said to Be Printed by Order of the Committee of the Councils; and Distributed among the Members of the Legislature* (Philadelphia: Zachariah Poulson, Jr., 1799), 2.

22. B. Henry Latrobe, *Remarks on the Address of the Committee of the Delaware and Schuylkill Canal Company to the Committee of the Senate and House of Representatives* (Philadelphia: Zachariah Poulson, Jr., [1799]), 17. These "remarks" took the form of a letter to Select Council member John Miller.

23. Benjamin Latrobe, *An Answer to the Joint Committee of the Select and Common Councils of Philadelphia, on the Subject of a Plan for Supplying the City with Water, &c.* [Philadelphia, 1799], 7.

24. City of Philadelphia, *Report on the Progress and State of the Water Works* (1799), 42, 16.

25. *Correspondence and Miscellaneous Papers of Benjamin Henry Latrobe*, ed. Van Horne and Formwalt, 1:109; David Freeman Hawke, *Nuts and Bolts of the Past: A History of American Technology, 1776–1860* (New York: Harper & Row, 1988), 66.

26. Eliza Cope Harrison, ed., *Philadelphia Merchant: The Diary of Thomas P. Cope, 1800–1851* (South Bend, IN: Gateway Editions, 1978), 56–57.

27. City of Philadelphia, *Report of the Committee Appointed by the Common Council, to Enquire into the State of the Water Works* (Philadelphia: William Duane, 1802), 45.

28. Pursell, *Early Stationary Steam Engines in America*, 31; Thompson Westcott, *A History of Philadelphia, from the Time of the First Settlements on the Delaware to the Consolidation of the City and Districts in 1854* (Philadelphia: A. C. Kline, 1867), vol. 4, chap. 427.

29. City of Philadelphia, *Report on the Progress and State of the Waterworks* (1799), 8; City of Philadelphia, *Report of the Committee for the Introduction of Wholesome Water into the City of Philadelphia* (Philadelphia: Zachariah Poulson, Jr., 1801), 13.

30. City of Philadelphia, *Report of the Watering Committee to the Select and Common Councils, November 1, 1803* (Philadelphia: William Duane, 1803), 4.

31. *Aurora*, January 13, 1802.

32. *United States Gazette* (Philadelphia), August 10, 1805; Eleanor A. Maass, "A Public Watchdog: Thomas Pym Cope and the Philadelphia Waterworks," *Proceedings of the American Philosophical Society* 125, no. 2 (April 30, 1981): 136; Talbot Hamlin, *Benjamin Henry Latrobe* (New York: Oxford University Press, 1955), 161–64; Nelson Manfred Blake, *Water for the Cities: A History of the Urban Water Supply Problem in the United States* (Syracuse, NY: Syracuse University Press, 1956), 39–40.

33. City of Philadelphia, *Report of the Committee for the Introduction of Wholesome Water* (1801), 12–13.

34. *Aurora*, October, 22 and 26, 1801. On top of everything else, Latrobe faced the loss and confusion caused by the theft in 1800 of a large amount of money and many important papers. Hamlin, *Benjamin Henry Latrobe*, 135.

35. Letter of Benjamin Henry Latrobe to Robert Goodloe Harper, March 28, 1803, in *Correspondence and Miscellaneous Papers of Benjamin Henry Latrobe*, ed. Van Horne and Formwalt, 1:266.

36. City of Philadelphia, *Report of the Watering Committee, to the Select and Common Councils, Read January 12, 1832* (Philadelphia: Lydia R. Bailey, 1832), 28.

37. Ibid., 25, 28.

38. *Aurora*, February 27, 1801; *United States Gazette*, December 14, 1801, May 10, 1806. For a similar story, see *United States Gazette*, February 11, 1805.

39. Henry Simpson, *The Lives of Eminent Philadelphians, Now Deceased* (Philadelphia: William Brotherhead, 1859), 431–36.

40. City of Philadelphia, *Report of the Watering Committee, upon the Present State of the Works for Supplying the City with Water, and the Several Other Plans Proposed for that Purpose, May 2, 1812* (Philadelphia: Jane Aitken, 1812), 3.

41. See City of Philadelphia, Department for Supplying the City with Water, *History of the Works, and Annual Report of the Chief Engineer of the Water Department of the City of*

Philadelphia (Philadelphia, C. E. Chichester, 1860), 15; *The Philadelphia Water Department: An Historical Perspective* [Philadelphia, 1990(?)], 6.

42. *Aurora*, August 13, 1812.

43. *Aurora*, November 5, 1812. Charles V. Hagner, writing in 1869, identified the person who signed himself "A Citizen" in the 1812 letter as Erskine Hazard, who had a financial interest in his proposal, since he was a partner of mill owners Josiah White and Joseph Gillingham, with whom the city would need to negotiate for rights to pursue the plan that Hazard had proposed. Hagner, *Early History of the Falls of Schuylkill, Manayunk, Schuylkill and Lehigh Navigation Companies, Fairmount Waterworks, Etc.* (Philadelphia: Claxton, Remsen & Haffelfinger, 1869), 47.

44. City of Philadelphia, *Report of the Watering Committee, to the Select and Common Councils* (1832), 25. Evans and Latrobe had previously been involved in a dispute over steam-engine design. See Hamlin, *Benjamin Henry Latrobe*, 326.

45. The city had experimented with iron pipe as early as 1804. The large-scale replacement of wood with iron began in 1820. The city continued to use wood in decreasing amounts before ceasing altogether in 1832. City of Philadelphia, *Annual Report of the Chief Engineer of the Water Department of the City of Philadelphia for the Year 1875* (Philadelphia: E. C. Markley & Son, 1875), 42, 48.

46. City of Philadelphia, *Report of the Watering Committee of the Agreements with the Schuylkill Navigation Company and White and Gillingham, Relating to the Water Power of the River Schuylkill* (Philadelphia: W. Fry, 1819) (see n. 43, above).

47. Joseph Lewis, who as chair of the Watering Committee had originally negotiated the agreement between the city and the canal, later became president of the Schuylkill Navigation Company. In 1832 he declared that the company's volume of business required the construction of another lock next to the dam, and that it would build this lock immediately. Philadelphia objected both to the proposal itself and to Lewis's assumption that the company did not need the city's approval. Lewis and a crew of company employees appeared at the dam with a constable, evicted the bewildered toll-keeper, and began the work of tearing down the tollhouse and erecting the additional lock. The city tried but failed to obtain either legal or financial satisfaction. During one episode in this strange drama, the long-suffering Graff, forced to venture out on a cold, wet evening to try to secure the works, contracted a respiratory ailment that laid him up for months and weakened his health. See City of Philadelphia, *Correspondence of the Watering Committee with the Schuylkill Navigation Company, in Relation to the Fair Mount Water Works* (Philadelphia: Lydia R. Bailey, 1833).

48. [City of Philadelphia, *Report of the Watering Committee . . . Relative to the Fair Mount Water Works* (Philadelphia: Lydia R. Bailey, 1824)], 3. The cover of the copy consulted is missing. It is owned by the Historical Society of Pennsylvania but held by the Library Company of Philadelphia.

49. City of Philadelphia, *Report of the Watering Committee, to the Select and Common Councils, Read in Select Council, January 8, 1824* (Philadelphia: Lydia R. Bailey, 1824), 10.

50. City of Philadelphia, *Report of the Watering Committee to the Select and Common Councils, Read February 10, 1831* (Philadelphia: Lydia R. Bailey, 1831), 5.

51. City of Philadelphia, *Annual Report of the Chief Engineer of the Water Department* (1875), 42–43, 48.

52. I am grateful to Jeffrey Ray, Senior Curator of the Philadelphia History Museum at the Atwater Kent, which owns the cup, for providing the information on this artifact. See also Simpson, *The Lives of Eminent Philadelphians*, 436.

53. Thomas Ewbank, *A Descriptive and Historical Account of Hydraulic and Other Machines for Raising Water, Ancient and Modern* (London: Tilt and Bogue, 1842), 301. Jane Mork Gibson describes the period 1830 to 1850 as "the golden age at Fairmount Waterworks," when "receipts were well over expenditures, the waterwheels were operating efficiently, and the public was enthusiastically responsive to the well-designed buildings and the picturesque setting." Gibson, "The Fairmount Waterworks," *Bulletin* (Philadelphia Museum of Art) 84, nos. 360–361 (Summer 1988): 28.

54. City of Philadelphia, *History of the Works*, 16–17.

55. [Nathaniel J. Bradlee], *History of the Introduction of Pure Water into the City of Boston, with a Description of Its Cochituate Water Works* (Boston: Alfred Mudge & Son, 1868), 3. Under its first corporate charter of 1822, the mayor and the members of the bicameral city legislature were elected directly. The eight members of the Board of Aldermen were chosen at large, while each of the city's twelve wards elected four representatives to the Common Council. As of 1821, all adult male residents of the state could vote if they had paid any state or county tax assessed within the previous two years. Keyssar, *The Right to Vote*, table A.9, p. 352. On Quincy's activist mayoralty—he served three two-year terms, from 1823 to 1829—see Matthew H. Crocker, *The Magic of the Many: Josiah Quincy and the Rise of Mass Politics in Boston, 1800–1830* (Amherst: University of Massachusetts Press, 1999). Quincy (who shared the name of his father and grandfather) will subsequently be referred to as Josiah Quincy Sr. to distinguish him from his son and subsequent Boston mayor from 1845 to 1849, Josiah Quincy Jr. Josiah Quincy Jr.'s grandson, Josiah Quincy III, was the city's mayor from 1896 to 1899.

56. "An Address to the Board of Aldermen, and Members of the Common Council of Boston, on the Organization of the City Government, January 2, 1826," in City of Boston, *The Inaugural Addresses of the Mayors of Boston*, 1:52–54.

57. Letter of John Collins Warren to Josiah Quincy, November 25, 1825, John Collins Warren Papers, Massachusetts Historical Society.

58. [Bradlee], *History of the Introduction of Pure Water into the City of Boston*, 31.

59. City of Boston, *Soft Water*, City Document No. 25, 1839 [Boston, 1839], 15.

60. Lemuel Shattuck, *Report to the Committee of the City Council Appointed to Obtain the Census of Boston for the Year 1845, Embracing Collateral Facts and Statistical Researches Illustrating the History and Condition of the Population, and their Means of Progress and Prosperity* (Boston: John H. Eastburn, 1846), 56, 26, 57.

61. Commonwealth of Massachusetts, *Proceedings before a Joint Special Committee of the Massachusetts Legislature, upon the Petition of the City of Boston, for Leave to Introduce a Supply of Pure Water into that City, from Long Pond, February and March, 1845* (Boston: John H. Eastburn, 1845), 125–26.

62. Ibid., 34. Shattuck's pamphlet was *Letter from Lemuel Shattuck, in Answer to Interrogatories of J. Preston, in Relation to the Introduction of Water into the City of Boston* (Boston: Samuel N. Dickinson, 1845).

63. Commonwealth of Massachusetts, *Proceedings before a Joint Special Committee of the Massachusetts Legislature* (1845), 69–70.

64. William J. Hubbard, *Argument on Behalf of Joseph Tilden and Others, Remonstrants, on the Hearing of the Mayor of the City of Boston, on Behalf of the City Council, for a Grant of the Requisite Powers to Construct an Aqueduct from Long Pond to the City, before a Joint Special Committee of the Massachusetts Legislature, March 6, 1845* (Boston: T. R. Marvin, 1845), 52.

65. Whether as family members or servants, women bore the major household responsibility of obtaining water from whatever source was available and affordable, which might be some distance from where they lived, and then disposing of it after it was used. Even piped water needed to taken from the faucet to its particular application, which might involve carrying it upstairs (and, subsequently, downstairs and outside). See Christine Stansell, *City of Women: Sex and Class in New York, 1789–1860* (New York: Knopf, 1986), 157–61. For similar conditions in England, see Elizabeth Wilson, *The Sphinx in the City: Urban Life, the Control of Disorder, and Women* (Berkeley: University of California Press, 1991), 36–37.

66. *Daily Advertiser and Patriot* (Boston), May 19, 1845. Shakum Pond is in Framingham.

67. *Daily Advertiser*, May 16, 1845.

68. *Daily Advertiser*, May 15, 1845.

69. *Daily Advertiser*, May 14, 1845.

70. *Daily Advertiser*, April 8, 1845.

71. *Daily Advertiser*, May 16, 1845.

72. *Address of the Faneuil Hall Committee, on the Project of a Supply of Pure Water for the City of Boston, May 5, 1845* (Boston: W. W. Clapp & Son, 1845), 3, 30–31.

73. *Daily Advertiser*, May 19, 1845.

74. *Daily Evening Transcript* (Boston), May 20, 1845.

75. *Daily Advertiser*, May 22, 1845.

76. *Daily Advertiser*, July 25, 1845

77. For the record of this meeting, see Union Water Convention, *Proceedings of the Union Water Convention, Concerning the Conflagration at South Boston, on the Morning of the 14th of September, 1845* (Boston: Eastburn's Press, 1845).

78. Blake, *Water for the Cities*, 207–8.

79. City of Boston, *Report of the Commissioners Appointed by Authority of the City Council, to Examine the Sources from Which a Supply of Pure Water May Be Obtained for the City of Boston* (Boston: J. H. Eastburn, 1845), 126.

80. [Bradlee], *History of the Introduction of Pure Water into the City of Boston*, 57. The new act still called for the concurrent election of commissioners by the City Council, but it made them more tightly subject to ordinances, rules, and regulations set by this body. Their service was limited to three years (unless the construction was completed sooner), and they could be removed by a two-thirds majority of the council, which would also determine their salaries. The commissioners were obligated to issue a written report every six months, more frequently if formally asked to do so. Commonwealth of Massachusetts, *An Act for Supplying the City of Boston with Pure Water* [Boston, 1846], 3–4.

81. [Bradlee], *History of the Introduction of Pure Water into the City of Boston*, 95–108, 110, 118.

82. Ibid., 104–5, 100, 110.

83. Ibid., 85.

84. City of Boston, *Water Ordinance*, City Document No. 63, 1849 [Boston, 1849].

85. City of Boston, *Report of Committee on Finance*, City Document No. 21, June 11, 1846 [Boston, 1846], 2.

86. Charles Phillips Huse, *The Financial History of Boston from May 1, 1822, to January 31, 1909* (Cambridge, MA: Harvard University Press, 1916), 82, 100.

87. City of Boston, *Final Water Report*, City Document No. 3, 1850 [Boston, 1850], 7–8.

88. Louis P. Cain examines the several implications of this in *Sanitation Strategy for*

a Lakefront Metropolis: The Case of Chicago (DeKalb: Northern Illinois University Press, 1978). See also James C. O'Connell, "Technology and Pollution: Chicago's Water Policy, 1833–1930" (Ph.D. diss., University of Chicago, 1980); Betsy Thomas Mendelsohn, "Law and Chicago's Waters, 1820–1920" (Ph.D. diss., University of Chicago, 1999); and Harold L. Platt, *Shock Cities: The Environmental Transformation and Reform of Manchester and Chicago* (Chicago: University of Chicago Press, 2005).

89. A. T. Andreas, *History of Chicago, from the Earliest Period to the Present Time*, vol. 1, *Ending with the Year 1857* (Chicago: A. T. Andreas Co., 1884), 185.

90. Ibid., 1:185–86; City of Chicago, Bureau of Engineering, *A Century of Progress in Water Works: Chicago, 1833–1933* (Chicago: Department of Public Works, 1933), 13; O'Connell, "Technology and Pollution," 10–11; Platt, *Shock Cities*, 108.

91. Andreas, *History of Chicago*, 1:186.

92. For a graphic representation of this, see Platt, *Shock Cities*, fig. 4.2, p. 144.

93. *Chicago Tribune*, December 28, 1850.

94. Mendelsohn, "Law and Chicago's Waters," 126.

95. At this time, Illinois granted the vote to all white male citizens who had lived in the state at least one year. Keyssar, *The Right to Vote*, table A9, p. 352.

96. City of Chicago, *Report of the Water Commissioners of the City of Chicago, Made to the Common Council, December 8, 1851: Together with the Act of Incorporation, and a Statement of the Financial Condition of the City, November 16, 1851* (Chicago: Seaton & Peck, 1851), 4–5, 10; Andreas, *History of Chicago*, 1:186–87.

97. City of Chicago, *Eighth Annual Report of the Board of Public Works to the Common Council of the City of Chicago, for the Municipal Fiscal Year Ending March 31st, 1869* (Chicago: Jameson & Morse, 1869), 61; City of Chicago, *Third Semi-Annual Report of the Water Commissioners, of the City of Chicago, to the Common Council of the City of Chicago* (Chicago: Democrat Office, 1853), 24.

98. City of Chicago, *Seventh Semi-Annual Report of the Board of Water Commissioners* (1854), 39–40.

99. City of Chicago, *Thirteenth Semi-Annual Report of the Board of Water Commissioners, to the Common Council of the City of Chicago, January 1st, 1858* (Chicago: Jameson & Morse, 1858), 47.

100. City of Chicago, *Fifteenth Semi-Annual Report of the Board of Water Commissioners, to the Common Council of the City of Chicago, January 1st, 1859* (Chicago: William H. Rand, 1859), 3.

101. As Platt points out, the situation was complicated and made more urgent by the financial downturn of 1857 and a serious fire the same year. *Shock Cities*, 112, 132.

102. *Chicago Tribune*, April 10, 1855.

103. This section read in part: "The citizens have for several times during the winter, been annoyed by the presence of small fish in the pipes, sometimes in great numbers." City of Chicago, *Seventeenth Semi-Annual Report of the Board of Water Commissioners to the Common Council of the City of Chicago, January 1st, 1860* (Chicago: Dunlop, Sewell & Spalding, 1860), 7. William Bross—who arrived in Chicago in 1846 and used his position as owner or part-owner of a succession of newspapers (including the *Tribune*) to promote Chicago enthusiastically—recalled the surprises that awaited anyone who took city water. Since "it was impossible to keep the young fish out of the reservoir," Bross wrote, "it was no uncommon thing to find the unwelcome fry sporting in one's wash-bowl, or dead and stuck in the faucets." If they found their way to a hot-water heater, "they would get stewed up into a

very nauseous fish chowder." Bross, "What I Remember of Early Chicago," in *Reminiscences of Chicago during the Forties and Fifties*, ed. Mabel McIlvaine (Chicago: R. R. Donnelley & Sons, 1913), 7, 6. Bross delivered the text of this essay as a talk in 1876.

104. *Chicago Tribune*, February 17, 1862.

105. Louis P. Cain describes the project as "the first comprehensive sewerage system undertaken by any major city in the United States." Cain, "Raising and Watering a City: Ellis Sylvester Chesbrough and Chicago's First Sanitation System," *Technology and Culture* 13, no. 3 (July 1973): 357. See also Platt, *Shock Cities*, 121–22. Chesbrough (1813–1886) began his engineering career as a teenager, working on the Baltimore and Ohio Railroad. He was employed by a series of other lines before becoming superintendent of construction of the Charleston and Cincinnati Railroad. Chesbrough was a leader in the organization of his profession in America. In 1869 he became president of the Civil Engineers Club of the Northwest (which in 1880 was renamed the Western Society of Engineers), and in 1877–78 was president of the American Society of Civil Engineers, which had been founded in 1852.

106. City of Chicago, *First Annual Report of the Board of Public Works to the Common Council of the City of Chicago, January 1st, 1862* (Chicago: S. P. Rounds, 1862), 6; City of Chicago, *Second Annual Report of the Board of Public Works to the Common Council of the City of Chicago, April 1st, 1863* (Chicago: Tribune Book and Job Printing Office, 1863), 38–54; A. T. Andreas, *History of Chicago, from the Earliest Period to the Present Time*, vol. 2, *From 1857 until the Fire of 1871* (Chicago: A. T. Andreas Co., 1885), 66.

107. City of Chicago, *First Annual Report of the Board of Public Works* (1862), 5.

108. City of Chicago, *Second Annual Report of the Board of Public Works* (1863), 7, 4.

109. City of Chicago, *First Annual Report of the Department of Public Works, to the City Council of the City of Chicago, for the Fiscal Year Ending December 31, 1876* (Chicago: Clark & Edwards, 1877), 85.

110. For predictions that the tunnel plan could not be accomplished and would not work, see the letters from John Gage and from "A Tax-Payer" that were published in the *Chicago Tribune* on November 20 and December 4, 1863. Chesbrough's confidence in the project generally and in his cost estimates specifically were based on his knowledge of other recent hydraulic engineering projects, including the Croton Aqueduct and the Thames Tunnel, not to mention the Cochituate Aqueduct. See Cain, "Raising and Watering a City," 368–69.

111. City of Chicago, *Eighth Annual Report of the Board of Public Works* (1869), 97–106.

112. *Chicago Tribune*, August 25, 1864.

113. Andreas, *History of Chicago*, 2:67; City of Chicago, *First Annual Report of the Department of Public Works* (1877), 95–100.

114. City of Chicago, *Sixth Annual Report of the Board of Public Works to the Common Council of the City of Chicago, for the Municipal Fiscal Year Ending March 31st, 1867* (Chicago: Republican Book and Job Office, 1867), 45.

115. City of Chicago, *Eighth Annual Report of the Board of Public Works* (1869), 71.

116. As Chicago's Board of Public Works explained in 1868, "Other cities, at vast expense, construct reservoirs in their midst and store up water for emergencies. The city of Chicago has the water at its door, but it must store up the power by which it is to be elevated." City of Chicago, *Seventh Annual Report of the Board of Public Works to the Common Council of the City of Chicago, for the Municipal Fiscal Year Ending March 31st, 1868* (Chicago: Jameson & Morse, 1869), 4.

117. The standpipe was 138 feet high. Much of the information here is drawn from

the annual reports of the Board of Public Works from 1861 to 1867; Andreas, *History of Chicago*, 2:66–67; and City of Chicago, *A Century of Progress in Water Works*, 19.

118. Andreas, *History of Chicago*, 2:69.

119. In late June 1842, four months before the opening of New York's Croton works, its aqueduct was filled to a depth of eighteen inches, and a small crew journeyed through this subterranean river in the *Croton Maid*, a flat-bottomed vessel constructed for this task. For an account, see Charles King, *A Memoir of the Construction, Cost, and Capacity of the Croton Aqueduct, Compiled from Official Documents* (New York: Charles King, 1843), 195–96. See also Dell Upton, *Another City: Urban Life and Urban Spaces in the New American Republic* (New Haven, CT: Yale University Press, 2008), 310.

120. *Chicago Tribune*, March 24, 1867; *Chicago Republican*, March 24, 1867. The practice of conducting such ceremonial tours of completed projects continued. In July 1874, shortly after the second Lake Michigan tunnel was finished, "a party of city officials, prominent residents, and visitors from abroad, were afforded, through the courtesy of the contractors, Messrs. Steele & McMahon, the novel pleasure of a trip through the tunnel to the crib." City of Chicago, *Fourteenth Annual Report of the Board of Public Works to the Common Council of the City of Chicago, for the Municipal Fiscal Year Ending March 31, 1875* (Chicago: J. S. Thompson & Co., 1875), 4.

121. City of Chicago, *Sixth Annual Report of the Board of Public Works* (1867), 4. As Platt points out, the anxiety was due partly to the serious cholera visitation of the year before, and the rejoicing was based on the hope that the new supply would provide better protection in the future. Platt, *Shock Cities*, 135, 144.

122. City of Chicago, *Eighth Annual Report of the Board of Public Works* (1869), 5.

123. *Chicago Tribune*, December 11, 1867. Chesbrough's salary figure is taken from City of Chicago, *Sixth Annual Report of the Board of Public Works* (1867), 63.

124. City of Chicago, *First Annual Report of the Department of Public Works* (1877), 63; Andreas, *History of Chicago*, 2:70.

125. Siegfried Giedion writes, "During the second half of the century, cities everywhere were to be supplied with water throughout. Running water entered first the basement, then the storeys, and finally each apartment. Words are too static. Only a moving picture could portray water's advance through the organism of the city, its leap to the higher levels, its distribution to the kitchen and ultimately to the bath." Giedion, *Mechanization Takes Command: A Contribution to Anonymous History* (New York: Norton, 1969), 684–85. This was originally published by Oxford University Press in 1948.

126. *The Great Chicago Lake Tunnel* (Chicago: Jack Wing, 1867), 26.

127. City of Chicago, *Thirteenth Semi-Annual Report of the Board of Water Commissioners* (1858), 4.

128. City of Chicago, *Seventh Semi-Annual Report of the Board of Water Commissioners* (1854), 33; City of Chicago, *Eighth Annual Report of the Board of Public Works* (1869), 64; *Chicago Tribune*, August 19, 1869. The *Tribune*, calling this "a calamity which will be felt very deeply by a large portion of the community," proposed better enforcement of the ordinance, as well as the arrest and prosecution of the captain responsible, "who wantonly endangered the safety, health and comfort of one hundred thousand people. If he were sent to prison fifteen years and his vessel confiscated it would not be adequate compensation for what he has done."

129. Wesley G. Skogan, *Chicago since 1840: A Time-Series Data Handbook* (Urbana: University of Illinois Institute of Government and Public Affairs, 1976), 8; City of Chicago,

Report of the Water Commissioners of the City of Chicago (1851), 47; City of Chicago, *Fifteenth Annual Report of the Board of Public Works to the Common Council of the City of Chicago, for the Municipal Fiscal Year Ending December 31st, 1875* (Chicago: J. S. Thompson & Co., 1876), 10.

130. City of Chicago, *Eleventh Annual Report of the Board of Public Works to the Common Council of the City of Chicago, for the Municipal Fiscal Year Ending March 31st, 1872* ([Chicago]: D. & C. H. Blakely, 1872), 5–6, 10.

Chapter Three

1. As Ann Durkin Keating writes, "The home had changed irrevocably from an independent unit to a part of numerous service systems connecting it with the outside world." Keating, *Building Chicago: Suburban Developers and the Creation of a Divided Metropolis* (Columbus: Ohio State University Press, 1988), 58. The same could be said of a commercial building that took city water.

2. Walter Channing, *A Plea for Pure Water: Being a Letter to Henry Williams, Esq., by Walter Channing: With an Address "To the Citizens of Boston," by Mr. H. Williams* [Boston, 1844], 25.

3. *Daily Advertiser and Patriot* (Boston), November 22, 1847. In the sermon he delivered for Thanksgiving Day in 1846, Congregationalist minister Nehemiah Adams predicted that the arrival of the water would create a new and more perfect Boston. Adams hoped that the waterworks project "may help to turn our adoring and grateful thoughts . . . to Him 'that made heaven, and earth, and the sea, and the fountains of waters.'" Long Pond's largess would "smile in every household" and bless man and beast in common. "The rich will be as grateful as the poor, and the poor will not envy the rich." Then Adams presented a glorious vision of how the arrival of the water would unify the populace: "The distant lake will pour itself abroad upon a hundred thousand grateful hearts; it will stimulate its unseen sources to supply the demands upon its benevolence; and the whole structure, so complicated in its preparation and composed of such various materials, will become one great monument of divine goodness, and of human energy." Nehemiah Adams, *The Song of the Well: A Discourse on the Expected Supply of Water in Boston, Preached to the Congregation in Essex Street, Boston, on Thanksgiving Day, November 26, 1846* (Boston: William D. Ticknor & Co., 1847), 11–12.

4. City of Philadelphia, *Report of the Watering Committee, Read in Select Council, November the Twelfth, 1818* (Philadelphia: William Fry, 1818), 5. Three years later, when the committee decided to switch from wood to iron pipes, it admitted that it felt "great reluctance" in asking for "additional sums to perfect a system which has already cost so much to the public," but since "the public interest requires it, they conceive it their duty to take the responsibility of recommending measures, however costly they may prove in their execution." City of Philadelphia, *Report of the Watering Committee to the Select and Common Councils, Read January 18, 1821* (Philadelphia: Lydia R. Bailey, 1821), addendum.

5. [Nathaniel J. Bradlee], *History of the Introduction of Pure Water into the City of Boston, with a Description of Its Cochituate Water Works* (Boston: Alfred Mudge & Son, 1868), 30.

6. City of Chicago, *Report of the Water Commissioners of the City of Chicago, Made to the Common Council, December 8, 1851: Together with the Act of Incorporation, and a Statement of the Financial Condition of the City, November 16, 1851* (Chicago: Seaton & Peck, 1851), 18. Similar language characterized discussions of subsequent improvements. An account of the building of the new lake tunnel described the condition of the city before this project: "And

Chicago was great, prosperous and happy, with the sole exception that it lacked the one great essential to vitality, to life itself and the life of its citizens—PURE WATER." *The Great Chicago Lake Tunnel* (Chicago: Jack Wing, 1867), 8.

7. Betsy Thomas Mendelsohn writes: "Surface water by its nature is a public good, a substance everyone uses from which no one can be excluded. Water problems are communicated between many people and therefore cannot be solved by individuals. Water is also a natural resource vital to every individual and the community as a whole. Making rules to share a public good such as water is the province of government, since individuals have neither the power nor knowledge to police everyone's use of it." Mendelsohn, "Law and Chicago's Waters, 1820–1920" (Ph.D. diss., University of Chicago, 1999), 116.

8. Channing, *A Plea for Pure Water*, 27.

9. Martin V. Melosi, *The Sanitary City: Urban Infrastructure in America from Colonial Times to the Present* (Baltimore: Johns Hopkins University Press, 2000), table 4.3, p. 74; Richardson Dilworth, *The Urban Origins of Suburban Autonomy* (Cambridge, MA: Harvard University Press, 2005), 12–13; Nelson Manfred Blake, *Water for the Cities: A History of the Urban Water Supply Problem in the United States* (Syracuse, NY: Syracuse University Press, 1956), 267.

10. For a brief discussion of early public water sources, see Blake, *Water for the Cities*, 12–13.

11. [Bradlee], *History of the Introduction of Pure Water*, 4.

12. There were exceptions, as when local government stepped in to provide services that were essential to both the individual and public good but that no private business could or would undertake. Among these services were the regulation of markets and the punishment of criminals. In the decades before the Revolution, Philadelphia assumed ownership of some private wells and made them available for public use. Its 1790 charter granted it the power to levy taxes for lighting, policing, watering, pitching, paving, and cleansing the streets. Theodore Thayer, "Town into City," in *Philadelphia: A 300-Year History*, ed. Russell F. Weigley (New York: Norton, 1982), 69; Richard G. Miller, "The Federal City," in *Philadelphia*, ed. Weigley, 166–67. The Select and Common Councils would continue to use this authority in passing numerous ordinances for specific projects. See for example, "An Ordinance, for the Improvement and Protection of the Footways within the City of Philadelphia," *Aurora and General Advertiser* (Philadelphia), October 8, 1811.

Sam Bass Warner Jr. examines Philadelphia as what he calls "the private city" in urbanizing America, a community that saw as its main common purpose to "keep the peace among individual money-makers, and, if possible, help to create an open and thriving setting where each citizen would have some substantial opportunity to prosper." Warner, *The Private City: Philadelphia in Three Periods of Its Growth* (Philadelphia: University of Pennsylvania Press, 1968), 3–4. See also Maureen Ogle, "Water Supply, Waste Disposal, and the Culture of Privatism in the Mid-Nineteenth Century American City," *Journal of Urban History* 25 (March 1999): 321–47.

Until Boston and Chicago were incorporated as cities, semi-autonomous boards carried out similar responsibilities that were then assumed by committees of their new representative governments, as already had become the practice in Philadelphia. For the evolution of Boston's city government in its first century, see John Koren, *Boston, 1822 to 1922: The Story of Its Government and Principal Activities during One Hundred Years* (Boston: City of Boston Printing Department, 1923). On Chicago, see Ann Durkin Keating, "The Expansion of City Government," in *Building Chicago*, 33–50. See also Dilworth, *The Urban Origins of*

Suburban Autonomy, 11–35; Hendrik Hartog, *Public Property and Private Power: The Corporation of the City of New York in American Law, 1730–1870* (Chapel Hill: University of North Carolina Press, 1983); Philip J. Ethington, *The Public City: The Political Construction of Urban Life in San Francisco, 1850–1900* (New York: Cambridge University Press, 1994); William J. Novak, *The People's Welfare: Law and Regulation in Nineteenth-Century America* (Chapel Hill: University of North Carolina Press, 1996); and Robin L. Einhorn, *Property Rules: Political Economy in Chicago, 1833–1872* (Chicago: University of Chicago Press, 1991).

13. Charles Sellers points out that in the period after the Revolution, America's earliest corporations were chartered mainly to serve purposes clearly in the public interest. Sellers argues that private enterprise was afforded remarkable encouragement and support by the legal system. Sellers, *The Market Revolution: Jacksonian America, 1815–1846* (New York: Oxford University Press, 1991), 45. As Jon C. Teaford explains in relation to cities in particular, "the ideal of regulated concord was yielding to an ideal of open competition." Teaford, *The Municipal Revolution in America: Origins of Modern Urban Government, 1650–1825* (Chicago: University of Chicago Press, 1975), 100.

14. City of Boston, "The Mayor's Address to the City Council of Boston, February 27, 1845," in City of Boston, *The Inaugural Addresses of the Mayors of Boston* (Boston: Rockwell & Churchill, 1894), 1:317.

15. City of Philadelphia, *Report to the Select and Common Councils, on the Progress and State of the Water Works, on the 24th of November, 1799* (Philadelphia: Zachariah Poulson, Jr., 1799), 5.

16. Commonwealth of Massachusetts, *Proceedings before a Joint Special Committee of the Massachusetts Legislature, upon the Petition of the City of Boston, for Leave to Introduce a Supply of Pure Water into that City, from Long Pond, February and March, 1845* (Boston: John H. Eastburn, 1845), 114.

17. Mendelsohn cites Common Council reports going back to the mid-1840s that complained about the inadequacy of the Chicago Hydraulic Company's supply. Mendelsohn, "Law and Chicago's Waters," 124.

18. On the machinations of the Manhattan Company, see Blake, "The Strange Birth of the Manhattan Company" and "New York Wrestles with the Water Problem," in *Water for the Cities*, 44–62, 100–120; as well as Gerard T. Koeppel, *Water for Gotham: A History* (Princeton, NJ: Princeton University Press, 2000). See also Blake, "Rise of the Private Water Companies," in *Water for the Cities*, 63–77.

19. Blake, *Water for the Cities*, 77; City of Philadelphia, *Report of the Joint Committee of the Select and Common Councils on the Subject of Bringing Water to the City* (Philadelphia: Zachariah Poulson, Jr., 1798), 6.

20. [Bradlee], *History of the Introduction of Pure Water into the City of Boston*, 6.

21. Lewis, quoted in Blake, *Water for the Cities*, 175.

22. [Bradlee], *History of the Introduction of Pure Water into the City of Boston*, 6–7.

23. City of Boston, *Soft Water*, City Document No. 25, 1839 [Boston, 1839], 14–15.

24. City of Boston, *Report [of the Standing Committee on the Introduction of Pure and Soft Water into the City]*, *January 29, 1838* [Boston, 1838], 2.

25. Channing, *A Plea for Pure Water*, 24. In his response that seconded Channing, Henry Williams dismissed as "miserable and weak" the assumption that giving the project to a private corporation was preferable. "I would as soon countenance the farming out [of] the pure air of heaven, of placing it in the hands of a few men among us to be stintedly doled out *at a*

price, to the numerous members of the human family." Channing, *A Plea for Pure Water*, 35.
Those with more expert knowledge, who recognized that apparently different water issues
were interconnected, saw this as a strong reason for public control.

26. Commonwealth of Massachusetts, *Proceedings before a Joint Special Committee of the
Massachusetts Legislature* (1845), 103, 2, 4, 8–9, 110.

27. Delaware and Schuylkill Canal Company, *Address of the Committee of the Delaware
and Schuylkill Canal Company, to the Committees of the House of Representatives, on the Memorial of Said Company* (Philadelphia: John Ormrod, [1799]), 4, 7, 12, 14. The cornerstone of
the 552-foot so-called Permanent Bridge that crossed the river at Market Street was laid on
October 13, 1800. Edgar P. Richardson, "The Athens of America (1800–1825)," in *Philadelphia*, ed. Weigley, 232–33; Koren, *Boston, 1822 to 1922*, 175.

28. Schuylkill Navigation Company, *Report of the President and Managers of the Schuylkill
Navigation Company, to the Stockholders, January 4, 1836* (Philadelphia: James Kay, Jr.,
1836), 22. A letter published in 1817 in the *United States Gazette* from "A Friend to the
Poor" had also suggested that the Schuylkill Navigation Company could help "contrive
employment for the poor." *United States Gazette* (Philadelphia), July 15, 1817.

29. "Temperance," *Hints to the Honest Taxpayers of Boston* [Boston, 1845], 1; "Prudence,"
Another View on the Subject of Water [Boston, 1845], 2.

30. *Daily Evening Transcript* (Boston), February 20, 1845.

31. The Boston Aqueduct Company had expressed concern about this possibility as early
as the 1830s, when the company formally asked the Common Council just what Boston's intentions were in regard to building a public water system. Before they invested more money
in the company, its owners wished to know if they faced "overwhelming competition with
the city." Boston Aqueduct Corporation, *Memorial* [Boston, 1838], 2.

32. [Lucius M. Sargent], *Boston Aqueduct and the City of Boston* (Boston: Dutton & Wentworth, [1849]), 3–5. While the Boston Aqueduct claimed that it was worth $200,000, it
finally settled for just over $45,000. City of Boston, *Report from the Cochituate Water Board
Stating that They Have Completed the Purchase of the Property of the Aqueduct Corporation, &c.*,
City Document No. 46, June 16, 1851 [Boston, 1851].

33. Chicago Hydraulic Company, *To the Honorable Mayor and Council of the City of Chicago*
[Chicago, 1852], 1–2, 6–7. The city countered by citing the Supreme Court decision in
Charles River Bridge v. Warren Bridge (1837), arguing that the Chicago Hydraulic Company's
charter did not grant it exclusive rights to provide the city with water. After turning down
an earlier buyout offer of $30,000, the Chicago Hydraulic Company ultimately received
$15,000. City of Chicago, *Third Semi-Annual Report of the Water Commissioners, of the City
of Chicago, to the Common Council of Chicago* (Chicago: Democrat Office, 1853), 6; City of Chicago, *Seventh Semi-Annual Report of the Board of Water Commissioners of the City of Chicago
to the Common Council of the City of Chicago, December 18, 1854* (Chicago: Democratic Press
Steam Print, 1855), 11.

34. *Address of the Faneuil Hall Committee, on the Project of a Supply of Pure Water for the
City of Boston, May 5, 1845* (Boston: W. W. Clapp & Son, 1845), 16, 31–32.

35. John Winthrop, "A Model of Christian Charity," in *The American Puritans: Their Prose
and Poetry*, ed. Perry Miller (New York: Anchor Books, 1956), 83.

36. City of Philadelphia, *Report of the Committee Appointed by the Common Council, to
Enquire into the State of the Water Works* (Philadelphia: William Duane, 1802), 76; City of
Philadelphia, *Report of the Watering Committee to the Select and Common Councils, Novem-
ber 1, 1803* (Philadelphia: William Duane, 1803), 18–19; Letter of Benjamin Henry Latrobe

to Robert Goodloe Harper, March 28, 1803, in *The Correspondence and Miscellaneous Papers of Benjamin Henry Latrobe*, ed. John C. Van Horne and Lee W. Formwalt (New Haven, CT: Yale University Press, 1984), 1:266.

37. City of Philadelphia, *Report to the Select and Common Councils, on the Progress and State of the Water Works* (1799), 28.

38. Not everyone shared this view. A letter published in the *Aurora* from three Philadelphians who dubbed themselves "Common Sense," "Correct Taste," and "Sound Policy" protested against those "who gave us the *dark shadow of the water-house*, in the very middle of the best avenue, to the best city in America, at an extra expence [*sic*] of *forty thousand dollars*, in defiance of all taste, all feeling, and of all economy." Three weeks later "Penn" sent a series of angry letters to the paper criticizing public structures in the city. He called the engine house a "monument of extravagance." *Aurora*, March 7 and 30, 1810.

39. Thompson Westcott, *A History of Philadelphia, from the Time of the First Settlements on the Delaware to the Consolidation of the City and Districts in 1854* (Philadelphia: A. C. Kline, 1867), vol. 4, chap. 454. Rush and his model became the subject of several paintings by Philadelphia painter Thomas Eakins early and late in his career. See *William Rush Carving His Allegorical Figure of the Schuylkill* (Philadelphia Museum of Art, 1876–77), *William Rush Carving His Allegorical Figure of the Schuylkill River* (Brooklyn Museum, 1908), and *William Rush and His Model* (Honolulu Academy of Arts, 1908).

40. [Bradlee], *History of the Introduction of Pure Water into the City of Boston*, 252; *Daily Evening Transcript*, November 17, 1849. The reservoir stood until 1880 in the block bounded by Derne, Temple, Mount Vernon, and Hancock Streets. Its dimensions were approximately 200 feet on each side, and its capacity was almost 2.7 million gallons. The much larger (approximately 7.5 million gallons) South Boston Telegraph Hill Reservoir was situated in what became Thomas Park.

41. City of Chicago, *Report of the Water Commissioners* (1851), 15, 14. When the reservoirs were almost finished, the commissioners declared, "They are substantial structures of a style of work suitable for their use, and of such agreeable appearance as to be ornaments in their localities." City of Chicago, *Fifteenth Semi-Annual Report of the Board of Water Commissioners, to the Common Council of the City of Chicago, January 1st, 1859* (Chicago: William H. Rand, 1859), 4–5.

42. City of Chicago, *Eighth Annual Report of the Board of Public Works to the Common Council of the City of Chicago, for the Municipal Fiscal Year Ending March 31st, 1869* (Chicago: Jameson & Morse, 1869), 74. Cregier was Chicago waterworks engineer from 1853 to 1879 before succeeding Chesbrough as city engineer from 1879 to 1882, at which point he became commissioner of public works for four years. He served as mayor of Chicago from 1889 to 1891.

43. Philadelphia waterworks art and artifacts inspired a 1988 exhibition, titled "The Fairmount Waterworks, 1812–1911," at the Philadelphia Museum of Art, which stands on the site where the Fairmount reservoir was once located. For selected illustrations of the works on display as well as a checklist of the entire exhibition, see *Bulletin* (Philadelphia Museum of Art) 84 (Summer 1988).

44. "To the taste of persons of the present generation it seems unusually chaste in design," Thompson Westcott wrote, "yet it was denounced when first erected as immodest." Westcott, *A History of Philadelphia*, vol. 4, chap. 454.

45. "Review of the Second Annual Exhibition," *Port Folio*, 8, no. 1 (July 1812): 24.

46. After the formal celebration was over, "the several corps marched to their several

places of entertainment." *Aurora*, July 6, 1803. Philadelphia Federalists usually repaired to sites within the city, while Democrats preferred its outskirts, including spots near the future location of the waterworks at Fairmount.

47. "All in all," Mary P. Ryan explains, "the nineteenth-century city was the site of a major expansion in the number and variety of American ceremonies." Ryan, *Women in Public: Between Banners and Ballots, 1825–1880* (Baltimore: Johns Hopkins University Press, 1990), 22. These might be recurring rituals, whether nationally observed holidays like July 4th or more local events.

48. Quoted in Westcott, *A History of Philadelphia*, vol. 4, chap. 427.

49. *Daily Advertiser*, August 21 and 22, 1846; *Daily Evening Transcript*, August 26, 1846; [Bradlee], *History of the Introduction of Pure Water into the City of Boston*, 97–98.

50. On the New York waterworks parade, see Charles King, *A Memoir of the Construction, Cost, and Capacity of the Croton Aqueduct, Compiled from Official Documents* (New York: Charles King, 1843), 225–306.

51. Other participants in the Boston parade included the Overseers of the Poor, the Society of the Cincinnati, newspaper editors, the president of Harvard, associations of skilled workers and merchants (the Boston Charitable Association of Master Tailors "displayed a representation of Adam and Eve, as a specimen of the human race before the invention of their craft"), the Ancient Order of Druids, the Council of the Star in the East, and various merchant societies; "the old Palanquin," an ornate sedan chair supported "by six stout negro bearers, dressed in white oriental costume, with white turbans," and occupied by "a fair young boy reclining in oriental style"; and a fully rigged model of "the famous and fortunate privateer, the *Grand Turk*," mounted on a carriage "drawn by four fine horses." City of Boston, *Celebration of the Introduction of the Water of Cochituate Lake into the City of Boston, October 25, 1848* (Boston: J. H. Eastburn, 1848), 14.

52. Ibid., 7, 5.

53. Ibid., 20.

54. Geo. Schnapp, *Cochituate Grand Quick Step* (Boston: Stephen W. Marsh, 1849). Library of Congress American Memory, http://memory.loc.gov (accessed April 9, 2011).

55. City of Boston, *Celebration of the Introduction of the Water of Cochituate Lake into the City of Boston* (1848), 22.

56. *Daily Advertiser*, October 31, 1848.

57. City of Boston, *Celebration of the Introduction of the Water of Cochituate Lake into the City of Boston* (1848), 30.

58. Ibid., 45.

59. *Daily Evening Transcript*, October 26, 1848; City of Boston, *Celebration of the Introduction of the Water of Cochituate Lake into the City of Boston* (1848), 44–45.

60. City of Boston, *Celebration of the Introduction of the Water of Cochituate Lake into the City of Boston* (1848), 45.

61. *Daily Evening Transcript*, October 26, 1848.

62. *Daily Advertiser*, October 25, 1848.

63. Boston's water celebration, and especially the parade, was, as David Waldstreicher says of the Grand Federal Procession in Philadelphia sixty years earlier, a purposeful exercise in "stylized order." Waldstreicher, *In the Midst of Perpetual Fetes: The Making of American Nationalism, 1776–1820* (Chapel Hill: University of North Carolina Press, 1997), 105. See also Mary P. Ryan, *Civic Wars: Democracy and Public Life in the American City during the Nineteenth Century* (Berkeley: University of California Press, 1997), 59–60; and David

Glassberg, "Civic Celebrations and the Invention of the Urban Public," *Mid-America* 82, nos. 1–2 (Summer 2000): 147–72. When these ceremonies were occasioned by a local achievement, they were double-layered celebrations of city life. Marchers and spectators gathered together to represent the urban collective in honor of achievements that this collective had accomplished.

64. City of Boston, *Celebration of the Introduction of the Water of Cochituate Lake into the City of Boston* (1848), 46–48.

65. Ibid., 10–11.

66. *Chicago Tribune*, March 18, 1864; A. T. Andreas, *History of Chicago, from the Earliest Period to the Present Time*, vol. 2, *From 1857 until the Fire of 1871* (Chicago: A. T. Andreas Co., 1885), 67.

67. *Chicago Tribune*, July 25, 1865.

68. There had been at least one official inspection of the work-in-progress. Participants included the Board of Public Works and most members of the Common Council. For an account, see *Chicago Tribune*, December 2, 1866.

69. *Chicago Tribune*, December 7, 1866; Andreas, *History of Chicago*, 2:67–68. These two sources report the wording of the speeches with some slight differences.

70. *Chicago Tribune*, March 26, 1867; *Chicago Republican*, March 26, 1867; *Chicago Evening Journal*, March 26, 1867. For another description of the celebration, see Harold L. Platt, *Shock Cities: The Environmental Transformation and Reform of Manchester and Chicago* (Chicago: University of Chicago Press, 2005), 135–37.

71. [Sargent], *Boston Aqueduct and the City of Boston*, 33.

72. "A Selfish Taxpayer" objected to the suggestion by Nathan Hale, owner and editor of the pro–Long Pond *Daily Advertiser*, that the matter be put to a popular vote. According to "A Selfish Taxpayer," people like Hale had used "highly improper means . . . to inflame the public mind," creating "a water mania" that was further fueled by the desire of "a very numerous body of artisans, who hope, in one way and another, to be employed in its execution," but "of whom many are utterly ignorant of the real merits of the question, and care nothing for the consequences." In short, "the people" might be far worse caretakers of the public interest than presumably wise and disinterested individuals like himself. The excitable electorate forced candidates for public office, whose "solemn duty" was to consider such issues "*in the most calm and unbiased manner*," to pander to public opinion rather than advance the actual public good. When this happened, "the public functionary surrenders the exercise of his reasoning powers." "A Selfish Taxpayer," *Thoughts about Water* [Boston, 1844], 8, 14, 15.

73. William J. Hubbard, *Argument on Behalf of Joseph Tilden and Others, Remonstrants, on the Hearing of the Mayor of the City of Boston, on Behalf of the City Council, for a Grant of the Requisite Powers to Construct an Aqueduct from Long Pond to the City, before a Joint Special Committee to the Massachusetts Legislature, March 6, 1845* (Boston: T. R. Marvin, 1845), 52.

74. Henry B. Rogers, *Remarks on the Present Project of the City Government, for Supplying the Inhabitants of Boston with Pure Soft Water* (Boston: S. N. Dickinson & Co., 1845), 4–5.

75. Jonathan Preston, in *Letter from Lemuel Shattuck, in Answer to Interrogatories of J. Preston, in Relation to the Introduction of Water into the City of Boston* (Boston: Samuel N. Dickinson, 1845), 6. While Shattuck is not listed as one of the investors in Spot Pond, one of his relatives, Boston physician George Cheyne Shattuck, was.

76. The anonymous author of the pamphlet *How Shall We Vote on the Water Act?* (the Massachusetts Historical Society identifies this person as Lemuel Shattuck) likewise

implied that he had the good of most Bostonians in mind when he said that shouldering the costs of a public waterworks would be bad for the city. *How Shall We Vote on the Water Act?* [Boston, 1845], 23. Nelson Manfred Blake quotes a letter that appeared in the Boston *Daily Evening Transcript* on September 3, 1844, in which "Anti-Humbug" observed that the rich were leaving Boston to avoid its taxes, so that the poor would end up paying for the waterworks. Blake, *Water for the Cities*, 193–94.

77. C. K. Yearley examines the reliance of antebellum governments, including urban governments in particular, on real estate taxes. He also discusses late nineteenth- and early twentieth-century efforts to reform this system. See Yearley, *The Money Machines: the Breakdown and Reform of Governmental and Party Finance in the North, 1860–1920* (Albany: State University of New York Press, 1970).

78. Commonwealth of Massachusetts, *Proceedings before a Joint Special Committee of the Massachusetts Legislature* (1845), 12. Derby was referring to the cost overruns of the Croton system. He claimed that he was not opposed to building a new system, but that the city could find a cheaper and closer source than Long Pond. With the exception of one year, 1858, when the poll tax was $2.10, it stayed at $1.50 through the period covered in this history. For a fuller discussion, see Charles Phillips Huse, *The Financial History of Boston from May 1, 1822, to January 31, 1909* (Cambridge, MA: Harvard University Press, 1916), 88–89.

79. Commonwealth of Massachusetts, *Proceedings before a Joint Special Committee of the Massachusetts Legislature* (1845), 7–8, 110.

80. Hubbard, *Argument on Behalf of Joseph Tilden and Others*, 9; [John H. Wilkins], *Arguments and Statements Addressed to the Members of the Legislature, in Relation to the Petition of the City of Boston for Power to Bring into the City the Water of Long Pond. By a Remonstrant* (Boston: Freeman and Bolles, 1845), 5.

81. "A Selfish Taxpayer," *Thoughts about Water*, 14. Boston mayors dealt with their limited authority in different ways. The first Mayor Quincy candidly stated that he would use the office as aggressively as he could. Aware that committees of the City Council possessed more power than the mayor, Quincy enhanced his influence over civic affairs by designating himself chair of key committees. If the powers he assumed appeared to be "too great for any individual," he stated in his first inaugural address, "let it be remembered, that they are necessary to attain the great objects of health, comfort, and safety to the city," notably city water. The people would see "the advantage of a vigorous and faithful administration." "Inaugural Address of Josiah Quincy, Delivered May 1, 1823," in City of Boston, *The Inaugural Addresses of the Mayors of Boston*, 1:14–15.

82. *How Shall We Vote on the Water Act?* (1845), 16, 24.

83. *Boston to the Rescue!!* [Boston, 1845].

84. Andrew W. Schocket points out that the Watering Committee was the sole subcommittee of the Councils to enjoy "a separate budget, authority to enter into contracts, and the ability to hire and fire its own employees," including its own staff of revenue collectors, giving it "almost total control over its own cash flow." Schocket writes, "With its hands on hundreds of thousands of municipal dollars in building and maintenance of the nation's first grand municipal infrastructure project, the Watering Committee had become the behemoth of Philadelphia politics." Schocket adds: "It [the Watering Committee] issued its own printed annual reports and kept its own records separate from the city councils'. It owned land, paid its own bills, and ran a growing bureaucracy, running up substantial debts in doing so." Schocket, *Founding Corporate Power in Early National Philadelphia* (DeKalb: Northern Illinois University Press, 2007), 127. See also Edward P. Allinson and Boies Penrose, "The

City Government of Philadelphia," in *Johns Hopkins University Studies in Historical and Political Science*, vol. 5: *Municipal Government and Politics* (Baltimore: N. Murray, 1887), 37–38.

Sam Bass Warner Jr. views the accomplishments of the Watering Committee through the 1830s as "a tribute to [Philadelphia's] old merchant-led committee system of government," which had a high regard for the needs of the community that was being superseded by the much more selfish ethos of "the private city," which put the needs and desires of the entrepreneurial individual ahead of the good of the collective. Warner, *The Private City*, 102. Schocket's and Warner's views are not necessarily in conflict. The committee did do remarkable work in overseeing the construction of the first major waterworks in America, and its members over the years included many civic-minded men. But they did command a great deal of power and public resources, which incited indignation and suspicion, especially in the troubled early years of the works.

85. *How Shall We Vote on the Water Act?* (1845), 20.

86. City of Philadelphia, *Report to the Select and Common Councils, on the Progress and State of the Water Works* (1799), 43–44.

87. In 1853 the first Chicago water commissioners made a similar plea. As a response to complaints, they said that they asked "no improper protection" and they did not "seek to avoid any scrutiny however searching," but they demanded that public disapproval of their decisions be based on "correct information" or "a proper consideration of the subject under consideration," as opposed to "the clamor of selfish or vicious individuals who are influenced by other motives than a regard for the public good," among whom they listed people who wanted their jobs or engineers and contractors who had been fired. The committee was especially angry with members of the Common Council, who questioned the commissioners' actions and who were all too ready to credit complaints of fraud, corruption, and incompetence. *Chicago Tribune*, October 26, 1853.

88. *Daily Advertiser*, June 19, 1845.

89. *Daily Advertiser*, July 25, 1845.

90. *Daily Advertiser*, June 17, 1845.

91. City of Chicago, *Report of the Water Commissioners* (1851), 15.

92. Edwin R. A. Seligman, *Essays in Taxation*, 8th ed. (London: Macmillan & Co., 1915), 415–16. Unlike a tax—which, in Seligman's words, "imposes a common burden" in the form of a broadly applied recurring charge, the proceeds of which the government can apply in any number of ways toward what it defines as the common good—a special assessment is a onetime payment for the fulfillment of a particular and measurable private request. A special assessment "can never be progressive," Seligman explains, but is to be "proportional to benefits." While it is not intended to be redistributive, many things that were paid for with special assessments—such as new or upgraded streets, alleys, or sidewalks—were public works, open to and presumably serving the general public interest. On the extensive use of special assessments in Chicago, see Einhorn, *Property Rules*.

The reports of the Board of Public Works for several years in the 1860s and 1870s list no special assessments for items relating to water service, and when such assessments do appear in other years they are trivial compared to those devoted to other infrastructure improvements and to the overall operating budget of the water system as a whole. For example, the board's report for the fiscal year ending March 31, 1869, lists special assessments of $376,111 for street openings and widenings, $27,967 for alley openings and widenings, $1,756,776 for wooden block pavement, $169,225 for miscellaneous street and alley improvements, $27,707 for the erection of lampposts, and only $8,903 for the construction of water service pipes and $13,551 for private drains. In that same year, the income from water rents (called "water tax" in the report) was $420,657, while the expenditures on new construction for the works were $335,472. Other expenses and repairs amounted to $137,933, and interest on the water loans was $171,762. City of Chicago, *Eighth Annual Report of the Board of Public Works* (1869), 25–31, 147–160. Two years later, the last fiscal year before the fire of 1871, total income from special assessments was $2,359,836, with only $137,498 related to water service pipes and $50,103 to street drains. City of Chicago, *Tenth Annual Report of the Board of Public Works to the Common Council of the City of Chicago, for the Municipal Fiscal Year Ending March 31, 1871* (Cambridge, MA: University Press, 1871), 34–37.

93. B. Henry Latrobe, *View of the Practicability and Means of Supplying the City of Philadelphia with Wholesome Water, in a Letter to John Miller, Esquire, from B. Henry Latrobe, Engineer, December 29, 1798* (Philadelphia: Zachariah Poulson, Jr., 1799), 12–13.

94. Einhorn, *Property Rules*, 133.

95. City of Philadelphia, *An Ordinance Providing for the Raising of a Sum of Money for Supplying the City of Philadelphia with Wholesome Water* (Philadelphia: Zachariah Poulson, Jr., 1799), 6.

96. City of Philadelphia, *Report of the Committee for the Introduction of Wholesome Water into the City of Philadelphia* (Philadelphia: Zachariah Poulson, Jr., 1801), 7.

97. City of Chicago, *Third Annual Report of the Board of Public Works, to the Common Council of the City of Chicago, April 1ˢᵗ, 1864* (Chicago: Jameson & Morse, 1864), 5.

98. The water commissioners justified this policy by pointing to their balance sheets. As they explained in their first report, Chicago received $1,375 per mile of pipe, while Boston took in $2,605 and New York $2,733. The problem was Chicago's lower average population density. "Probably no large city covers so much ground in proportion to its population as does Chicago," the board explained. City of Chicago, *First Annual Report of the Board of Public Works to the Common Council of the City of Chicago, January 1st, 1862* (Chicago: S. P. Rounds, 1862), 51. Three years later the board emphasized how much less Chicago was making per mile of pipe than five other cities (though it acknowledged that Philadelphia was doing worse), and in 1874 its members were still repeating this point. City of Chicago, *Fourth Annual Report of the Board of Public Works to the Common Council of the City of Chicago, April 1st, 1865* (Chicago: George H. Fergus, 1865), 7–8. Boston, with an equivalent length of supply and distributing mains (about 135 miles versus Chicago's 132), received almost $400,000 in income, compared to Chicago's $225,000. Cincinnati, with much less pipe, enjoyed higher revenue than Chicago. New York's 295-mile system generated the most dollars (just over $3,000) per mile. In the 1874 report, the board wrote, "In estimating the extent, the variety, the speed, and the value of public improvements carried on by the Board, and the difficulties to be constantly overcome, the peculiar conditions under which the city of Chicago has been built should be taken into account. The older cities developed according to a fixed ratio, and grew compactly. The development of Chicago has been so swift that few years have furnished reliable standards for subsequent ones, and in no large city of the world is the ratio of area to population so great." The board was not so much bragging as offering an explanation for not being able to keep up with the city it served. In its view, "a consideration of this vast territory, and of the great net-work of improvements which minutely cover it, will sufficiently account for the difficulties experienced in prosecuting public works at the same pace which marks private enterprise." City of Chicago, *Thirteenth Annual Report*

of the Board of Public Works to the Common Council of the City of Chicago, for the Municipal Fiscal Year Ending March 31, 1874 (Chicago: J. S. Thompson & Co., 1874), 24–25.

99. City of Chicago, Fourth Annual Report of the Board of Public Works (1865), 7–8.

100. City of Chicago, Ninth Annual Report of the Board of Public Works, to the Common Council of the City of Chicago, for the Municipal Fiscal Year Ending March 31, 1870 (Chicago: Lakeshore Press, 1870), 7.

101. Daily Advertiser, March 22, 1845.

102. Commonwealth of Massachusetts, Proceedings before a Joint Special Committee of the Massachusetts Legislature (1845), 8.

103. Ibid., 114–15. The water eventually reached East Boston via a roundabout route through Charlestown and Chelsea. See [Bradlee], History of the Introduction of Pure Water into the City of Boston, 259–60. For a short account of the work while it was in progress, see the Daily Advertiser of June 11, 1850. For the engineering challenges and proposed solutions regarding East Boston, see City of Boston, Report on the Petition of S. W. Hall, That the Water from Lake Cochituate, May Be Carried to East Boston, City Document No. 22, May 11, 1848 [Boston, 1848]; City of Boston, Report on Water, City Document No. 24, April 19, 1849 [Boston, 1849]; and City of Boston, Engineer's Report in Relation to Taking Water to East Boston, City Document No. 41, July 21, 1849 [Boston, 1849].

104. Aurora, January 20, 1802.

105. City of Boston, Soft Water, City Document No. 19, May 2, 1839 [Boston, 1839], 13–15.

106. Commonwealth of Massachusetts, Proceedings before a Joint Special Committee of the Massachusetts Legislature (1845), 10.

107. City of Chicago, Report of the Water Commissioners (1851), 6–7; City of Chicago, Seventh Semi-Annual Report of the Board of Water Commissioners (1854), 5.

108. City of Chicago, Seventh Semi-Annual Report of the Board of Water Commissioners (1854), 11. Newspapers subsequently carried notices that told owners or occupants of buildings "situated on lots adjoining any street avenue or alley through which the Distributing Water Pipes of the Chicago Water Works are laid" that they were being charged for the use of water. See, for example, the Chicago Tribune of April 19 and 21, 1862.

109. Letter of Benjamin Henry Latrobe to Robert Goodloe Harper, March 28, 1803, in Correspondence and Miscellaneous Papers of Benjamin Henry Latrobe, ed. Van Horne and Formwalt, 1:267. Latrobe was offering advice on how Baltimore, where Harper lived, should proceed.

110. City of Chicago, Seventh Semi-Annual Report of the Board of Water Commissioners (1854), 10.

111. City of Chicago, Ninth Semi-Annual Report of the Board of Water Commissioners to the Common Council of the City of Chicago: Together with Act of Incorporation and Amendments Thereto. Also the Ordinances of the City in Relation to the Chicago Hydraulic Company, January 1st, 1856 (Chicago: Democratic Press Printing House, 1856), 15. Walker was soon able to report that the city's hydrants functioned better, with the notable exception of those that had been tapped "by the water cartmen [who were evidently selling water they took for free from the public hydrants], street sprinklers, and for building purposes." City of Chicago, Eleventh Semi-Annual Report of the Board of Water Commissioners of the City of Chicago to the Common Council of the City of Chicago, January 1st, 1857 (Chicago: Jameson & Morse, 1857), 12.

112. [Bradlee], *History of the Introduction of Pure Water into the City of Boston*, 128. The winter of 1868–69 proved particularly hard on Chicago hydrants. "It is hoped that the system already commenced of endeavoring to make air-tight chambers around the hydrants, by means of boards and tan bark, will not only reduce the expenses of this department of the Works," City Engineer Ellis S. Chesbrough stated, "but will diminish the cases of freezing, and do away with the manure nuisance. It will probably never be safe to neglect frequent examinations during and immediately after very cold weather, in localities peculiarly exposed." City of Chicago, *Tenth Annual Report of the Board of Public Works* (1871), 85.

113. City of Chicago, *Ninth Semi-Annual Report of the Board of Water* (1856), 61–63; Keating, *Building Chicago*, 40; James C. O'Connell, "Technology and Pollution: Chicago's Water Policy, 1833–1930" (Ph.D. diss., University of Chicago, 1980), 16. Keating states that by 1856 a majority of homes had been connected directly to the system. She offers a listing of dwellings with and without water in the late 1850s, based on the reports of the water commissioners, which indicates that in 1858 76.1 percent of the buildings were on the water grid. Because the city was growing so fast, the percentage was lower than the previous two years despite a very large increase in the number of dwellings connected (from 5,640 in 1857 to 7,777 in 1858). Keating, *Building Chicago*, 39, 190.

114. City of Boston, *Report of the Cochituate Water Board, on the Subject of a New Main Pipe, from Brookline to Boston, 1857*, City Document No. 50, August 13, 1857 [Boston, 1857], 14. Eight years later the board still believed that meters were "impracticable," since the cheapest price for a dependable one was $50, "and as out of the 27,000 water takers, at least 13,000 are at a rate of $12 per year, or less, and most of them at $6, it will be seen that the interest and depreciation on a metre, would nearly equal the bill." City of Boston, *Communication of the Cochituate Water Board to the City Council, 1865*, City Document No. 76, October 5, 1865 [Boston, 1865], 7.

115. City of Chicago, *Thirteenth Annual Report of the Board of Public Works* (1874), 4. An 1869 editorial on "The Water Supply" in the *Tribune* read, "The introduction of metres is about as just a means of reaching this abuse [i.e., the waste of water] as any other." *Chicago Tribune*, September 5, 1869.

116. According to the website *Measuring Worth*, the equivalent worth in 2010 of $5 in 1802 was $105, but in terms of affordability, it was comparable to almost $3,000. *Measuring Worth*, http://www.measuringworth.com/uscompare/ (accessed December 25, 2011).

117. City of Philadelphia, *Report of the Committee Appointed by the Common Council, to Enquire into the State of the Water Works* (1802), 50–51; City of Philadelphia, *Report of the Joint Committee, Appointed by the Select and Common Councils for the Purpose of Superintending and Directing the Water Works* (Philadelphia: Robert Cochran, 1802), 23.

118. City of Philadelphia, *Report of the Watering Committee to the Select and Common Councils, November 2, 1809* (Philadelphia: Jane Aitken, 1809), 16.

119. City of Philadelphia, *Report of the Watering Committee, to the Select and Common Councils, Read January 12, 1826* (Philadelphia: Lydia R. Bailey, 1826), 19.

120. City of Philadelphia, *Report of the Watering Committee, to the Select and Common Councils, Read February 11, 1830* (Philadelphia: Lydia R. Bailey, 1830), 4.

121. Philadelphia Water Department, *Water Rates of the City of Philadelphia* (Philadelphia, 1859). The equivalent worth in 2010 of $5 in 1859 was $135. *Measuring Worth*, http://www.measuringworth.com/uscompare/ (accessed December 25, 2011).

122. City of Boston, *An Ordinance in Relation to Water Rates* [Boston, 1850], 6.

123. City of Boston, *City Solicitor's Opinion upon the Right to Give Water to the Children's Friend Society*, City Document No. 13, May 2, 1850 [Boston, 1850], 4, 7.

124. According to the hospital's trustees, the city had encouraged them to take city water with assurances that the cost would not be burdensome. While the hospital had accepted previous increases "without a murmur," even though the charges seemed "unnecessarily high," the latest hike would prove "a serious and very onerous draft upon its funds." The trustees also pointed out that in treating so many poor patients at its own expense, the hospital was freeing the city from the considerable expense of doing so. Meanwhile, it was dependent on charitable contributions, which provided far less than its annual expenses. City of Boston, *Report of the Committee on Water on the Petition of the Massachusetts General Hospital for Reduction of Water Rates*, City Document No. 47 [Boston, 1863], 3–6.

125. The city solicitor explained that the members of the City Council, as trustees of income from water rents, were legally bound to deposit this income in a sinking fund for the redemption of the water debt, "and they can no more appropriate the funds thus receivable to objects foreign to the trust, however charitable and useful, than other trustees can divert their trust funds to other purposes than those for which they are bound to hold them." City of Boston, *Report of the Committee on Water on the Petition of the Massachusetts General Hospital for Reduction of Water Rates* (1863), 6–10. While he did not do so, the city solicitor might have offered the hospital the limited solace that had been conveyed to the Children's Friend Society. He told the society that the great regret the government felt because it could not grant water rent relief was diminished by the knowledge that the "excellence" of such organizations "so strongly commends [the Children's Friend Society] to the individuals of whom cities and towns are composed, that they cannot languish for want of support in a generous community like ours." In other words, private individuals dedicated to the public good would voluntarily step up where the public itself legally could not. City of Boston, *City Solicitor's Opinion upon the Right to Give Water to the Children's Friend Society* (1850), 8.

126. City of Chicago, *Fourteenth Annual Report of the Board of Public Works to the Common Council of the City of Chicago, for the Municipal Fiscal Year Ending March 31, 1875* (Chicago: J. S. Thompson & Co., 1875), 8, 23, 7.

127. City of Philadelphia, *Report of the Committee for the Introduction of Wholesome Water* (1801), 7.

128. Lyman estimated, based on figures from London, that a family would use 200 gallons a day. It was usually assumed that the size of a family was around six people. But 200 gallons proved to be considerably lower than actual use. Lyman's estimate also did not take into account commercial consumption. Lyman added, "The only general remark to be made, appears to be this,—that the supply should be so abundant, or rather bountiful, as to allow the people to use it, at discretion, for every purpose of comfort, cleanliness, and convenience." Theodore Lyman Jr., *Communication to the City Council, on the Subject of Introducing Water into the City* (Boston: J. H. Eastburn, 1834), 13.

129. Capt. [Frederick] Marryat, *A Diary in America, with Remarks on Its Institutions* (New York: Wm. H. Colyer, 1839), 1:79.

130. Walter Channing, *An Address on the Prevention of Pauperism* (Boston: Office of the Christian World, 1843), 71.

131. *Daily Advertiser*, November 18, 1848. "Wantonly and willfully, and uselessly to squander it," he added, "is ingratitude and crime."

132. Nelson Manfred Blake states that the city of Philadelphia tried to explain this to

the public by doing the math, pointing out that if a single house ran water for a half hour a day, this would end up costing the city $7.56 a year, or about 150 percent of what it was receiving in water rent. An 1806 ordinance levied a $5 fine for running water to no useful purpose, but, as noted, laws like this were generally hard to enforce. Blake, *Water for the Cities*, 42–43.

133. City of Boston, *Report of the Cochituate Water Board on Proposed Amendment of Water Rates, 1865*, City Document No. 35 [Boston, 1865], 8. See also City of Boston, *Report on Rates for Measured Water, 1865*, City Document No. 69, September 21, 1865 [Boston, 1865].

134. City of Philadelphia, *Report of the Committee Appointed by the Common Council, to Enquire into the State of the Water Works* (1802), 76; City of Philadelphia, *Report of the Watering Committee to the Select and Common Councils* (1803), 4. The committee was hopeful about a plan to replace hydrants with a system by which Philadelphians would pump water from cisterns buried below the frost line, but the problem persisted. Boston's first water commissioners decided, after consulting with their counterparts in Philadelphia and New York, that hydrants would easily break and, even if they functioned properly, would become a public nuisance, "in consequence of boys collecting around them, and throwing water about; and in the consequence of the mud and dirt, which it is almost impossible to prevent from collecting about them." [Bradlee], *History of the Introduction of Pure Water into the City of Boston*, 128.

135. Letter of James Vanuxem to Robert Wharton, November 27, 1814, Watering Committee Correspondence, Historical Society of Pennsylvania Miscellaneous Collection [0425], the Historical Society of Pennsylvania.

136. City of Boston, *Water Works*, City Document No. 32 [Boston, 1850], 10. The Boston works also had to deal with numerous leaks.

137. Wetmore used the wine gallon, a British measure that is equivalent to the modern U.S. gallon. England had adopted the larger imperial gallon in 1826.

138. City of Boston, *Report of the Cochituate Water Board on the Daily Consumption of Water in This City*, City Document No. 51 [Boston, 1852], 3–5. Up to this point, Wetmore pointed out, the only water-related penal offenses were "illegal and malicious diversion of the water" and "injury to the works," but not "malicious waste." The board could shut off a user, plus issue a fine of $2 on top of the forfeiture of the most recent payment (service was billed in advance), but since it did not make widely known such terminations of service, "all benefit of the example, as the means of deterring others, is lost by the want of publicity."

139. [Bradlee], *History of the Introduction of Pure Water into the City of Boston*, 197–98.

140. One reason to which Boston officials attributed the waste was the high quality of the sewerage system, which, in contrast to some other cities (they cited Philadelphia), removed users' possible concerns about disposing of all the water they consumed. Good sewers, combined with fixed rates for water, also reduced the incentive to fix a leaky faucet or toilet. City of Boston, *Water Rates*, City Document No. 74, November 28, 1853 [Boston, 1853], 4–7. In 1854 Wetmore informed Mayor Jerome V. C. Smith that the "exorbitant consumption of water in the city" made it "incumbent on all City officers, and especially the Water Board[,] to detect and prevent all unnecessary waste." At the same time, he singled out urinals in which water ran constantly, "particularly those belonging to the City Buildings," as a "great source of waste," and declared it "my duty thereupon to direct the supply to these places to be at once cut off." City of Boston, *An Ordinance in Addition to an Ordinance Relating to the Water Works*, City Document No. 19 [Boston, 1854], 3. In 1862 Wetmore's successor, Ebenezer Johnson, estimated that a supposedly self-regulating hop-

per closet used almost ten times more water than one with a pan that required emptying, and he called for a high annual premium for every fixture of this kind that a water customer owned. He also addressed the concomitant problem of disposal of soiled water that now overflowed private waste receptacles and by this point challenged the city's sewerage system. City of Boston, *Report of the Water Registrar on Waste of Water by Hopper Closets, City Document No. 11, January 27, 1862* [Boston, 1862]. On the special burden on sewerage caused by the water closet, see Joel A. Tarr, "Building the Urban Infrastructure in the Nineteenth Century: An Introduction," *Essays in Public Works History* 14 (December 1985): 68; and Tarr, *The Search for the Ultimate Sink: Urban Pollution in Historical Perspective* (Akron, OH: University of Akron Press, 1996), 114–16.

141. "Inaugural Address to the Aldermen and Common Council by Benjamin Seaver, Mayor of the City of Boston, January, 1852," in City of Boston, *The Inaugural Addresses of the Mayors of Boston* (Boston: Rockwell & Churchill, 1894), 2:22–24.

142. City of Boston, *An Ordinance in Addition to an Ordinance Relating to the Water Works* (1854), 3.

143. City of Boston, *Report of the Cochituate Water Board, on the Subject of a New Main Pipe, from Brookline to Boston, 1857*, 6–7, 13–14. According to Charles Phillips Huse, in 1860 the city of New York, which had four times the population of Boston, was using only a third more water. Huse, *Financial History of Boston*, 141.

144. City of Chicago, *Eighth Annual Report of the Board of Public Works* (1869), 62.

145. City of Chicago, *Ninth Semi-Annual Report of the Board of Water Commissioners* (1856), 63–64.

146. City of Chicago, *Seventeenth Semi-Annual Report of the Board of Water Commissioners to the Common Council of the City of Chicago, January 1st, 1860* (Chicago: Dunlop, Sewell & Spalding, 1860), 5.

147. City of Chicago, *Rules and Regulations, Board of Public Works, Water Department 1869* (Chicago: Jameson & Morse, 1869), 8; City of Chicago, *Ninth Annual Report of the Board of Public Works* (1870), 122–23.

148. City of Chicago, *Seventh Annual Report of the Board of Public Works to the Common Council of the City of Chicago, for the Municipal Fiscal Year Ending March 31st, 1868* (Chicago: Jameson & Morse, 1869), 3–4.

149. City of Chicago, *Ninth Annual Report of the Board of Public Works* (1870), 87.

150. City of Chicago, *Third Annual Report of the Department of Public Works to the City Council of the City of Chicago, for the Fiscal Year Ending December 31, 1878* (Chicago: Clark & Edwards, 1879), 53. For a chart of the city's per capita consumption, average daily pumpage, population, and mileage of pipe in service from just before 1855 to 1935, see City of Chicago, Bureau of Engineering, *A Century of Progress in Water Works: Chicago, 1833–1933* (Chicago: Department of Public Works, 1933), 25. For statistics on usage in other cities, see Tarr, *The Search for the Ultimate Sink*, 10.

151. City of Boston, *Report of the Cochituate Water Board on Proposed Amendment of Water Rates, 1865*, 4.

152. *Chicago Tribune*, June 27, 1870.

153. City of Chicago, *Ninth Annual Report of the Board of Public Works* (1870), 87–88.

154. *Chicago Tribune*, June 5, 1871. The paper complained the next day that it was bad policy to expand the city's mains without increasing the supply.

155. City of Chicago, *Fourteenth Annual Report of the Board of Public Works* (1875), 14.

156. City of Chicago, *Fifteenth Annual Report of the Board of Public Works to the Common*

Council of the City of Chicago, for the Municipal Fiscal Year Ending December 31st, 1875 (Chicago: J. S. Thompson & Co., 1876), 11–12.

157. City of Boston, *Communication of the Cochituate Water Board to the City Council, 1865*, 7.

158. See the discussion of this painting, its origins, and its composition, in Anneliese Harding, *John Lewis Krimmel: Genre Artist of the Early Republic* (Winterthur, DE: Henry Francis DuPont Winterthur Museum, 1994), 166–71.

159. Harding points out that Krimmel's sketches for the painting indicate that it was inspired in part by Andrew Jackson's exploits in the Seminole War of 1818 and his visit to Philadelphia in February 1819. Ibid., 169.

160. *Daily Advertiser* and Wharton proclamation, cited by J. Thomas Scharf and Thompson Westcott, *History of Philadelphia, 1609–1884* (Philadelphia: L. H. Everts and Co., 1884), 3:1844. Scharf and Westcott state that "the liquor-sellers and gamblers removed to Bush Hill, where they soon became more objectionable than ever." See also Susan G. Davis, *Parades and Power: Street Theatre in Nineteenth-Century Philadelphia* (Philadelphia: Temple University Press, 1986), 42.

161. Edwin Wolf II, *Philadelphia: Portrait of an American City*, new and enl. ed. (Philadelphia: Camino Books, 1990), 159.

162. Gary B. Nash contends that the earlier painting depicts Centre Square as less rowdy than it probably was even then. He states that one of the signs of increasing social divisions was that the white working class had taken over July Fourth celebrations, driving away wealthier Philadelphians and most African Americans—the former by choice, the latter to avoid race-based violence. As for the standing African American boy and couple in the earlier painting (he does not mention the African American boy under the drink seller's table), Nash writes, "It is possible that Krimmel painted the picture just before the incident where white ruffians, driving black Philadelphians from Centre Square celebrations, initiated a 'whites only' policy regulating freedom celebrations." Nash, *First City: Philadelphia and the Forging of Historical Memory* (Philadelphia: University of Pennsylvania Press, 2002), 117. See also Simon P. Newman, *Parades and the Politics of the Street: Festive Culture in the Early American Republic* (Philadelphia: University of Pennsylvania Press, 1997), 87, 103.

Nash discusses how New Year's Day, the date of the prohibition of the slave trade in 1808 and of Haitian independence in 1804, became "the black Fourth of July" in Philadelphia. Nash, *Forging Freedom: The Formation of Philadelphia's Black Community, 1720–1840* (Cambridge, MA: Harvard University Press, 1988), 177, 189. On the city's race riots of the 1830s, see 273–77.

163. *Daily Evening Transcript*, October 23, 1848.

164. In the widely distributed panoramic lithograph of the crowd on the Common, Webster stands in stern profile in the left foreground, with most of the other spectators and the fountain far behind him. A nearby figure who resembles Henry Wadsworth Longfellow converses with some ladies in attendance. For the identification of Webster and Longfellow, see Samuel Barber, *Boston Common: A Diary of Notable Events, Incidents, and Neighboring Occurrences* (Boston: Christopher Publishing House, 1916), 177.

165. Wentworth, quoted in the *Chicago Tribune*, October 28, 1853. Sounding very much like those in Boston who wanted to follow the judgment of men of "wealth" and "competence" in regard to water and all other significant matters, the *Tribune* described the current commissioners as beyond reproach because they were "practical businessmen, . . . not politi-

cians." They were ideally suited for the job since "their interests are intimately identified with as rapid a prosecution of these works as is compatible with economy, and the welfare of the city." As men of "large property," their investment in Chicago "will be affected advantageously if they prosecute their trust with fidelity, and injuriously if they do not." These individuals were innocent of "the wasteful extravagance and wholesale robbery that had characterized similar enterprises previously, when they were managed by politicians" who were interested in "fattening upon the public treasury by a species of peculation not half as honorable as highway robbery." *Chicago Tribune*, October 19, 1853. Five days later, the three commissioners sent a letter to the mayor and Common Council. They stated that while they respected "popular opinion" when it was based on "correct information," they "have not felt it our duty to be controlled by the clamor of selfish or vicious individuals who are influenced by other motives than a regard for the public good." This letter, dated October 24, was published in the *Chicago Tribune* on October 26, 1853.

166. The *Tribune* endorsed George W. Dole (who won) over Eighth Ward alderman Samuel Ashton. The paper praised longtime resident Dole as "a prominent and successful merchant and an honorable and valuable citizen" who from humble beginnings had "toiled upward, until by industry, energy and probity he has won the esteem and confidence of all who have known him," earning a fortune without "blot or taint" on his character. Ashton, who would later become a major figure in Chicago politics and a justice of the Illinois Supreme Court, was in the *Tribune*'s estimation a mediocre lawyer, "better known as a small politician than anything else," lacking in the experience in managing large projects and big budgets that the job required. *Chicago Tribune*, May 2, 1856.

167. The paper expressed great distress four years later at the prospect that internal divisions in the Republican Party might enable an unacceptable candidate to take office, as had happened in the most recent Sewerage Commission election. With the term of a Republican water commissioner coming to an end, the paper asked, "Who wants to see an Irish Democrat in his place, and another in the place of the next one of our men who goes out? Who wants to put these two great public trusts [i.e., water and sewerage], upon which the daily comfort and the health of every citizen depend, into the field of partisan strife, and make them the servants of the Democrats rather than of the people?" *Chicago Tribune*, February 29, 1860. The *Tribune* did not acknowledge that the matter was already "in the field of partisan strife," but instead presented its position as impartial and devoted only to the public good.

Chapter Four

1. See Robert Pogue Harrison, *Forests: In the Shadow of Civilization* (Chicago: University of Chicago Press, 1992), 2. On conceptions of the nineteenth-century American city in relation to nature, see, among other studies, Peter J. Schmitt, *Back to Nature: The Arcadian Myth in Urban America* (New York: Oxford University Press, 1969); Thomas Bender, *Toward an Urban Vision: Ideas and Institutions in Nineteenth-Century America* (Lexington: University Press of Kentucky, 1975); David Schuyler, *The New Urban Landscape: The Redefinition of City Form in Nineteenth-Century America* (Baltimore: Johns Hopkins University Press, 1986); and Stanley K. Schultz, *Constructing Urban Culture: American Cities and City Planning, 1800–1920* (Philadelphia: Temple University Press, 1989). For more theoretical analyses, with specific reference to water, see Maria Kaika, *City of Flows: Modernity, Nature, and the City* (New York: Routledge, 2005); and Erik Swyngedouw, *Social Power and the Urbanization of Water* (New York: Oxford University Press, 2004).

2. B. Henry Latrobe, *View of the Practicability and Means of Supplying the City of Philadelphia with Wholesome Water, in a Letter to John Miller, Esquire, from B. Henry Latrobe, Engineer, December 29, 1798* (Philadelphia: Zachariah Poulson, Jr., 1799), 4, 6, 7.

3. For a description of the considerable changes made in and around Long Pond, see [Nathaniel J. Bradlee], *History of the Introduction of Pure Water into the City of Boston, with a Description of Its Cochituate Water Works* (Boston: Alfred Mudge & Son, 1868), 235–38. As Nancy S. Seasholes explains in detail, most of Boston consists of made land. Seasholes, *Gaining Ground: A History of Landmaking in Boston* (Cambridge, MA: MIT Press, 2003), appen. 1, p. 422.

4. Jean-Pierre Goubert writes, "In the nineteenth century, water, an essentially free gift from God or nature, became an industrial product manufactured by man." Goubert, *The Conquest of Water: The Advent of Health in the Industrial Age*, trans. Andrew Wilson (Princeton, NJ: Princeton University Press, 1989), 169. In addition, entrepreneurs harvested ice from frozen ponds in winter, stored it in special buildings, and then cut it into blocks for sale to those who wished to enjoy the unnatural pleasure of a chilled drink in the heat of summer. Factory owners viewed moving water from naturally occurring or artificial mill streams and ponds as what Theodore Steinberg describes as "a set of 'inputs'" powering the industrial economy, which revolved around, as the mechanistic metaphor would have it, urban "hubs." Steinberg, *Nature Incorporated: Industrialization and the Waters of New England* (New York: Cambridge University Press, 1991), 11.

Some historians point to the assumption that man can conquer and control nature as the imaginative basis of such otherwise disparate ways of understanding the world as those underlying the Enlightenment, industrialization, capitalism, and totalitarianism. They see each of these as attempts at mastery that alienate humans from the natural world and one another. Both David Harvey and Donald Worster cite Marx's assertion that the rise of Enlightenment scientific thought and capitalism stripped nature of power, authority, and meaning beyond its usefulness to the humans who subdued it. Harvey, *Social Justice and the City* (Baltimore: Johns Hopkins University Press, 1973), 215; Worster, *Rivers of Empire: Water, Aridity, and the Growth of the American West* (New York: Pantheon Books, 1985), 26. In a related point, Karl Wittfogel asserts that "nature changes profoundly whenever man, in response to simple or complex historical causes, profoundly changes his technical equipment, his social organization, and his world outlook." Wittfogel, *Oriental Despotism: A Comparative Study of Total Power* (New Haven, CT: Yale University Press, 1957), 11.

5. David Nye writes, "Most nineteenth-century Americans believed in a deceptively simple story in which the natural world was incomplete and awaited fulfillment through human intervention." Nye, *America as Second Creation: Technology and Narratives of New Beginnings* (Cambridge, MA: MIT Press, 2003), 9.

6. Goubert, who evocatively describes underground urban water infrastructure as "a subterranean universe . . . a sort of invisible, upside-down city," that displays the "conquest" of water, also points out that this achievement made humans all the more dependent on it: "Water thus conquered man in a triumph linked to increasing industrialization and an economy that devoured water." He immediately adds, "Nevertheless, it has retained its ancient power." Goubert, *The Conquest of Water*, 68, 25.

7. Samuel Breck, *Sketch of the Internal Improvements Already Made by Pennsylvania: With Observations upon Her Physical and Fiscal Means for Their Extension; Particularly as They Have Reference to the Future Growth and Prosperity of Philadelphia* (Philadelphia: J. Maxwell, 1818), 30.

8. Ralph Waldo Emerson, *Nature*, in *Emerson: Essays and Lectures*, ed. Joel Porte (New York: Library of America, 1983), 8.

9. "An Address to the Board of Aldermen, and the Members of the Common Council, of Boston, on the Organization of City Government, May 1, 1824," in City of Boston, *The Inaugural Addresses of the Mayors of Boston* (Boston: Rockwell & Churchill, 1894), 1:27–28.

10. *Address of the Faneuil Hall Committee, on the Project of a Supply of Pure Water for the City of Boston, May 5, 1845* (Boston: W. W. Clapp & Son, 1845), 23.

11. Emerson, *Nature*, 24.

12. *Daily Advertiser and Patriot* (Boston), August 22, 1846.

13. Ibid.

14. Ibid. The free translation did not stop here. Speaking for the water commissioners, T. B. Curtis said that the literal meaning of the lake's Indian name was "great, eminent, clear, sweet, and plenty of it."

15. Nehemiah Adams, *The Song of the Well: A Discourse on the Expected Supply of Water in Boston, Preached to the Congregation in Essex Street, Boston, on Thanksgiving Day, November 26, 1846* (Boston: William D. Ticknor & Co., 1847), 2, 5.

16. [Bradlee], *History of the Introduction of Pure Water into the City of Boston*, 95–97, 105.

17. Adams continued: "That purpose has long been concealed and delayed. The snows of myriads of winters, 'likewise the small rain and the great rain of His strength' [Job 37:6], have descended into it; secret springs have been contributing to its depth and amplitude; it has been full, year after year; but still the purpose of God with regard to it has not been made manifest. The leaves of uncounted autumns have fallen round it; the nuts and berries in successive harvests have perished there; by it the fowls of heaven have had their habitation, which sing among the branches; the monthly 'changes of the watery star' have passed over it age after age; but not till the present year has it begun to serve any purpose commensurate with its value." Adams, *The Song of the Well*, 5. This passage anticipated the language of a contemporary with whom Adams otherwise had little in common, Henry David Thoreau, who would write in *Walden*, "A lake is the landscape's most beautiful and expressive feature. It is earth's eye; looking into which the beholder measures the depth of his own nature." Thoreau, *Walden*, in *Henry David Thoreau*, ed. Robert F. Sayre (New York: Library of America, 1985), 471. *Walden* was first published in 1854.

18. Adams, *The Song of the Well*, 6–7.

19. City of Boston, *Celebration of the Introduction of the Water of Cochituate Lake into the City of Boston, October 25, 1848* (Boston: J. H. Eastburn, 1848), 5.

20. Ibid., 21–22.

21. Ibid., 25, 29.

22. See Exodus 17:6: "Behold, I will stand before thee there upon the rock in Horeb; and thou shalt smite the rock, and there shall come water out of it, that the people may drink. And Moses did so in the sight of the elders of Israel." The comparison with Moses also implied that God granted Adams a prospective vision of the new Canaan that was a well-watered Boston but did not permit him to live to dwell within it.

23. City of Boston, *Celebration of the Introduction of Water of Cochituate Lake* (1848), 43.

24. Holmes voiced this observation with pride and approval, expressing contempt for what he called "little toad-eating cities." Oliver Wendell Holmes, *The Autocrat of the Breakfast-Table* (Boston: Houghton Mifflin, 1892), 127. This book was first published in 1858.

25. *Daily Advertiser*, August 22, 1846.

26. Like Quincy, Chickering saw the inflow of rural population and goods as both the

cause and the effect of the growth of Boston. As he succinctly put it, "The city is derived from the country," whether one was talking about those "attracted thither by the hope of employment, of trade, of pleasure or of support," or "the materials with which the edifices are reared, the wharves and sewers are constructed, and the streets covered, the merchandise to be sold and distributed in all directions where it may be wanted by the people of the surrounding country." Cities were where "genius and invention are stimulated to higher attainments in the means of administering to the refinement and luxuries of mankind." City of Boston, *Report of the Committee Appointed by the City Council; and also a Comparative View of the Population of Boston in 1850, with the Births, Marriages, and Deaths, in 1849 and 1850, by Jesse Chickering*, City Document No. 60 (Boston: J. H. Eastburn, 1851), 50–51.

27. Lemuel Shattuck, *Report to the Committee of the City Council Appointed to Obtain the Census of Boston for the Year 1845, Embracing Collateral Facts and Statistical Researches Illustrating the History and Condition of the Population, and Their Means of Progress and Prosperity* (Boston: John H. Eastburn, 1846), 37.

28. City of Boston, *Comparative View of the Population of Boston in 1850*, 57; Jesse Chickering, *Immigration into the United States* (Boston: Charles C. Little and James Brown, 1848), 34.

29. Henry Brown, *Present and Future Prospects of Chicago: An Address Delivered before the Chicago Lyceum, January 20, 1846* (Chicago: Fergus Printing Co., 1876), 6, 10. See Psalms 50:10: "For every beast of the forest is mine, and the cattle upon a thousand hills." On the relationship between nature and Chicago's built environment, see Jonathan J. Keyes, "Urbs in Horto: Chicago and Nature, 1833–1874" (Ph.D. diss., University of Chicago, 2003); and Joshua A. T. Salzmann, "Safe Harbor: Chicago's Waterfront and the Political Economy of the Built Environment, 1847–1918" (Ph.D. diss., University of Illinois at Chicago, 2008).

30. John S. Wright, *Chicago: Past, Present, Future*, 2nd ed. (Chicago: Horton & Leonard, 1870), 77, 34, ii, xii–xiii, xlii.

31. John B. Rice, quoted in A. T. Andreas, *History of Chicago, from the Earliest Period until the Present Time*, vol. 2, *From 1857 until the Fire of 1871* (Chicago: A. T. Andreas Co., 1885), 68.

32. *Chicago Republican*, March 26, 1867.

33. Acknowledging Marx and Hegel, William Cronon defines "second nature" as "the artificial nature that people erect atop first nature." His broader point is that what we understand as city and country are mutually interdependent creations of each other. "Both are second nature to us." Cronon, *Nature's Metropolis: Chicago and the Great West* (New York: Norton, 1991), xvii, 384–85.

34. Emerson, *Nature*, 8.

35. As Cronon points out, the distinction between first and second nature is inherently problematic, especially once cities appeared. This is because an interdependent economy of producers, manufacturers, transporters, marketers, and customers transformed the "natural" hinterland that raised crops and livestock and mined coal and minerals at the same time it created and continuously developed the cities where these were processed into goods to be traded and consumed. In addition, any discussion of the distinction between first and second nature is complicated by the fact that "nature" itself is a concept invented by humans. Cronon, *Nature's Metropolis*, xvii. Raymond Williams similarly states that "the idea of nature contains, though often unnoticed, an extraordinary amount of human history." Man and nature have become so implicated in each other that "the idea of nature is

the idea of man, and this not only generally, or in ultimate ways, but the idea of man in society, indeed the ideas of kinds of societies." Williams continues, "We have mixed our labour with the earth, our forces with its forces too deeply to be able to draw back and separate each other out." Williams, "Ideas of Nature," in *Problems in Materialism and Culture: Selected Essays* (London: Verso, 1980), 67, 83. Lawrence Buell writes, "Humanity *qua* geographer is *Homo faber* [man the maker], the environment's constructor, and the sense of place is necessarily always a social product and not simply what is 'there.'" Buell, *The Environmental Imagination: Thoreau, Nature Writing, and the Formation of American Culture* (Cambridge, MA: Belknap Press of Harvard University Press, 1995), 77. This is a central argument in Simon Schama's *Landscape and Memory* (New York: Knopf, 1995).

36. Thoreau, *Walden*, 395, 399.

37. "Be it life or death, we crave only reality," Thoreau continued. "If we are really dying, let us hear the rattle in our throats and feel cold in the extremities; if we are alive, let us go about our business." Ibid., 400.

38. Ibid.

39. Thompson Westcott, *A History of Philadelphia, from the Time of the First Settlements on the Delaware to the Consolidation of the City and Districts in 1854* (Philadelphia: A. C. Kline, 1867), vol. 4, chaps. 486, 492.

40. George Perkins Marsh, *Man and Nature; or, Physical Geography as Modified by Human Action*, ed. David Lowenthal (Cambridge, MA: Belknap Press of Harvard University Press, 1965), 39–40, 12–13. *Man and Nature* was originally published in 1864. A revised version, titled *The Earth as Modified by Human Action*, was published in 1874.

41. Ibid., 36.

42. Even before the appearance of humans, Marsh explained, nature was in a state of dynamic balance more delicate than people understood, and people were all too ready to upset it. He hoped "to suggest the possibility and the importance of the restoration" of what he called nature's "disturbed harmonies," as well as "the material improvement of waste and exhausted regions." Marsh wished to remind humans that the rest of the creation is, like them, "nourished at the table of bounteous nature." They must realize the consequences of how much they altered the planet by the way they chose to live. Ibid., v–vii, 3, 11, 35.

43. Ibid., 42, 308–9.

44. City of Philadelphia, *Annual Report of the Chief Engineer of the Water Department of the City of Philadelphia, for the Year 1875* (Philadelphia: E. C. Markley & Son, 1876), 9–10.

45. City of Boston, *The Sewerage of Boston: A Report by a Commission, Consisting of E. S. Chesbrough, C.E., Moses Lane, C.E., Charles F. Folsom, M.D.* (Boston: Rockwell & Churchill, 1876), 14.

46. J. H. Rauch, M.D., *The Sanitary Problems of Chicago, Past and Present* (Cambridge, MA: Riverside Press, 1879), 1. This was reprinted from vol. 4 of *Transactions of the American Public Health Association*. In her study of Chicago's water history from the early nineteenth to the early twentieth century, Betsy Thomas Mendelsohn challenges what she describes as "the prevailing historical interpretation that Chicago triumphed over its low, marshy city site." She argues that "the knowledge of hydrology . . . was ignored by local politicians when they built a sewerage and drinking water system contrary to their sanitary engineer's recommendation," that is, a sewerage system that could readily pollute the water supply. Mendelsohn maintains that "public problems linked to water were difficult to solve because water was a misunderstood environmental factor in the calculus of social and natural ills."

She adds, "These problems persisted in part because they depended on environmental water factors new to cities." Mendelsohn, "Law and Chicago's Waters, 1820–1920" (Ph.D. diss., University of Chicago, 1999), 9, 120.

47. As Raymond Williams points out, nostalgia "is universal and persistent." Williams, *The Country and the City* (New York: Oxford University Press, 1973), 12. Lawrence Buell characterizes pastoralism as "the cosmopolitan's recurring daydream." Buell, *New England Literary Culture: From Revolution through Renaissance* (New York: Cambridge University Press, 1986), 86.

48. Holmes, *The Autocrat of the Breakfast-Table*, 200. The quotation is from the opening words of the first of "The Olympian Odes" by the Greek lyric poet Pindar (ca. 522–ca. 438 B.C.).

49. Samuel Woodworth, *The Poetical Works of Samuel Woodworth* (New York: Charles Scribner, 1861), 1:27.

50. Ibid., 1:31–32.

51. The Erie Canal happily united the Hudson River and Lake Erie, Woodworth writes, "by weaving a ligament nought can disjoin," with the result that "regions once dreary, are now smiling cheery." Ibid., 1:247, 251–52, 271.

52. Bostonius, *Voice in the City; to Water Drinkers, a Recitative Poem* (Boston: Published by the Author, 1839), 4, 18.

53. Bostonius's recitative is representative of other responses to contemporary change by local authors. In antebellum Boston, which was the literary and intellectual capital of America as well as of New England at this time, many writers based in the city and the region voiced an appreciation of the countryside in ways that entailed what seems to have been a willful obliviousness of the fact that Massachusetts was quickly becoming the most industrialized state in the nation. Lawrence Buell notes that, with few exceptions until at least mid-century, the work of the best of these writers ignored the dramatic alterations—urbanization central among them—taking place before their very eyes and in their own lives. Buell adds that numerous works written at what he calls "the 'subliterary' level" did deal with the new urban subject "in such urban modes as Dickensian gothic." Buell, *New England Literary Culture*, 300–301. On the continuing imaginative presence of the pastoral in industrializing and urbanizing America, see Leo Marx, *The Machine in the Garden: Technology and the Pastoral Ideal in America* (New York: Oxford University Press, 1964).

54. Penn laid out these specifications in his instructions as governor of the colony, dated September 30, 1681. *The Papers of William Penn*, ed. Mary Maples Dunn, Richard S. Dunn et al. (Philadelphia: University of Pennsylvania Press, 1982), 2:119–21.

55. As Mary Maples Dunn and Richard S. Dunn explain, Philadelphia would still have "far more generous city limits than in any other early American town, with enough expansion room to permit many years of orderly future growth." Dunn and Dunn, "The Founding 1681–1701," in *Philadelphia: A 300-Year History*, ed. Russell F. Weigley (New York: Norton, 1982), 7.

56. Ibid., 7.

57. On Holme's grid and Philadelphia's street plan, see Dell Upton, *Another City: Urban Life and Urban Spaces in the New American Republic* (New Haven, CT: Yale University Press, 2008), 113–14, 124, 139.

58. Penn, *Papers of William Penn*, 2:197–99; Dunn and Dunn, "The Founding," 7–8, 15–16. Holme's grid and the squares are evident in fig. 2.1.

59. City of Philadelphia, *Report of the Committee for the Introduction of Wholesome Water into the City of Philadelphia* (Philadelphia: Zachariah Poulson, Jr., 1801), 12.

60. City of Philadelphia, *Report of the Watering Committee to the Select and Common Councils, November 1, 1803* (Philadelphia: William Duane, 1803), 18. In the 1820s and 1830s, Philadelphia similarly improved its other public squares, putting in dozens of varieties of trees and plants. Other Philadelphians who were similarly honored by having a public square named after them were scholar and statesman James Logan, astronomer and scientific instrument maker David Rittenhouse, and Benjamin Franklin, as well as George Washington, who lived in the city during his presidency. A notice in the *United States Gazette* on September 16, 1815, indicates that property owners in the vicinity of the southeast (Washington) square agreed to pay a special assessment to install a culvert in it. On the development of Washington Square, see Jon C. Teaford, *The Municipal Revolution in America: Origins of Modern Urban Government, 1650–1825* (Chicago: University of Chicago Press, 1975), 106–7.

61. In discussing its plans for improving the setting, the Watering Committee had noted that it "will present uniformity in taste as well as utility." City of Philadelphia, *Report of the Watering Committee, to the Select and Common Councils, Read January 12, 1832* (Philadelphia: Lydia R. Bailey, 1832), 5.

62. Westcott, *A History of Philadelphia*, vol. 4, chap. 511. See also Henri Marceau, *William Rush, 1756–1833: The First Native American Sculptor* (Philadelphia, 1937), 26–29; Linda Bantel, "Catalogue," in *William Rush, American Sculptor* (Philadelphia: Pennsylvania Academy of the Fine Arts, 1982), 114–17.

63. Commissioners of Fairmount Park, *First Annual Report of the Commissioners of Fairmount Park* (Philadelphia: King & Baird, 1869), 7.

64. Fairmount Park Commission, *Laws, Ordinances and Regulations Relating to Fairmount Park and Other Parks Under the Control of the Fairmount Park Commission* (Philadelphia: Fairmount Park Commission, 1933), 5.

65. *A Description of the Boston Water Works, Embracing All the Reservoirs, Bridges, Gates, Pipe Chambers, and Other Objects of Interest, from Lake Cochituate to the City of Boston* (Boston: Geo. R. Holbrook & Co., 1848), 27–28.

66. City of Chicago, *Eleventh Semi-Annual Report of the Board of Water Commissioners of the City of Chicago to the Common Council of the City of Chicago, January 1st, 1857* (Chicago: Jameson & Morse, 1857), 12.

67. *Chicago Tribune*, January 14, 1867.

68. Westcott, *A History of Philadelphia*, vol. 4, chap. 454.

69. "Suspension Bridge at Fairmount," *Gleason's Pictorial Drawing-Room Companion* 1, no. 15 (August 9, 1851): 232.

70. James Hosmer Penniman, *Philadelphia in the Early Eighteen Hundreds* (Philadelphia: St. Stephen's Church, 1923), 43, cited in Andrew M. Schocket, *Founding Corporate Power in Early National Philadelphia* (DeKalb: Northern Illinois University Press, 2007), 130–31.

71. On Lafayette in Philadelphia, see John Russell Young, ed., *Memorial History of the City of Philadelphia from Its First Settlement to the Year 1895* (New York: New-York History Co., 1895), 453–55.

72. Frances Trollope, *Domestic Manners of the Americans* (London: Whittaker, Treacher, & Co., 1832), 1:211–12.

73. Charles Dickens, *American Notes for General Circulation* (New York: D. Appleton & Co., 1868), 43.

74. Samuel L. Clemens to Orion and Henry Clemens, October 26–28, 1853, in University of California, Mark Twain Project Online, http://www.marktwainproject.org/xtf/view?docId=letters/UCCL00002.xml;style=letter;brand=mtp (accessed January 8, 2010). The pine eventually rotted as a result of exposure to the weather, and the city replaced the statue with a bronze reproduction. The head of the original figure is all that remains, and it is now in the collections of the Pennsylvania Academy of the Fine Arts, while the bronze reproduction is in the Philadelphia Museum of Art.

75. J. C. Wild, *Panorama and Views of Philadelphia, and Its Vicinity, Embracing a Collection of Twenty Views, Drawn on Stone, by J. C. Wild, from His Own Sketches and Paintings, with Poetical Illustrations of Each Subject, by Andrew Mc'Makin* (Philadelphia: J. B. Chevalier, 1838), illustration 1, "Fairmount, from the Basin."

76. An illustrated guide to Chicago published in 1867 stated that the waterworks at Chicago Avenue would be a great attraction, but mainly as an engineering achievement, not as an appealing natural space. The guide read, "These buildings are but a mile from the heart of the city, and associated as they will be with this wonder of the age, will always prove an object of great interest to visitors, as they are of pride to the people of Chicago." James W. Sheahan, *Chicago Illustrated, 1830–1866* (Chicago: Jevne & Almini, [1867]), 58.

77. *The Great Chicago Lake Tunnel* (Chicago: Jack Wing, 1867), 30–39. *Chicago Tribune*, November 30, 1865. In July 1842 former New York mayor Philip Hone and his wife took their carriage uptown to see the new reservoirs then filling with Croton water. Hone noted in his diary what an attraction the reservoirs had become: "There were a great number of visitors at this place,—pedestrians, horsemen, railroad travellers, and those who, like myself, came in their own carriages (which, if they had no more right than me to do, was very reprehensible),—for it has become a fashionable place of resort; and well it may, for it is well worth seeing." Hone, *The Diary of Philip Hone, 1828–1851*, ed. Bayard Tuckerman (New York: Dodd, Mead & Co., 1889), 2:137.

78. Philadelphia's Vauxhall Gardens, which took its name from its London model, charged the high admission price of fifty cents for an evening visit, during which one could enjoy the enchantment of three thousand lamps, "arranged in the Venicean [sic] style." *United States Gazette* (Philadelphia), April 30, 1817. A longtime resident reminisced about Vauxhall's "numerous graveled walks bordered by beautiful flower beds, fanciful embankments, and a great variety of native trees and plants, which altogether rendered this spot a little paradise." Frank Colliger, "Recollections of a Cinquegenarian," in *A Collection of Miscellanies: Scraps Illustrating the History of the City of Philadelphia, in the "Olden Time," Collected & Arranged, with Notes &c., by C.A.P*, ed. Charles A. Poulson, 8:8. This is a series of scrapbooks of clippings, collected by Poulson, in the collections of the Library Company of Philadelphia. Vauxhall also hosted displays of fireworks and, in later years, balloon ascensions. See J. Thomas Scharf and Thompson Westcott, *History of Philadelphia, 1609–1884* (Philadelphia: L. H. Everts & Co., 1884), 2:958–59, 1052. For a fuller description of Vauxhall and other Philadelphia public gardens, see John F. Watson, *Annals of Philadelphia, and Pennsylvania, in the Olden Time* (Philadelphia: Leary, Stuart & Co., 1909), 3:400–404.

79. These organizations included the Pennsylvania Horticultural Society, founded in 1827, and the Massachusetts Horticultural Society, organized two years later. Both attracted a mix of businessmen, gentlemen, farmers, and scientists, many of whom were extensively involved in civic affairs. The societies emphasized education and research

directed toward the improvement of local horticulture as well as the display of both native species and exotic imports. As Alexander Von Hoffman explains, such groups suited "the urban gentlemen's patriotic aspirations" in proving that their city was a civilized place. Von Hoffman, *Local Attachments: The Making of an American Urban Neighborhood, 1850 to 1920* (Baltimore: Johns Hopkins University Press, 1994), 267. See also Tamara Plakins Thornton, *Cultivating Gentlemen: The Meaning of Country Life among the Boston Elite, 1785–1860* (New Haven, CT: Yale University Press, 1989).

80. The result was a remarkable intersection of science, capitalism, and mastery of nature until it was lost to a fire. Nathaniel Shurtleff, *A Topographical History of Boston* (Boston: Alfred Mudge & Son, 1871), 355–67; Marshall Pinckney Wilder, "The Horticulture of Boston and Vicinity," in *The Memorial History of Boston, including Suffolk County, Massachusetts*, vol. 4, *1630–1880*, ed. Justin Winsor (Boston: William D. Ticknor & Co., 1886), 614–17; Walter Muir Whitehill, *Boston: A Topographical History*, 2nd ed. (Cambridge, MA.: Belknap Press of Harvard University Press, 1968), 142–45; *The Public Garden, Boston* (Boston: Friends of the Public Garden and Common, 1988).

81. *Chicago Tribune*, August 31, 1860.

82. Chicago's Dr. John H. Rauch favored burying the dead outside the city limits in garden cemeteries primarily as a sanitary measure, but he also spoke of the spiritual value of these retreats. "In the slumbering silence that pervades these Cities of the Dead," Rauch wrote, "far different feelings are awakened from those suggested by the busy turmoil of the living. The uncertainty of life, the vanity of human pursuits—wealth, power, ambition, conquest—for which we toil and struggle, all are forcibly impressed upon the contemplative mind." Rauch, *Intramural Interments in Populous Cities* (Chicago: Tribune Book and Job Printers), 45.

83. Herman Melville, *Moby-Dick*, 150th Anniversary Edition, ed. Herschel Parker and Harrison Hayford (New York: Norton, 2002), 19.

84. In the two hundred years between the time the Puritans founded the Massachusetts Bay Colony and the water debates of the 1830s and 1840s, Bostonians put the Common's approximately fifty acres to numerous uses. As its name suggests, it mainly served at first as a common pasture, but it was also the site of public hangings and of a slave market. The British troops who fought in the Battle of Lexington drilled here. As Boston urbanized, the Common became more and more a place where a person could escape from the city around it, even if the "nature" it contained was increasingly one that Bostonians shaped themselves. Though the last cattle were not removed until the mid-1820s, by this time the city had installed many of the Common's "natural" features. Beginning in the eighteenth century, it was appended, trimmed, and sculpted, and lawns, trees, and flowers were planted by both official and self-appointed landscapers. The city also added malls for civilized promenades. In 1836 an iron fence around the Common succeeded a wooden one. Shurtleff, *A Topographical History of Boston*, 295–354; Edward L. Bynner, "Topography and Landmarks of the Colonial Period," in *Memorial History of Boston*, ed. Winsor, 1:552–54; Edward Stanwood, "Topography and Landmarks of the Last Hundred Years," in *Memorial History of Boston*, ed. Winsor, 4:60–63; Whitehill, *Boston*, 4, 35.

85. City of Chicago, *Twelfth Annual Report of the Board of Public Works to the Common Council of the City of Chicago, for the Municipal Fiscal Year Ending March 31, 1873* (Chicago: National Printing Co., 1873), 44–45.

86. William Wrighte, *Grotesque Architecture; or, Rural Amusement* (London: J. Taylor, 1802). Among other designs, Wrighte offered plans for a "grotto in a modern architectonic

style, ornamented with jet d'eaux, sea weeds, looking-glass, fountains, and other grotesque decorations," and an open Chinese grotto at the head of a grand canal (4–6).

87. Latrobe explained further, "The air produced by the agitation of water is of the purest kind, and the sudden evaporation of water, scattered through the air, absorbs astonishing quantities of heat,—or to use the common phrase, creates a great degree of cold." Latrobe, *View of the Practicability and Means of Supplying the City of Philadelphia with Wholesome Water*, 11, 18.

88. City of Boston, *Soft Water*, City Document No. 19, May 2, 1839 [Boston, 1839], 10.

89. City of Philadelphia, *Report of the Committee Appointed by the Common Council, to Enquire into the State of the Water Works* (Philadelphia: William Duane, 1802), 52–54.

90. *Daily Evening Transcript* (Boston), May 26, 1849.

91. G. B. Farnam, *Description of Farnam's Patent Hydraulic Apparatus, for Raising Water* (New York: B. R. Barlow, 1851), 69. The first edition appeared in 1845. By 1853 there was an enlarged fourth edition, published by E. N. Grossman.

92. City of Philadelphia, *Report of the Committee Appointed by the Common Council, to Enquire into the State of the Water Works* (1802), 52–54. As a subscriber to multiple shares of the water loan, Sansom received a $20 discount on his water bill for three years.

93. City of Philadelphia, Department for Supplying the City with Water, *History of the Works, and Annual Report of the Chief Engineer of the Water Department of the City of Philadelphia* (Philadelphia: C. E. Chichester, 1860), 51. The basic charges ranged from $6 a year for a sixteenth-inch jet to $32 for a half-inch one.

94. City of Boston, *Water Commissioners Report, in Relation to Fountains*, City Document No. 53, November 1, 1849 [Boston, 1849], 3–7.

95. City of Boston, *Water*, City Document No. 24 [Boston, 1852], 3–11; [Bradlee], *History of the Introduction of Pure Water into the City of Boston*, 144–48. The proprietors claimed that Mayor Josiah Quincy Jr. had promised them free water and that the installation of the fountain at a cost of $1,300 to them in fact raised their property taxes, gave pleasure to others in the vicinity, and would be helpful in case of fire. According to Bradlee, the City Council recommended that the water be supplied at a nominal rate, and the charge was eventually reduced to $25 a year, with limits on the hours the fountain could be operated.

96. City of Chicago, *Eleventh Semi-Annual Report of the Board of Water Commissioners* (1857), 35.

97. City of Boston, *Water Commissioners Report, in Relation to Fountains* (1849), 4.

98. Walter Channing, *A Plea for Pure Water: Being a Letter to Henry Williams, Esq., by Walter Channing: With an Address "To the Citizens of Boston," by Mr. H. Williams* [Boston, 1844], 27.

99. "Address of the Mayor to the City Council of Boston, January 4, 1847," in City of Boston, *Inaugural Addresses of the Mayor of Boston*, 1:348.

100. *Daily Advertiser*, October 29, 1848.

101. *Gleason's Pictorial Drawing-Room Companion*, 1, no. 13 (July 26, 1851): 201.

102. Geo. Schnapp, *Cochituate Grand Quick Step* (Boston: Stephen W. Marsh, 1849). Library of Congress American Memory, http://memory.loc (accessed April 9, 2011).

103. *Gleason's Pictorial Drawing-Room Companion* 1, no. 13 (July 26, 1851): 201.

104. Fredrika Bremer, *The Homes of the New World: Impressions of America*, trans. Mary Howitt (New York: Harper & Brothers, 1858), 1:197.

105. *Daily Advertiser*, October 31, 1848.

106. Ibid.

107. *Daily Evening Transcript*, November 6, 1848.

108. The original 1851 charter of the Chicago City Water Company empowered the commissioners "to construct fountains in the public squares, or such other public grounds of said city as they shall deem expedient." The Board of Public Works took pride in the "large and substantial fountain" that it had placed in front of its main building on Chicago Avenue. This fountain, along with new sod, shrubs, and evergreens, the commissioners stated, "make this portion of the premises a tasteful and agreeable resort, and harmonize with the character of the buildings." City of Chicago, *Report of the Water Commissioners of the City of Chicago, Made to the Common Council, December 8, 1851: Together with the Act of Incorporation, and a Statement of the Financial Condition of the City, November 16, 1851* (Chicago: Seaton & Peck, 1851), 4; City of Chicago, *Fourteenth Annual Report of the Board of Public Works to the Common Council of the City of Chicago, for the Municipal Fiscal Year Ending March 31, 1875* (Chicago: J. S. Thompson & Co., 1875), 35. By 1877, however, DeWitt Cregier, then engineer of the Pumping Station, stated that the narrow path past the fountain was "illy-adapted to a thoroughfare, and consequently there are frequent collisions" among pleasure vehicles bound for the newly completed Lake Shore Drive north of the works. "Besides," he added, "horses do not appreciate the beauties of the fountain, which they are obliged to pass, much to the danger and annoyance of their drivers." City of Chicago, *First Annual Report of the Department of Public Works, to the City Council of the City of Chicago, for the Fiscal Year Ending December 31, 1876* (Chicago: Clark & Edwards, 1877), 39.

109. On the restoration and rededication of the fountain after years of neglect, see "Ether Monument gets a fresh start in the Public Garden," *Boston Globe*, September 27, 2006, http://www.boston.com/news/globe/city_region/breaking_news/2006/09/the _worlds_only.html (accessed November 5, 2010).

110. [Bradlee], *History of the Introduction of Pure Water into the City of Boston*, 263–65.

111. *Daily Evening Transcript*, October 16, 1848; Walter Channing, M. D., *A Treatise on Etherization in Childbirth, Illustrated by Five Hundred and Eighty-One Cases* (Boston: William D. Ticknor & Co., 1848).

Chapter Five

1. W. M. Cornell, M.D., "Utility of Bathing," *Boston Medical and Surgical Journal* 32, no. 16 (May 21, 1845): 309.

2. Emile Zola famously described Les Halles as "the belly of Paris" in *Le Ventre de Paris* (1873). Graeme Davison dates the city-as-body metaphor to at least the time of Aristotle. Davison, "The City as a Natural System: Theories of Urban Society in Early Nineteenth-Century Britain," in *The Pursuit of Urban History*, ed. Derek Fraser and Anthony Sutcliffe (London: Edward Arnold, 1983), 349. See also Richard Sennett, *Flesh and Stone: The Body and the City in Western Civilization* (New York: Norton, 1994).

3. "The people of Boston came in a body and threw themselves at the feet of the Legislature, asking a simple boon," announced city attorney Charles Warren in asking the General Court to authorize the city to build a public waterworks. Commonwealth of Massachusetts, *Proceedings before a Joint Special Committee of the Massachusetts Legislature, for Leave to Introduce a Supply of Pure Water into That City, from Long Pond* (Boston: John H. Eastburn, 1845), 9.

4. Carl Sandburg, "Chicago," in *The Complete Poems of Carl Sandburg*, ed. Archibald MacLeish (New York: Harcourt Brace, 1970), 3.

5. J. C. Wild, *Panorama and Views of Philadelphia, and Its Vicinity, Embracing a Collection of*

Twenty Views, Drawn on Stone, by J. C. Wild, from His Own Sketches and Paintings, with Poetical Illustrations of Each Subject, by Andrew Mc'Makin (Philadelphia: J. B. Chevalier, 1838), view 1: "Fairmount, from the Basin." Sewerage networks have likewise been compared to excretory systems.

6. "Through *living* in the city—through everyday experience in and of the material world of buildings, spaces, and people—American urbanites developed active senses of themselves and individuals and as members of a new republican society," Dell Upton explains. Upton, *Another City: Urban Life and Urban Spaces in the New American Republic* (New Haven, CT: Yale University Press, 2008), 1. David Harvey writes, "Capitalism these last two hundred years has produced, through its dominant form of urbanization, not only a 'second nature' of built environments even harder to transform than the virgin nature of frontier regions years ago, but also an urbanized human nature, endowed with a very specific sense of time, space, and money as sources of social power and with sophisticated abilities and strategies to win back from one corner of urban life what may be lost in another." Harvey, *The Urban Experience* (Baltimore: Johns Hopkins University Press, 1989), 199. And, as Donald Worster points out, perhaps the most profound question Karl Wittfogel asks in his *Oriental Despotism* is, "How, in the remaking of nature, do we remake ourselves?" Worster, *Rivers of Empire: Water, Aridity, and the Growth of the American West* (New York: Pantheon, 1985), 30.

7. Thomas Jefferson, *Notes on the State of Virginia*, ed. William Peden (Chapel Hill: University of North Carolina Press, 1954), 165. Barbara Gutmann Rosenkrantz points out that the view of disease as something unnatural and foreign to America has long been a popular one in this country. Rosencrantz, *Public Health and the State: Changing Views in Massachusetts, 1842–1936* (Cambridge, MA: Harvard University Press, 1972), 1, 2.

8. On cholera and social thought, see Charles E. Rosenberg, *The Cholera Years: The United States in 1832, 1849, and 1866* (Chicago: University of Chicago Press, 1962).

9. The word "sanitary" dates to 1838, "sanitarian" to 1859. *Merriam-Webster's Collegiate Dictionary*, 11th ed. (Springfield IL: Merriam-Webster, 2003), 1101.

10. Charles Caldwell, M.D., *Thoughts on Quarantine and Other Sanitary Systems* (Boston: Marsh, Capen & Lyon, 1834), 19, 64. Merchant and Harvard benefactor Ward Nicholas Boylston endowed the Boylston Prize. His purpose was to encourage research in chemistry, anatomy, and medicine. Contestants submitted their entries anonymously, and the winner received a cash award of $100 or a gold medal. I am grateful to Jack Eckert, Reference Librarian, Center for the History of Medicine, Francis A. Countway Library of Medicine at Harvard, for this information.

11. *Chicago Tribune*, May 8, 1855.

12. While sanitary reformers viewed the cleansing power of water as the great liberator from urban filth and the maladies it caused, they supported additional measures to improve the city's cleanliness. One of the ways in which city governments expanded their activities and authority in the first half of the nineteenth century was by assuming fuller responsibility for carting away solid waste and by passing many more health regulations.

13. *United States Gazette* (Philadelphia), December 28, 1804.

14. "First Report of the Committee on Public Hygiene of the American Medical Association," *Transactions of the American Medical Association* 2 (1849): 437.

15. City of Boston, *Report of the Committee of Internal Health on the Asiatic Cholera, Together with a Report of the City Physician on the Cholera Hospital* (Boston: J. H. Eastburn, 1849), 7, 6.

16. Lemuel Shattuck et al., *Report of the Sanitary Commission of Massachusetts, 1850* (Cambridge, MA: Harvard University Press, 1948), 152–53. The full title of the original edition, published by Dutton & Wentworth in 1850, is *Report of a General Plan for the Promotion of Public and Personal Health, devised, prepared and recommended by the Commissioners Appointed under a Resolve of the Legislature of Massachusetts, Relating to a Sanitary Survey of the State, presented April 25, 1850.*

17. During the period in which Philadelphia, Boston, and Chicago built their first reliable central waterworks systems, miasmists debated the causes of diseases and their spread with contagionists (also called importationists), who maintained that many maladies passed directly from one infected body to another. Contagionists held that commercial cities were particularly vulnerable to disease because they were places where so many people arrived every day, and where large, mobile, and shifting populations lived and worked in such close proximity. Both theories, each of which dates to antiquity, found plenty of advocates in urbanizing America, though the miasmist theory dominated through much of the nineteenth century, until it was superseded by the germ theory during the century's final decades. The disputes between champions of the opposing views were sometimes fierce and heavily inflected by local politics, notably in Philadelphia during repeated assaults by yellow fever at the turn of the nineteenth century. The two theories overlapped in some respects more than many who advocated one or the other admitted. Whatever their differences, however, miasmists and contagionists agreed that the city was an unhealthy place to live and that an improved water supply would be a great benefit. See Simon Finger, *The Contagious City: The Politics of Public Health in Early Philadelphia* (Ithaca, NY: Cornell University Press, 2012).

18. Writing in his 1845 census of Boston, Shattuck said that "the *density of a population* is a matter deserving consideration" because of its clear correlation with rates of sickness and death. City of Boston, *Report to the Committee of the City Council Appointed to Obtain the Census of Boston for the Year 1845, Embracing Collateral Facts and Statistical Researches Illustrating the History and Condition of the Population, and Their Means of Progress and Prosperity* (Boston: John H. Eastburn, 1846), 31. Sixteen years later a committee (including former mayor Josiah Quincy Jr.) appointed by the Boston Sanitary Association found that rising population density "is attended with an increased mortality, especially of children—and a lessened longevity of the whole." In spite of this, the committee noted, "There is a constant tendency of population to gather in compact masses and live in cities." Although these cities could be said to be "the destroyers of families and of generations," they "have increased rapidly, while the country districts have increased much more slowly and some are stationary." *Memorial of the Boston Sanitary Association to the Legislature of Massachusetts, Asking for the Establishment of a Board of Health and of Vital Statistics* (Boston: State Printing Office, 1861), 21, 26.

19. Josiah Quincy Sr., "An Address to the Board of Aldermen, and Members of the Common Council, of Boston, on the Organization of the City Government, at Faneuil Hall, May 1, 1824," and "Address to the Board of Aldermen, of the City of Boston, Jan. 3, 1829, on Taking Final Leave," in City of Boston, *The Inaugural Addresses of the Mayors of Boston* (Boston: Rockwell & Churchill, 1894), 1:21, 103.

20. After Elizabeth Drinker, the wife of Philadelphia merchant and future member of the Watering Committee Henry Drinker, used the freestanding shower that her husband had constructed behind their home in 1798, she confided to her diary that this was the first time in almost thirty years her entire body had been wet all at once. See Richard L.

Bushman and Claudia L. Bushman, "The Early History of Cleanliness in America," *Journal of American History* 74, no. 4 (March 1988): 1214. See also Suellen Hoy, *Chasing Dirt: The American Pursuit of Cleanliness* (New York: Oxford University Press, 1995).

21. B. Henry Latrobe, *View of the Practicability and Means of Supplying the City of Philadelphia with Wholesome Water, in a Letter to John Miller, Esquire, from B. Henry Latrobe, Engineer, December 29, 1798* (Philadelphia: Zachariah Poulson, Jr., 1799), 19.

22. William A. Alcott, ed., *The Moral Reformer and Teacher on the Human Constitution* (Boston: Light and Horton, 1835), 1:11–12, 202. This book consists of issues of the magazine of the same name, bound together. Volume 2 appeared in 1836. According to John B. Blake, "Dr. John G. Coffin, fearing that many persons were injuring their health by improper or excessive bathing, offered lectures in June 1818 on how to bathe. In 1820 he got the Board of Health to endorse his views publicly, though it failed to adopt his suggested regulation prohibiting anyone from bathing for longer than two or three minutes at a time or more often than once every other day." Blake, *Public Health in the Town of Boston, 1630–1822* (Cambridge, MA: Harvard University Press, 1959), 239–40.

23. Caroline Gilman, *The Lady's Annual Register, and Housewife's Almanac, for 1840* (Boston: Otis, Broaders, & Co., 1840), 61.

24. *Godey's Lady's Book* 39 (July 1849): 8. For whatever reason, this comment appeared in the issue before the one containing the illustration.

25. City of Boston, *Soft Water*, City Document No. 19, May 2, 1839 [Boston, 1839], 9.

26. The rate of increase in fixtures per taker remained relatively modest at first, but it then rose faster than did the percentage of people installing service connections to their homes and businesses. The Philadelphia Watering Committee did not start counting bathtubs among its customers for about a decade after the opening of the Centre Square works, when it began assessing an extra charge for this amenity. The committee's 1813 report listed 172 baths among 2,426 domestic users. By 1826 fewer than twice that many customers had installed 587 baths, and by 1832 the number was up to 992. In 1849 there were 4,107 baths being used by close to 17,000 customers. City of Philadelphia, *Report of the Watering Committee, to the Select and Common Councils* (Philadelphia: Lydia R. Bailey, 1813), 16; City of Philadelphia, *Report of the Watering Committee, to the Select and Common Councils* (Philadelphia: Lydia R. Bailey, 1827), 24; City of Philadelphia, *Report of the Watering Committee, to the Select and Common Councils* (Philadelphia: Lydia R. Bailey, 1832), 40; Elizabeth M. Geffen, "Industrial Development and Social Crisis, 1841–1854," in *Philadelphia: A 300-Year History*, ed. Russell F. Weigley (New York: Norton, 1982), 317. In 1830 Philadelphia had a population of a little over 80,000, which rose to slightly more than 121,000 by 1850.

The Boston figures show how much more common fixtures were by mid-century, as water systems expanded and the manufacture of plumbing devices became a major industry. The number of sinks in the city increased from 19,287 to 31,098 between 1853 and 1860, and other fixtures spread even faster: wash hand basins, 3,149 to 10,141; bathing tubs, 1,838 to 3,910; water closets, 2,479 to 9,864; urinals, 218 to 1,070; wash tubs, 476 to 3,006. City of Boston, *Report of the Water Registrar on Waste of Water by Hopper Closets*, City Document No. 11, January 27, 1862 [Boston, 1862], 4, 8. As Richard and Claudia Bushman explain, however, bathtubs remained uncommon even among those who could afford them. They note that the proposed installation of a bathtub in the White House in 1851 "raised a furor over the unnecessary expense." Bushman and Bushman, "The Early History of Cleanliness in America," 1225–26. In the late 1860s, Boston began to install public urinals flushed with water from the Cochituate system. There were eleven of these by 1868. See [Nathaniel J. Bradlee], *History of the Introduction of Pure Water into the City of Boston, with a Description of Its Cochituate Water Works* (Boston: Alfred Mudge & Son, 1868), 290–92.

27. An 1853 report from the Cochituate Water Board, for example, includes an inventory of all the fixtures in Boston's leading hotels. The largest, the American House, which had 217 beds, provided its guests with 7 bathtubs, 16 water closets, 30 washtubs, 6 sinks, 9 basins, 3 urinals, and 1 fountain, for which it paid a total water bill of $375.50. City of Boston, *Water Rates*, City Document No. 74, November 28, 1853 [Boston, 1853], 8. Both Philadelphia and Boston charged $3 a year for a home tub, a few dollars more for every hotel tub.

28. Both domestic bathtubs and private bathhouses fed by very local sources existed in small numbers before the availability of running water. In the early 1790s, there was a bathhouse along the Schuylkill at Race Street, almost two miles from what was then the heart of Philadelphia. Bushman and Bushman, "The Early History of Cleanliness in America," 1215. Thompson Westcott, *A History of Philadelphia, from the Time of the First Settlements on the Delaware to the Consolidation of the City and Districts in 1854* (Philadelphia: A. C. Kline, 1867), vol. 4, chap. 428; *United States Gazette*, April 30, 1806. It is not clear whether this is the same bathhouse listed in the 1813 Philadelphia Watering Committee report.

29. Edwin Wolf II, *Philadelphia: Portrait of an American City*, new and enl. ed. (Philadelphia: Camino Books, 1990), 160; Nicholas B. Wainwright, "The Age of Nicholas Biddle, 1825–1841," in *Philadelphia: A 300-Year History*, ed. Weigley, 290, 317.

30. The 1850 United States census, the first to record country of birth, revealed that the percentage of the white foreign-born among the residents of Philadelphia County (which became coextensive with the city four years later) was 29.8. For Boston the figure was 34.1, for Chicago fully 52.3. U.S. Bureau of the Census, "Table 21. Nativity of the Population for the 25 Largest Urban Places and for Selected Counties: 1850," http://www.census.gov/population/www/documentation/twps0029/tab21.html (accessed September 30, 2009). There was a particularly large and sudden jump in the number of Irish Catholic immigrants in the 1840s, largely as a result of the potato famine. The Irish became an especially strong presence in Boston, much to the dismay of many native-born white Protestants. In his 1850 census of Boston, Jesse Chickering combined immigrants from Ireland and their American-born children under one heading in calculating their presence as almost 40 percent of the city's 138,788 residents, far outnumbering the second largest non-English group, Germans and their offspring, who made up less than 2 percent. While Boston's native population was, by Chickering's count, almost 97 percent New England–born, a little less than two-thirds of these people were born in Boston itself. City of Boston, *Report of the Committee Appointed by the City Council; and also a Comparative View of the Population of Boston in 1850, with the Births, Marriages, and Deaths, in 1849 and 1850, by Jesse Chickering*, City Document No. 60 (Boston: J. H. Eastburn, 1851), 39. See Peter R. Knights's detailed study of population trends in Boston in *The Plain People of Boston, 1830–1860: A Study in City Growth* (New York: Oxford University Press, 1971).

31. "Address Made to the City Council of Boston, January 5, 1835," in City of Boston, *Inaugural Addresses of the Mayors of Boston*, 1:196. The Massachusetts Medical Society's 1832 report on cholera attributed the fact that the disease hit certain portions of the population harder to the way these people conducted their lives, singling out "drunkards" and "the filthy and those whose habitations are crowded," as well as "those exhausted by

fatigue," "those who live on a poor diet," and "those who are depressed by anxiety or fear." *A Report on Spasmodic Cholera, Prepared by a Committee under the Direction of the Counsellors of the Massachusetts Medical Society* (Boston: Carter and Hendee, 1832), 38.

As Mary Poovey explains in her discussion of the British poor in the nineteenth century, "pauperism" was different from "poverty" in that it described a moral and physical condition rather than a strictly economic one. Poovey, *Making a Social Body: British Cultural Formation, 1850–1864* (Chicago: University of Chicago Press, 1995), 13. In the United States, which was distinguished by its heterogeneous ethnic mix, there was a tendency to assume that while some native-born might be impoverished, they were less likely than immigrants to be paupers. The "greater calamity" that attended foreign immigration, which Shattuck called a "monstrous evil," was that the native-born, "who mingle with these recipients of their bounty," often became "contaminated with diseases, and sicken, and die." As a result, "the physical and moral power of the living is depreciated, and the healthy, social and moral character we once enjoyed is liable to be forever lost. Pauperism, crime, disease and death, stare us in the face." Shattuck, *Report of the Sanitary Commission of Massachusetts, 1850*, 205.

There was evidence behind the perception that immigrants settled disproportionately in cities. Joseph P. Ferrie notes that in 1850, 83 percent of free male native-born Americans between 20 and 65 lived in rural places (defined by the U.S. Census as having a population of less than 2,500), while only 9 percent lived in cities with at least 10,000 people. For immigrants, the respective figures were 54 and 36 percent. In the same year, some 60 percent of Irish immigrants dwelt in eastern cities, a higher percentage than other immigrant groups. Ferrie, *Yankeys Now: Immigrants in the Antebellum United States, 1840–1860* (New York: Oxford University Press, 1999), 35, 66.

32. According to Shattuck, "In 1849 [the year of a major cholera outbreak] there died of cholera, in Boston, 707 persons, of whom 572, or 81 per cent., were foreigners; and 135, or 19 per cent., were Americans; 42 only were Bostonians." Of 5,079 people who died in Boston of all causes that year, 59 percent were foreigners. Shattuck, *Report of the Sanitary Commission of Massachusetts, 1850*, 204. In his 1850 census, Jesse Chickering quoted the city registrar, who held that "the unusual mortality" of 1849 not just from cholera but also spikes in scarlet fever and other diseases was "fully accounted for by the deplorable condition of emigrants from Europe, constantly arriving in a state of entire destitution and exhaustion,—fatally diseased themselves, and spreading sickness and death among their relatives and countrymen here." Chickering, *Comparative View of the Population of Boston in 1850*, 29. In 1834 the number of foreign-born in the Boston House of Industry was almost twice that of "Americans." John Koren, *Boston, 1822 to 1922: The Story of Its Government and Principal Activities during One Hundred Years* (Boston: City of Boston Printing Department, 1923), 138. See also Richard A. Meckel, "Immigration, Mortality, and Population Growth in Boston, 1840–1880," *Journal of Interdisciplinary History* 15, no. 3 (Winter, 1985): 393–417.

33. R. C. Waterston, *An Address on Pauperism, Its Extent, Causes, and the Best Means of Prevention; Delivered at the Church in Bowdoin Square, February 4, 1844* (Boston: Charles C. Little & James Brown, 1844), 21. This address was published by the Society for the Prevention of Pauperism, before whom it was first delivered at the Central Church. It was repeated, by request, at the Bowdoin Square Church. Boston mayor John P. Bigelow, who followed Josiah Quincy Jr., complained that European nations were exporting their mentally insane as well as their physically sick to the United States. "The immigration of such as these," Bigelow stated, "is a flagrant abuse of the hospitalities of a nation which welcomes to its shores

the honest and industrious of every land, who are capable of sustaining a proportionate share in developing the resources, and promoting the welfare of their adopted country." "Inaugural Address to the Aldermen and Common Council by John Prescott Bigelow, Mayor of the City of Boston, January 6, 1851," in City of Boston, *Inaugural Addresses of the Mayors of Boston*, 1:399.

34. Mrs. L. H. Sigourney, "The Harwoods," in *Water-Drops* (New York: Robert Carter, 1848), 169. *Water-Drops* is a thoroughly lugubrious collection of Sigourney's poems, stories, and essays, all touching on the horrors of alcohol.

35. Shattuck, *Report of the Sanitary Commission of Massachusetts, 1850*, 205. Observers used similar water metaphors to talk about movements of population generally. In a book warning of the consequences of the arrival of so many foreign-born newcomers, Chickering also bemoaned the movement of desirable native-born residents to more recently settled areas of the nation: "This current of emigration, flowing from almost every town and city, has continued without interruption for more than half a century, and conveyed to the west whatever improvements may have been discovered or adopted in the older parts of the country." Jesse Chickering, *Immigration into the United States* (Boston: Charles C. Little & James Brown, 1848), 2. Chickering was concerned about the flow of members of local families out of the city. He was horrified to discover that in the years 1849 and 1850, as the city's total population swelled with immigrants from Ireland, more than twice as many Boston natives died as gave birth to children. Even though the mortality rate of this group was lower than the average of all residents in the city, theirs was the only one of Chickering's demographic groups among whom the number of deaths exceeded the number of births. According to Chickering, some Bostonians retired from business while others pursued opportunities elsewhere. Among those who remained, Chickering listed the aged, the unadventurous, and those forced by circumstances to stay in Boston—none of whom, in his opinion, promised to assume positions of civic leadership. Chickering, *Comparative View of the Population of Boston in 1850*, 52, 34, 38. See chap. 4, n. 28.

36. *Daily Evening Transcript* (Boston), October 9, 1848.

37. Shattuck, *Report of the Sanitary Commission of Massachusetts, 1850*, 201.

38. Ibid., 206.

39. Rev. Walter Mears, *Water Supply of Our Great Cities* (Philadelphia: Water Committee of the Councils of Philadelphia, n.d.), 7. This booklet states that it is based on articles that appeared in late June and July 1866 in the *American Presbyterian*.

40. In this regard, see also Boston's 1849 *Report of the Committee of Internal Health on the Asiatic Cholera*. The report supplemented its verbal descriptions of terrible neighborhoods with unflinching line drawings of dark and damp living spaces and of noisome back alleys with rickety porches, tipped garbage bins, murky mud puddles, and crisscrossed clotheslines. The 182-page report not only reads but also looks like the first edition of Jacob Riis's *How the Other Half Lives* (1890), which featured drawings based on his famous photographs.

41. Walter Channing, *An Address on the Prevention of Pauperism* (Boston: Office of the Christian World, 1843), 21.

42. City of Boston, *Report of the Committee on the Expediency of Providing Better Tenements for the Poor* (Boston: Eastburn's Press, 1846), 11. Bowditch had a distinguished career in public health. His *Public Hygiene in America* was published in 1877.

43. Commonwealth of Massachusetts, *Statement of Evidence before the Committee of the Legislature, at the Session of 1839, on the Petition of the City of Boston, for the Introduction of Pure Soft Water* (Boston: John H. Eastburn, 1839), 25.

44. Ibid., 55.

45. Channing, *An Address on the Prevention of Pauperism*, 71.

46. Walter Channing, *A Plea for Pure Water: Being a Letter to Henry Williams, Esq., by Walter Channing: With an Address "To the Citizens of Boston," by Mr. H. Williams* [Boston, 1844], 27, 14.

47. *Daily Advertiser and Patriot* (Boston), May 10, 1845.

48. City of Boston, *Communication to the City Council, on the Subject of Introducing Water into the City* (Boston: J. H. Eastburn, 1834), 13.

49. Channing, *A Plea for Pure Water*, 11, 14–15.

50. Shattuck, *Report of the Sanitary Commission of Massachusetts, 1850*, 264.

51. Shattuck credited Liverpool with leading the way in this "new movement for the general and sanitary benefit of the poor" when it opened two such institutions in the 1840s. Ibid., 209. Marilyn Thornton Williams states that the first indoor public bath in England built at public expense was in Liverpool, but that it dated to 1828, not the 1840s. By 1900 more than two hundred British cities had such baths. Williams, *Washing "The Great Unwashed": Public Baths in Urban America, 1840–1920* (Columbus: Ohio State University Press, 1991), 7–8.

52. Shattuck, *Report of the Sanitary Commission of Massachusetts, 1850*, 209.

53. *Boston Surgical and Medical Journal* 32, no. 23 (July 16, 1845): 466–67.

54. *Daily Advertiser*, February 13, 1845.

55. Waterston, *An Address on Pauperism*, 31. There is no denying the conservative and nativist bias of such statements, or that elite support for health measures aimed at the poor and foreign-born, while intended to improve individual and public health, found another motive in the desire to defend and bolster "American" institutions, which were presumed to be superior to those of the newcomers. As Oscar Handlin notes, the Irish were understandably suspicious and fearful of many sanitary, educational, and benevolent institutions and programs, which they rightly interpreted as enforcing insensitive and even hostile policies obsessed with simultaneously isolating and "cleansing" them not of dirt but of their "dirty" culture. Handlin, *Boston's Immigrants, 1790–1880: A Study in Acculturation*, rev. and enl. ed. (Cambridge, MA: Belknap Press of Harvard University Press, 1991; originally published 1941), 161. The state Sanitary Commission, for example, singled out Irish wakes as "improper, and dangerous to the public health and to good morals." Shattuck, *Report of the Sanitary Commission of Massachusetts, 1850*, 195.

56. *Daily Evening Transcript*, October 28, 1848. On the call for public bathing facilities in Chicago, see Bessie Louise Pierce, *A History of Chicago*, vol. 2: *From Town to City 1848–1871* (New York: Knopf, 1937; reprint, Chicago: University of Chicago Press, 2007), 333.

57. *Daily Advertiser*, June 19, 1845.

58. Henry L. Hammond, *Memorial Sketch of Philo Carpenter* (Chicago: Fergus Publishing Co., 1888), 8, 15; *Chicago Democrat*, December 17, 1833.

59. Shattuck, *Report of the Sanitary Commission of Massachusetts, 1850*, 183–84. W. J. Rorabaugh maintains that in the first third of the nineteenth century, "the typical American annually drank more distilled liquor than at any other time in our history," and trade in this product was an important part of the development the national and local economies. Rorabaugh, *The Alcoholic Republic: An American Tradition* (New York: Oxford University Press, 1979), 7. Rorabaugh attributes this rise, which he says peaked in 1830, to prosperity, improved technology, and the availability of alcohol, regardless of legal restrictions.

60. The list was an assortment of things both physical and moral, "in fine every violation

of diet and regimen, and every immoral indulgence which impairs the vigor of the body or enervates the mind." It also included "profligacy," "immorality," and "sensual indulgence," as well as "fear," "excessive labor," "extreme fasting," "innutritious diet," and "want of sleep." City of Boston, *Opinion of the Consulting Physicians, on Cholera, 1866* (Boston, 1866), 6. On alcohol and the 1832 cholera epidemic, see n. 31.

61. The Boston-based Massachusetts Temperance Society maintained that though its members wished to confront the practical worldly problems attributable to drinking, which ranged from waste and inefficiency to the higher population of prisons and almshouses, they were motivated primarily by their "moral duty to remove from our city the occasions of this vast moral evil, and by kindness and all good means, make reformation as easy as it can be made, even to the most abandoned." Council of the Massachusetts Temperance Society, *Plain Facts, Addressed to the Inhabitants of Boston, on the City Expenses for the Support of Pauperism, Vice and Crime* (Boston: Ford & Damrell, 1834), iii–iv.

62. The many temperance songs in praise of water associated it with all that was best about unspoiled nature. The lyrics of a composition titled "Cold Water Chase," which appeared in an 1844 collection called *Boston Temperance Songster*, compared the condition of temperance to a "bright rosy morning" that "Peeps over the hills / With blushes adorning / The meadows and fields" in a place "Where pleasure, and vigor, / And health, all embrace." The tone of this genre, often somehow both cheerful and dreary, is related to but different from nostalgia, since the emphasis in the contrast between the city and the country is moral rather than temporal. Another selection, "Pure Water," which was included in the same publication two years later, deemed water "an emblem of virtue" and of "peace to the mind," a bearer of "beauties and comfort" that banishes misery from the home of the former drunkard. Robert K. Potter, *Boston Temperance Songster* (Boston: W. White, 1844), 13; (1846), 41–42.

63. Walter Channing, M.D., *Annual Address Delivered before the Massachusetts Temperance Society, May 29, 1836* (Boston: John Ford, 1836), 18–19.

64. Lucius M. Sargent, who would later protest in vain when the Cochituate system devalued his investment in Jamaica Pond, was a tireless spokesman for temperance in his many essays, lectures, and stories on the subject. In the headnote to a story entitled "An Irish Heart," Sargent stated that the mounting emigration from Ireland in the late 1840s of "drunken and depraved" newcomers "addicted notoriously to the free employment of spirituous liquor" was "a subject for grave and fearful reflection." Sargent, "An Irish Heart," in *The Temperance Tales* (Boston: W. J. Reynolds & Co., 1848), 1:177. Sargent raged against urban democracy, which gave irresponsible people the vote, dooming the prospects of passing laws outlawing alcohol. "SEEK NOT TO DESTROY THIS LOATHSOME IDOL BY THE AID OF LEGISLATORS, WHO ARE ELECTED BY THE WORSHIPPERS THEMSELVES," he warned in the published version of a speech he delivered to the Harvard Temperance Society. L. M. Sargent, *An Address Delivered before the Temperance Society of Harvard University, November 20, 1834* (Cambridge, MA: Metcalf, Torry & Ballou, 1834), 15. Regardless of the reasons why they drank, or whether and how drinking affected the quality of their citizenship, the presence of Irish immigrants evidently did increase the amount of drinking in Boston. Handlin states that by 1849 there were 1,200 "groggeries" operating in Boston, most of them serving mainly Irish customers in Irish neighborhoods. In addition, he points out, many Irish families made extra money selling gin. Handlin, *Boston's Immigrants*, 121.

65. In anticipation of July 4, 1842, the Massachusetts Temperance Union's annual *Temperance Almanac* (the organization also published the monthly *Temperance Journal*) urged

all "Soldiers" to "be on the ground bright and early for the Fourth," adding, "Don't let the enemy get the start on you." "Cold Water Army," *Temperance Almanac of the Massachusetts Temperance Union* 1, no . 4 (1842). The 1843 *Almanac* proudly recalled the "signal triumphs" of the previous year, including the exclusion of wine at dinners held by the Boston Latin School, the Massachusetts Medical Society, and the Association of Congregationalist Clergymen, as well as at festivities hosted by the Mechanic's Charitable Association and the Horticultural Society. "All of the public 4th of July celebrations, sundry political gatherings and military excursions, were conducted on cold water principles," the *Almanac* reported. *Temperance Almanac* 1, no. 5 (1843): inside front cover.

66. City of Boston, *Soft Water*(1839), 8–9.

67. Commonwealth of Massachusetts, *Statement of Evidence before the Committee of the Legislature, at the Session of 1839*, 52, 56.

68. On objections to water and other non-alcoholic beverages as being bad-tasting, expensive, lacking nutritional value, and harmful or even dangerous to one's health, see Rorabaugh, *The Alcoholic Republic*, 95–97.

69. Loammi Baldwin, Esq., Civil Engineer, *Report on Introducing Pure Water into the City of Boston*, 2nd ed., with Additions (Boston: Hilliard, Gray and Co., 1835), iv.

70. City of Boston, *Celebration of the Introduction of the Water of Cochituate Lake into the City of Boston, October 25, 1848* (Boston: J. H. Eastburn, 1848), 38. While Quincy was technically correct—Philadelphia had spent almost $900 for liquor while building the first Centre Square works and more in following years—it seems unlikely that the crews on the Boston works and other projects were teetotalers. On the inclusion of alcohol in Philadelphia's waterworks budget, see City of Philadelphia, *Report of the Committee for the Introduction of Wholesome Water into the City of Philadelphia* (Philadelphia: Zachariah Poulson, Jr., 1801), 13; City of Philadelphia, Department for Supplying the City with Water, *History of the Works, and Annual Report of the Chief Engineer of the Water Department of the City of Philadelphia* (Philadelphia: C. E. Chichester, 1860), 14. In a memoir, former Chicago mayor John Wentworth joked about the presence of liquor at the "shovel day" of the Illinois and Michigan Canal, which took place on July 4, 1836. Near the site of the ceremonies in the Bridgeport neighborhood was a spring, he explained, into which were thrown chopped lemons, "to make lemonade for the temperance people." Wentworth continued, "Then they spoiled the lemonade, by emptying into it a whole barrel of whisky, which so penetrated the fountain-head of the spring, that Bridgeport people feel the effects of it to this day!" Wentworth, *Early Chicago: A Lecture, Delivered before the Sunday Lecture Society, at McCormick Hall, on Sunday Afternoon, May 7th, 1876, by Hon. John Wentworth* (Chicago: Fergus Printing Co., 1876), 28.

71. "O That's the Drink for Me," in *Temperance Song Book of the Massachusetts Temperance Union* (Boston: Kidder & Wright, 1842), 20–21. The ribbon also featured an image of a man taking water from a well with a bucket, in spite of the fact that the new system had been built because Boston's wells had dried up or gone bad, and because getting water from a tap would be more convenient in any case. This reveals the appeal of the idea that the new system was bringing natural and healthy country water to the city. See fig. 5.2.

72. *Chicago Republican*, March 26, 1867.

73. *Chicago Tribune*, March 23 and 26, 1867.

74. The author bitterly attacked the champions of the aqueduct for being motivated by ambition. He specifically criticized Channing, though not by name: "In this stage of the affair, an eminent physician, within the high district, is besought to petition for water. The

memorial is so contrived as to operate upon the sensibilities and awaken the fears of every nervous woman in the city, who operates, in turn upon a father, or a brother, or a husband. Water, which has been drunken with perfect satisfaction, becomes suddenly offensive to the smell and taste, and old ladies put on their spectacles and look for eels." "A Selfish Taxpayer" accused some self-described "ardent friends of temperance" of hypocrisy, of being men who "prate of pure and soft water, while every syllable they utter is accompanied with the compound stench of brandy and tobacco." "A Selfish Taxpayer," *Thoughts about Water* [Boston, 1844], 15–16.

75. In an angry address in 1845, E. K. Whitaker took the wealthy and privileged to task for their moral lassitude in regard to total abstinence. Whitaker was chairman of the Washingtonians, the popular national organization of self-described reformed drinkers that drew members from across the classes. There was "more to dread in the wine cup in the hands of great and reverend men, than in the gay saloon of the rumseller," he observed. The throngs of common people who had rejected alcohol "have given a stern rebuke to the more highly educated classes, and have furnished sublime proof, to the world, that they can govern themselves." Whitaker even suggested that current social divisions might be dissolved and remade on the principles of recognizing "as men and as brothers" all abstainers regardless of background, including Roman Catholics. Even "the slave sons of Africa" would have "no separate seat in the gallery." *Proceedings and Address of the Washingtonian Mass Convention Held in the City of Boston* (Boston: New England Washingtonian Office, 1845), 13, 15.

76. Letter of Josiah Quincy Jr. to the Organizers of the Delegation of Congress, at the Revere House, Boston, March 11, 1848. Courtesy of the Trustees of the Boston Public Library/Rare Books. The organizers evidently decided to ban liquor officially but informed the mayor that some attendees would provide it on their own. Quincy replied on March 13 that this was "a subterfuge" in which he would not participate.

77. *Daily Advertiser*, November 18, 1848.

78. See Joseph R. Gusfield, "Temperance, Status Control, and Mobility," in *Ante-Bellum Reform*, ed. David Brion Davis (New York: Harper & Row, 1967), 120–39.

79. Alcott, *The Moral Reformer and Teacher on the Human Constitution*, 1:2.

80. Walter Channing, *Thoughts on the Origin, Nature, Principles and Prospects of the Temperance Reform* (Boston: Council of the Massachusetts Temperance Society, 1834), 15–16.

81. In one of the most familiar of the many images produced by the temperance movement, the 1846 Currier & Ives lithograph depicting the drunkard's progress, the first "harmless" social drink leads a person of apparently good character and the best of intentions successively to disease, loneliness, financial ruin, desperation, crime, and ignominious death. See the Library of Congress Prints and Photographs Online Catalog, http://loc.gov/pictures/resource/cph.3b53131/ (accessed April 16, 2011).

82. Penn's contemporary Gabriel Thomas reported in his firsthand account of Pennsylvania, "Not two mile[s] from the *Metropolis* are also *Purging Mineral-Waters* that pass by both siege and urine, all out as good as Epsom," the spa town near London whose water contained Epsom salt, which remains a popular home treatment for various ills. Gabriel Thomas, *An Historical and Geographical Account of the Province and Country of Pensilvania [sic]; and of West-New-Jersey in America* (London: A. Baldwin, 1698), 12.

83. In his pamphlet *An Inquiry into the Effects of Spiritous Liquors on the Human Body*, Rush appended his well-known "Moral and Physical Thermometer," an illustration that was much reproduced in other publications. It endorsed water as the guarantor of "Health, Wealth, Serenity of mind, Reputation, long life, and Happiness." Benjamin Rush, *An Inquiry*

into the Effects of Spiritous Liquors on the Human Body (Boston: Thomas and Andrews, 1790), 12. See fig. 5.3.

84. Benjamin Rush, *Directions for the Use of the Mineral Water and Cold Bath, at Harrogate, Near Philadelphia* (Philadelphia: Melchior Steiner, 1786). The proprietor of the spa asked Rush to write this pamphlet, which included a note stating that it was based on a paper that Rush had read thirteen years earlier before the American Philosophical Society. In an advertisement he placed in the Philadelphia *Aurora and General Advertiser* in 1795, the owner of Harrogate (who here spelled it "Harrowgate") referred to Rush's endorsement. He claimed his water also removed worms, relieved gout and rheumatism, and "in many instances cured ulcers and other eruptions of the skin." *Aurora*, May 26, 1795.

85. John Bell, *On Baths and Mineral Waters* (Philadelphia: Journal of Health and the Family Library of Health, 1831). Bell's other books included *A Treatise on Baths*, which went through multiple editions. Its full title is *A Treatise on Baths; Including Cold, Sea, Warm, Hot, Vapour, Gas, and Mud Baths; also, on the Watery Regimen, Hydropathy and Pulmonary Inhalation; with a Description of Bathing in Ancient and Modern Times* (Philadelphia: Barrington & Haswell, 1850).

86. Rush, *Directions for the Use of the Mineral Water and Cold Bath*, 9.

87. In addition to applauding bathing as the way to keep the body clean, Caroline Gilman's *Lady's Annual Register, and Housewife's Almanac* and Dr. William Alcott's *Moral Reformer and Teacher on the Human Constitution* both endorsed hydropathy. The latter book contained a testimonial from a man in his forties, who stated that a combination of "dyspepsia, inflammation of the lungs, severe and frequent colds, influenza &c." had rendered him an invalid through most of his life. Not anymore. A cold full-body bath every morning, followed by "friction with a coarse towel," completely revitalized him. He was now less sensitive to winter's chill and could do double the work he had accomplished previously. Hydropathy improved his sleep so markedly that he was virtually free of fatigue, and it had ruddied up his "sallow countenance." His triumphant self-diagnosis: "I am a healthy man." Alcott's book also included a section on "medicated baths." Alcott, *The Moral Reformer and Teacher on the Human Constitution*, 2:223, 1:205. See also Dr. J. R. Wells, M.D., *The Family Companion, Containing Many Hundred Rare and Useful Receipts, on Every Branch of Domestic Economy*, 10th ed. (Boston: Printed for the Author, 1847), 67.

88. O. S. Fowler, *Physiology, Animal and Mental: Applied to the Preservation and Restoration of Health of Body, and Power of Mind*, 5th ed. (New York: Fowlers & Wells, 1848), 303. The proprietors of the New York publishing house of Fowlers & Wells were Fowler, his brother Lorenzo, and S. R. Wells. The firm advertised a whole line of hydropathy books, in addition to multiple volumes on physiology, psychology, and mesmerism, as well as phrenology. In addition to Fowler, the company's author list included William Alcott, Joel Shew, and dietary reformer Sylvester Graham.

89. Joel Shew, M.D., *The Water-Cure Manual* (New York: La Morte Barney, Office of Water-Cure Journal, 1847), 138, 81–82.

90. Fowler, *Physiology, Animal and Mental*, 311.

91. Joel Shew dismissed the idea that springs and wells possessed "some mysterious power" that would lead some to name them after a patron saint. Shew, *The Water-Cure Manual*, 62.

92. William Swaim issued pamphlets replete with testimonials of how his panacea cured scrofula, ulcers, cutaneous (skin) afflictions, swelling, syphilis, and mercury poisoning, and how it "perfectly re-established" the health of a young woman who had reached the point

"when the knell of death would have been hailed with emotions of pleasure." Swaim, *The Case of Nancy Linton, Illustrative of the Efficacy of Swaim's Panacea* (Philadelphia: Clark & Raser, 1827), 4. See also Swaim, *A Treatise on Swaim's Panacea, Being a Recent Discovery for the Cure of Scrofula or King's Evil, Mercurial Disease, Deep-Seated Syphilis, Rheumatism, and All Disorders Arising from a Contaminated and Impure State of the Blood; with Cases Illustrating Its Success* (Philadelphia: J. Maxwell, 1825); and Swaim, *Plain and Practical Observations on Diseases Resulting from Worms: With Remarks upon the Utility of Swaim's Vermifuge in Cholera Morbus, Dysentery, and in All Other Disease Originating in Debility of the Digestive Organs* (Philadelphia: J. Harding, 1850).

93. Shew, *The Water-Cure Manual*, 14.

94. Alcott, *The Moral Reformer and Teacher on the Human Constitution*, 2:233. Alcott did warn, however, that unless one had a "vigorous" constitution, a cold bath in the evening or early morning might do more harm than good. O. S. Fowler also emphasized the value of "bathing, friction, and the healthy action of the skin." Fowler, *Physiology, Animal and Mental*, 303. Dr. John Collins Warren endorsed cold bathing and the shower bath "as very conducive to invigoration of the body, and to lessening the susceptibility to the injurious effects of cold on the surface of the skin." He added, "I would speak of the advantages of regular frictions over the whole surface, and especially the chest and neck, those parts which are constantly to be exposed to the air." Warren, *Physical Education and the Preservation of Health*, 2nd ed. (Boston: William D. Ticknor & Co., 1846), 49. The first edition appeared the year before.

95. Warren, *Physical Education and the Preservation of Health*, 5, 10.

96. The poem can be read as a critique of contemporary city life. The blacksmith sheds only "honest sweat"; he "looks the whole world in the face, / For he owes not any man." His "village," unlike the city, is outside the urban credit economy, free of any compromises to his self-sufficiency and integrity. There is a sacred cadence in the rhythm of his hammer, which Longfellow likens to the "sexton ringing the village bell." Henry Wadsworth Longfellow, "The Village Blacksmith," in *Ballads and Other Poems*, 3rd ed. (Cambridge, MA: John Owen, 1841), 99–101.

97. Warren, *Physical Education and the Preservation of Health*, 50. Noted New York health reformer Dr. John H. Griscom, writing in his sanitary survey of New York that was contemporary with the work of Shattuck (the two men corresponded with one another), asked anyone who needed proof of this to compare "the pale face of the city belle, or matron, after the long confinement of the winter and spring, with the same countenance in the fall, upon her return from a few weeks tour to the Springs and Niagara, and observe whether the return of the long absent rose upon the cheek, is not accompanied with a greater elasticity of frame, and a happier and stronger tone of mind." Griscom also offered the testimony of an attorney friend who said that those of his children who attended school in the country had never been sick, while the ones who did so in New York, "though in a comparatively salubrious position in the city," ran up substantial bills for doctors and medicine. Griscom, *The Sanitary Condition of the Laboring Population of New York* (New York: Harper & Brothers, 1845; reprint, New York: Arno and the New York Times, 1970), 12.

98. Warren, *Physical Education and the Preservation of Health*, 70, 71. Warren was not opposed to reading when it involved fine literature, but he observed that many people ruined their eyes by the time they reached middle age, "and when that period arrives, at which the active occupations are diminished, and the pleasures of literature are wanted in their place, the power of vision is so much impaired, as scarcely to be capable of employment for the most common purposes."

99. See M. L. Shew, *Water-Cure for Ladies: A Popular Work on the Health, Diet, and Regi-men of Females and Children, and the Prevention and Cure of Diseases; with a Full Account of the Processes of Water-Cure; Illustrated with Various Cases*, rev. ed. by Joel Shew (New York: Wiley and Putnam, 1844). People of both sexes, including famous men like Horace Greeley and Mark Twain, patronized water cures (among British patients were Alfred Tennyson, Thomas Carlyle, Charles Darwin, and Darwin critic Bishop Samuel Wilberforce), but hydropathy was especially popular among women, since the water cure paid special attention to women's medical concerns. Unlike conventional medicine, hydropathy also welcomed female prac-titioners, among them the radical feminist Mary Sargeant Gove Nichols. In addition to writ-ing a book on the subject, she and her husband operated a hydropathic training school in New York in the early 1850s that was open to female applicants.

Jane B. Donegan contends that "by emphasizing individual self-reliance and nature's curative powers hydropathy provided women with a viable alternative for meeting health care responsibilities at a time when orthodox therapeutics frequently were dangerous, distasteful, and ineffective." Basing her conclusions on personal accounts, Donegan states that "the hydropathic system, whether practiced at home or at a water-cure establishment, proved physiologically and psychologically rejuvenating for many weary, ailing middle- and upper-class women." Donegan, *"Hydropathic Highway to Health": Women and Water-Cure in Antebellum America* (New York: Greenwood Press, 1986), xiv, 192. Donegan pays extended attention to health matters related to bearing children. Susan E. Cayleff states, "By solicit-ing a female readership, rethinking the treatment of female diseases, urging women's active participation in home health care, and actively supporting the inclusion of female physi-cians, the water-cure movement appealed to women as the primary caretakers of others and fostered an extension of women's sphere of influence from the domestic into the informally political realm." Cayleff contends that "for its followers, hydropathy was a way of living; for the larger public, it offered an articulate critique of medical care and social relations." The water cure, according to Cayleff, fostered "an extension of woman's sphere of influence from the domestic into the informally political realm." Cayleff, *Wash and Be Healed: The Water-Cure Movement and Women's Health* (Philadelphia: Temple University Press, 1987), 18, 5. See also Cecilia Tichi, "Hot Springs: American Hygeia," in *Embodiment of a Nation: Human Form in American Places* (Cambridge, MA: Harvard University Press, 2001), 173–215.

Catharine Beecher and her sister Harriet Beecher Stowe's advice manual *The American Woman's Home* included a section on hydropathy. It was published in 1869, by which time the Beecher sisters, especially Catharine (as well as some Beecher brothers and Har-riet's husband, Calvin Stowe), had become firm believers in the healing powers of water. Cayleff, among other scholars, maintains that women patronized water cures as much or more because of the relationships they developed among themselves as for the therapy. As Cayleff points out, Susan B. Anthony and Clara Barton were among other promi-nent American women who took the water cure. Cayleff, *Wash and Be Healed*, 148–51. Catharine Beecher shared her enthusiasm for hydropathy in an article that appeared in the fall of 1846 in the weekly *New York Observer and Chronicle*, drawing a scathing and sarcastic response from the *Boston Medical and Surgical Journal*, which she answered in its pages. See "Miss Beecher on the Water Cure," *New York Observer and Chronicle* 24, no. 43 (October 24, 1848): 171; "The Water-Cure," *Boston Medical and Surgical Journal* 35, no. 18 (December 2, 1846): 349–56; *Boston Medical and Surgical Journal* 35, no. 22 (December 30, 1846): 450.

100. Geffen, "Industrial Development and Social Crisis, 1841–1854," 320. Geffen writes

that a committee of the Philadelphia County Medical Society, which had been organized three years earlier, reported in 1851 that despite the efforts of conventional doctors to regulate the profession, "in addition to 397 physicians in 'legitimate' practice, there were 42 homeopathists, 30 Thomsonians, 2 hydropaths, 32 advertising doctors, 37 druggist-physicians, and 42 'non-descripts.'" Thomsonians followed the methods of Samuel Thomson (1769–1843), whose cure was based on the use of cayenne peppers and hot baths (to raise body heat), as well as herbal laxatives and emetics.

101. *Chicago Tribune*, April 8, 1856; April 21, 1856; July 16, 1862; October 6, 1863; *The Water-Cure Journal and Herald of Reforms, Devoted to Physiology, Hydropathy, and the Laws of Life* 23, no. 5 (May 1857): 117.

102. Shew, *The Water-Cure Manual*, 290, 31–32.

103. Kneeland, cited in C. Farrar, *Letter upon the Water-Cure, Addressed to the Hon. Edward Everett* [Boston, 1851], 15. Farrar was a Maine water-cure entrepreneur who wrote to Everett as part of an effort to raise $50,000 to open his own establishment. Farrar claimed to hold the degree of "magister aquarum" from a water-cure institution. Farrar appended to his fulsome letter testimonials from Everett, Henry Wadsworth Longfellow, Calvin and Harriet Beecher Stowe, Lyman Beecher, and several physicians.

104. Kneeland, cited in ibid., 15.

105. "Chit-Chat upon Watering-Place Fashions," *Godey's Lady's Book* 41, no. 1 (July 1850): 62.

106. Mrs. E. Wellmont, "Chat about Watering Places," *Gleason's Pictorial Drawing-Room Companion* 1, no. 18 (August 30, 1851): 286. Joel Shew had spent the first part of his professional career at Lebanon Springs.

107. *Daily Advertiser*, August 22, 1846.

108. On bathing and privacy, primarily in France, see Jean-Pierre Goubert, *The Conquest of Water: The Advent of Health in the Industrial Age*, trans. Andrew Wilson (Princeton, NJ: Princeton University Press, 1989); and Alain Corbin, *The Foul and the Fragrant: Odor and the French Social Imagination*, trans. Miriam L. Kochan, Roy Porter, and Christopher Prendergast (Cambridge, MA: Harvard University Press, 1986). For a more general discussion, see Siegfried Giedion, *Mechanization Takes Command: A Contribution to Anonymous History* (New York: Norton, 1969), 628–712.

109. [Ellis Sylvester Chesbrough], *Chicago Sewerage. Report of the Results of Examinations Made in Relation to Sewerage in Several European Cities, in the Winter of 1856–7* (Chicago: Board of Sewerage Commissioners, 1858), 27.

110. *Chicago Tribune*, July 12, 1868.

111. Goubert writes, "Within a few generations, between 1830 and 1840, it had become the established practice to wash and attend to one's bodily functions in private. The result was the development of real control over one's body." Goubert, *The Conquest of Water*, 97. See also Maureen Ogle, *All the Modern Conveniences: American Household Plumbing, 1840–1890* (Baltimore: Johns Hopkins University Press, 1996).

Chapter Six

1. "Inaugural Address of Mayor James Curtiss," March 9, 1847, Chicago Public Library, http://www.chipublib.org/cplbooksmovies/cplarchive/mayors/curtiss_inaug01.php (accessed April 17, 2011).

2. Roxanne Panchasi uses a similar term in *Future Tense: The Culture of Anticipation in France between the Wars* (Ithaca, NY: Cornell University Press, 2009).

3. Copp's Hill in the North End and Fort Hill in the old South End underwent similar surgery. See Walter Muir Whitehill, "Cutting Down the Hills to Fill the Coves," in *Boston: A Topographical History*, 2nd ed. (Cambridge, MA.: Belknap Press of Harvard University Press, 1968), 72–94.

4. Gary B. Nash, *First City: Philadelphia and the Forging of Historical Memory* (Philadelphia: University of Pennsylvania Press, 2002), 2, 8.

5. *Aurora and General Advertiser* (Philadelphia), December 6, 1805. If the *Aurora* seemed to accept Philadelphia's being surpassed by New York, it bristled at the political agreement by which Washington, D.C., succeeded Philadelphia as the national capital: "It [Philadelphia] must also yield metropolitan precedence to the doubtful policy of a *seat of government, far removed from the centre of wealth and population*, the pendulum of national activity, which must long vibrate (perhaps for ever) between Baltimore, Philadelphia, and New-York; a chain of commercial cities, whose vigorous [*sic*] in pulse is already accelerated by the bold ramification of turnpikes and canals." Two years later, "A Pennsylvanian," writing to the *United States Gazette*, strongly advised residents of the state to join the effort to move the route of a proposed national road farther north through Pennsylvania and toward Philadelphia, since at present "it is as completely calculated to carry the whole trade of the south western states, into Baltimore, as any road could possibly be." Gaining this trade was essential, since "New York must, and will be the mart of the north western part of the U [*sic*] States. Nature has placed her without a competitor for this trade on account of her inland navigation. A single glance at the U. States map, will clearly demonstrate the truth of this observation." *United States Gazette* (Philadelphia), December 22, 1807.

6. In his contribution to Justin Winsor's 250th anniversary history of the city, Charles Francis Adams Jr. presented a candid analysis of how circumstances both within and beyond its control caused Boston to fall behind in competition between American cities for commercial preeminence. Adams, "The Canal and Railroad Enterprise," in *The Memorial History of Boston, including Suffolk County, Massachusetts, 1630–1880*, ed. Justin Winsor (Boston: William D. Ticknor & Co., 1880), 4:111–50. At Boston's 250th anniversary celebration in 1880, Robert Winthrop, a descendant of Puritan leader John Winthrop, recalled the 200th anniversary ceremonies fifty years earlier, when Winthrop had the privilege of dining with Josiah Quincy Sr. and "so many more of the illustrious men who were the pride and glory of the Commonwealth in those days." Winthrop admitted that Boston had been reduced in its relative rank among American cities, but he found consolation in the fact that "there is enough left this day for us to contemplate with gratitude and pride. It has been from the first a city set on a hill,—yes, on three hills. It has never been hid. It never can be hid. The hills on which it was built, and which gave it the designation which was changed for Boston, on the 17th of September, 1630, have been levelled and swept into the sea, and we, who knew them and played on them as boys, now look for them in vain. But Boston remains,—with a character all its own, with a history which can never be obliterated, and with a future, as we all hope and believe, not less prosperous or less glorious than its past." City of Boston, *Celebration of the Two Hundred and Fiftieth Anniversary of the Settlement of Boston, September 17, 1880* (Boston, 1880), 35–37.

7. "Inaugural Address of Mayor Walter S. Gurnee," March 9, 1852, Chicago Public Library, http://www.chipublib.org/cplbooksmovies/cplarchive/mayors/gurnee_inaug02.php (accessed April 17, 2011).

8. *Chicago Tribune*, August 6, 1867.

9. "An Address to the Board of Aldermen, and Members of the Common Council, of

Boston, on the Organization of the City Government, January 5, 1829," in City of Boston, *Inaugural Addresses of the Mayors of Boston* (Boston: Rockwell & Churchill, 1894), 1:124.

10. City of Boston, *Communication to the City Council on the Subject of Introducing Water into the City* (Boston: J. H. Eastburn, 1834), 21.

11. Ibid., 22. In 1843, when the proprietors of the Middlesex Canal wished to salvage their investment in this project, which was now devalued because of the construction of a new railroad, they recommended that the canal be permitted to amend its authorized purpose so that it might be considered as the source for Boston's water. The proprietors' immediate concerns were selfish, but their larger point was that it was dangerous to the city's future not to provide for a more ample water supply. They cited the 1835 report of the water commissioners of New York on the necessity of building a centralized urban water system. The commissioners had noted with shame that New York, which was by far the largest city in the nation, possessed a worse water supply than that of Philadelphia, of smaller towns up the Hudson, of European cities half New York's size, and of far-off Constantinople, where "even the Turk, performing the rites of his infidel religion, bathes . . . in waters brought from the mountains." Rather than accept this situation, the Middlesex Canal owners noted, New York had constructed the Croton works. Boston must now act likewise. The unhappy truth was clear: "We are now behind the age." *Historical Sketch of the Middlesex Canal, with Remarks for the Consideration of the Proprietors. By the Agent of the Corporation* (Boston: Samuel N. Dickinson, 1843), 53.

12. City of Boston, *Report of the Commissioners Appointed Under an Order of the City Council, March 16, 1837, to Devise a Plan for Supplying the City of Boston with Pure Water* (Boston: John H. Eastburn, 1837), 41. James Baldwin was the brother of Loammi Baldwin Jr., who prepared the city's 1834 report on water. Five Baldwin brothers in all followed their distinguished father, Loammi Baldwin Sr., a colonel in the American Revolution and one of the new nation's first civil engineers, into the profession.

13. City of Boston, *Water. Mr. Baldwin's Report*, City Document No. 5 (January 22, 1839) [Boston, 1839], 5. Seven years later, at the height of the city's water debates, Walter Channing concluded an account of how desperate Bostonians were for good water with the question, "Is it right that facts like these should exist in such a community as ours?" He was outraged that anyone would oppose a measure like the Long Pond aqueduct, "which makes a sure provision for the long future." Walter Channing, M.D., *Parliamentary Sketches and Water Statistics. Being Another Word Addressed to the Citizens of Boston, in Support of Supplying the City with the Pure Water of Long Pond* (Boston: Benjamin H. Greene, 1845), 25–26.

14. *Chicago Tribune*, February 21, 1862.

15. *Chicago Tribune*, September 8, 1864.

16. City of Chicago, *Sixth Annual Report of the Board of Public Works to the Common Council of the City of Chicago, for the Municipal Fiscal Year Ending March 31st, 1867* (Chicago: Republican Book and Job Office, 1867), 4.

17. City of Boston, *Communication to the City Council on the Subject of Introducing Water into the City* (1834), 22.

18. "Address of the Mayor to the City Council of Boston, January 5, 1846," in City of Boston, *Inaugural Addresses of the Mayors of Boston*, 1:326–28, 332.

19. Martin V. Melosi writes, "Water supply was the first important public utility in the United States and the first municipal service that demonstrated a city's commitment to growth." Melosi, *The Sanitary City: Urban Infrastructure in America from Colonial Times to the Present* (Baltimore: Johns Hopkins University Press, 2000), 119.

20. Nelson Manfred Blake characterizes the 1836 report submitted by engineer R. H. Eddy as reflecting "tight-fisted Yankee logic." Eddy criticized larger projects, like the one favored by Loammi Baldwin Jr. two years earlier when Baldwin endorsed Long Pond, the source that Boston would ultimately select. Blake states that Eddy, who preferred the smaller Spot Pond, opposed very ambitious plans because "they were based on estimates of future rather than present needs." He instead favored a more incremental approach. Blake, *Water for the Cities: A History of the Urban Water Supply Problem in the United States* (Syracuse, NY: Syracuse University Press, 1956), 179. See R. H. Eddy, *Report on the Introduction of Soft Water into the City of Boston* (Boston: John H. Eastburn, 1836).

21. As proof, Fletcher told the Massachusetts state legislature that investigations of other cities that had recently built waterworks revealed that "however many mistakes and delays had occurred in their construction, they were clung to by the people with such a force of attachment, that nothing would compel them to give them up." In particular, New York's far more expensive Croton Aqueduct "had approved itself to the most intelligent and the most wealthy inhabitants, and nothing would now induce them to part with its benefits." Commonwealth of Massachusetts, *Proceedings before a Joint Special Committee of the Massachusetts Legislature, upon the Petition of the City of Boston, for Leave to Introduce a Supply of Pure Water into that City, from Long Pond, February and March, 1845* (Boston: John H. Eastburn, 1845), 110, 104.

22. *Address of the Faneuil Hall Committee, on the Project of a Supply of Pure Water for the City of Boston, May 5, 1845* (Boston: W. W. Clapp & Son, 1845), 9.

23. City of Chicago, *Thirteenth Semi-Annual Report of the Board of Water Commissioners, to the Common Council of the City of Chicago, January 1st, 1858* (Chicago: Jameson & Morse, 1858), 47.

24. City of Chicago, *Ninth Annual Report of the Board of Public Works, to the Common Council of the City of Chicago, for the Municipal Fiscal Year Ending March 31, 1870* (Chicago: Lakeshore Press, 1870), 7. See chap. 3, n. 100.

25. City of Chicago, *Thirteenth Annual Report of the Board of Public Works to the Common Council of the City of Chicago, for the Municipal Fiscal Year Ending March 31, 1874* (Chicago: J. S. Thompson & Co., 1874), 3.

26. As Eric H. Monkkonen explains, "Like other economic actors, the city manipulates the time dimension by borrowing." "It is the long-term debt that funds the building of cities," Monkkonen writes. "It pays for the big stage on which the city's social, economic, and cultural life are all played out. Long-term debt funds the streets, buildings, and sewers: it is the literally the foundation of modern urban life." Monkkonen, *The Local State: Public Money and American Cities* (Stanford, CA: Stanford University Press, 1995), 1.

27. A. M. Hillhouse, *Municipal Bonds: A Century of Experience* (New York: Prentice-Hall, 1936), 31–46. This figure rose in uneven spurts, passing a billion dollars by 1900 and $15 billion in 1932, when the Depression caused major defaults. On the history of defaults up to this period, see ibid., 37–46. See also Paul Studensky, *Public Borrowing* (New York: National Municipal League, 1930), 5–7.

28. City of Philadelphia, *Report of the Joint-Committee of the Select and Common Councils, on the City Debts and Expenditures, and on the City Credits and Resources* (Philadelphia: Zachariah Poulson, Jr., 1801), 3–5.

29. City of Philadelphia, *Accounts of the Corporation of the City of Philadelphia, for the Year 1806* (Philadelphia: Jane Aitken, 1807), unpaginated. Another $1,700 went for boring machines, which may have been used for the waterworks. Other substantial expenses included

paving ($7,627), "cleansing the city generally" ($7,170), lighting and watching ($15,878), and salaries of the mayor and other officers ($7,025) Andrew M. Schocket graphs the enormous portion of annual expenditures devoted to the waterworks between 1799 and 1825 in *Founding Corporate Power in Early National Philadelphia* (DeKalb: Northern Illinois University Press, 2007), 127.

30. Charles Phillip Huse, *The Financial History of Boston from May 1, 1822, to January 31, 1909* (Cambridge, MA: Harvard University Press, 1916), 349, 351, 369, 371, 378, 380.

31. According to A. T. Andreas, "When any proposition was made to borrow money, the utmost consternation seems to have been created. Several town officials had even resigned rather than sanction such recklessness." Andreas, *History of Chicago, from the Earliest Period to the Present Time*, vol. 1, *Ending with the Year 1857* (Chicago: A. T. Andreas Co., 1884), 181.

32. Bessie Louise Pierce, *A History of Chicago*, vol. 2, *From Town to City 1848–1871* (New York: Knopf, 1937; reprint, Chicago: University of Chicago Press, 2007), 344, 350; City of Chicago, *Eighth Annual Report of the Board of Public Works to the Common Council of the City of Chicago, for the Municipal Fiscal Year Ending March 31st, 1869* (Chicago: Jameson & Morse, 1869), 11; City of Chicago, *Tenth Annual Report of the Board of Public Works to the Common Council of the City of Chicago, for the Municipal Fiscal Year Ending March 31st, 1871* (Cambridge, MA: University Press, 1871), 11–12. Chicago's emphasis on its sewerage infrastructure at this time was reflected in the fact that sewerage bonds represented almost 40 percent of the total debt. Water bonds ($4.82 million) were a close second, at 34 percent. Pierce, *A History of Chicago*, 2:350–51. For more on the finances of water provision, see Ann Durkin Keating, *Building Chicago: Suburban Developers and the Creation of a Divided Metropolis* (Columbus: Ohio State University Press, 1988), 39–40.

33. City of Philadelphia, *Report to the Select and Common Councils, on the Progress and State of the Water Works, On the 24th of November, 1799* (Philadelphia: Zachariah Poulson, Jr., 1799), 19.

34. Huse, *Financial History of Boston*, 58; Caleb Snow, M.D., *A History of Boston, the Metropolis of Massachusetts, from Its Origin to the Present Period; with Some Account of the Environs* (Boston: Abel Bowen, 1825), 371.

35. "An Address to the Board of Aldermen, and Members of the Common Council, of Boston, on the Organization of the City Government, at Faneuil Hall, May 1, 1824," in City of Boston, *Inaugural Addresses of the Mayors of Boston*, 1:26–27.

36. Walter Channing, *A Plea for Pure Water: Being a Letter to Henry Williams, Esq., by Walter Channing: With an Address "To the Citizens of Boston," by Mr. H. Williams* [Boston, 1844], 16–19, 38.

37. *Daily Evening Transcript* (Boston), May 27, 1845.

38. As Nelson Manfred Blake explains, Philadelphia tried to float its original 6 percent bonds when the national government was offering a higher rate. Blake, *Water for the Cities*, 33.

39. Andreas, *History of Chicago*, 1:181, 183; *Chicago Democrat*, April 23, 1852. For subsequent bond negotiations, see Andreas, *History of Chicago*, 1:187. By the spring of 1869, by which time the lake tunnel, the Pumping Station, and the Water Tower were nearly completed, the Water Department had outstanding bonds of a little more than $1 million at 6 percent, and almost $2 million at 7 percent. See City of Chicago, *Eighth Annual Report of the Board of Public Works* (1869), 145, 11.

40. Schocket, *Founding Corporate Power*, 121–25. Schocket discusses how political leaders used the sinking fund in a variety of ways besides reducing waterworks debt.

41. Hillhouse, *Municipal Bonds*, 39.

42. *Daily Advertiser and Patriot* (Boston), November 22, 1847; City of Boston, *Celebration of the Introduction of the Water of Cochituate Lake into the City of Boston, October 25, 1848* (Boston: J. H. Eastburn, 1848), 39–40; "Address of the Mayor to the City Council of Boston, January 4, 1847," in City of Boston, *Inaugural Addresses of the Mayors of Boston*, 1:337–40. See also City of Boston, *Water Loan. Report of Finance Committee*, City Document No. 67, December 27, 1852 [Boston, 1852], for the announcement of a twenty-year loan of £400,000 at 4.5 percent from the London firm of Baring Brothers.

43. "Temperance," *Hints to the Honest Taxpayers of Boston* [Boston, 1845], 3. A letter in the *Daily Evening Transcript* of May 15, 1844, maintained that Boston would never have more than 200,000 people, and therefore Spot Pond would be completely adequate as a source.

44. *Letter from Lemuel Shattuck, in Answer to Interrogatories of J. Preston, in Relation to the Introduction of Water into the City of Boston* (Boston: Samuel N. Dickinson, 1845), 15.

45. [Nathaniel J. Bradlee], *History of the Introduction of Pure Water into the City of Boston, with a Description of Its Cochituate Water Works* (Boston: Alfred Mudge & Son, 1868), 31–32.

46. "Address of the Mayor to the City Council of Boston, January 6, 1840," and "Address of the Mayor of [*sic*] the City Council of Boston, January 4, 1841," in City of Boston, *Inaugural Addresses of the Mayors of Boston*, 1:250, 271; Josiah Quincy, *The History of Harvard University* (Cambridge, MA: J. Owen, 1840), 2:362. While Quincy was willing to put Boston under debt obligations for major improvements, even as mayor he advocated financial discipline, including provision for paying off bonds and general deficit reduction. "An Address to the Board of Aldermen, and Members of the Common Council, of Boston, on the Organization of City Government, January 2, 1826," in City of Boston, *Inaugural Addresses of the Mayors of Boston*, 1:65–66. The size of Boston's sinking fund doubled to almost $300,000 in the fiscal year ending April 30, 1827, and the funded debt, which had climbed the previous year from $218,000 to $712,000, modestly declined or stayed virtually level for the next five years. Huse, *Financial History of Boston*, 378.

47. William J. Hubbard, *Arguments on Behalf of Joseph Tilden and Others, Remonstrants, on the Hearing of the Mayor of the City of Boston, on Behalf of the City Council, for a Grant of the Requisite Powers to Construct an Aqueduct from Long Pond to the City, before a Joint Special Committee of the Massachusetts Legislature, March 6, 1845* (Boston: T. R. Marvin, 1845), 4.

48. "Inaugural Address of Mayor James Woodworth," March 14, 1848, Chicago Public Library, http://www.chipublib.org/cplbooksmovies/cplarchive/mayors/woodworth_inaug01.php (accessed April 17, 2011).

49. "Inaugural Address of Mayor James Curtiss," March 12, 1850, Chicago Public Library, http://www.chipublib.org/cplbooksmovies/cplarchive/mayors/curtiss_inaug02.php (accessed April 17, 2011). In 1851 future mayor John Wentworth opposed the publicly owned Chicago City Hydraulic Company on the grounds that it would force the city to go deeply into debt. Pierce, *A History of Chicago*, 2:200.

50. City of Boston, *Report [of the Standing Committee on the Introduction of Pure and Soft Water into the City], January 29, 1838* [Boston, 1838], 7.

51. "Inaugural Address of Mayor Walter S. Gurnee," March 9, 1852, Chicago Public Library, http://www.chipublib.org/cplbooksmovies/cplarchive/mayors/gurnee_inaug02.php (accessed April 17, 2011).

52. "Inaugural Address of Mayor John Wentworth," March 10, 1857, Chicago Public Library, http://www.chipublib.org/cplbooksmovies/cplarchive/mayors/wentworth_inaug

.php (accessed April 17, 2011). On Chicago's funded debt from the 1850s to the early 1870s, see Pierce, *A History of Chicago*, 2:349–52.

53. *Daily Evening Transcript*, October 31, 1848.

54. The mayor of Charlestown in turn toasted the two mayors Quincy, calling them "a very useful family for Boston—the father gave her a market—the son pure water; the one meat—the other drink." *Daily Advertiser*, November 27, 1849.

55. "An Address to the Board of Aldermen, and Members of the Common Council, of Boston, on the Organization of the City Government, at Faneuil Hall, May 1, 1824," in City of Boston, *Inaugural Addresses of the Mayors of Boston*, 1:27.

56. *Address of the Faneuil Hall Committee, on the Project of a Supply of Pure Water for the City of Boston* (1845), 6, 8.

57. "Address of the Mayor to the City Council of Boston, January 4, 1847," in City of Boston, *Inaugural Addresses of the Mayors of Boston*, 1:331.

58. City of Philadelphia, *Report of the Joint-Committee of the Select and Common Councils, on the City Debts and Expenditures, and on the City Credits and Resources* (1801), 2.

59. The first read, "A loud call from the people for pure water! Cry heard by Hon. Josiah Quincy, Mayor, 1825," and on the last was inscribed, "Ground broken at Cochituate Lake by the Hon. Josiah Quincy, Jr., August 20th, 1846." City of Boston, *Celebration of the Introduction of the Water of Cochituate Lake into the City of Boston* (1848), 5–6.

60. "Chicago was a long time since voted the eighth wonder of the world," the story began, "and none have disputed it for years past, though there be some who have striven to ridicule the claims put forth, not so much by the people of this city as by those who have come here to look and to wonder at our growing greatness. Chicago is a conglomerate wonder, because she is made up of wonders—full of intensified vitality, which is always moving, and always moving ahead." *Chicago Tribune*, November 28, 1866.

61. *Chicago Tribune*, December 7, 1866.

62. City of Chicago, *Thirteenth Annual Report of the Board of Public Works to the Common Council of the City of Chicago* (1874), 3.

63. *Address of the Faneuil Hall Committee, on the Project of a Supply of Pure Water for the City of Boston* (1845), 8. By 1850, five years after the committee made this statement, the population of Liverpool was about 250,000, St. Petersburg's was almost 490,000, Vienna just above 550,000, and Paris around a million. Boston's population was just under 137,000. It was the third biggest city in the United States, behind New York, which had just over half a million, and Baltimore, just under 170,000. Philadelphia was fourth, with slightly more than 121,000, though this is somewhat misleading since it was surrounded by other leading urban centers, including Spring Garden, Northern Liberties, and Kensington, that it would annex four years later. The consolidation made it the nation's second biggest city.

64. *Daily Advertiser*, May 19, 1845. The fifty-one-mile Marseille Aqueduct was a remarkably complex engineering achievement that included multiple bridges and dozens of tunnels.

65. Walter Channing, *An Address on the Prevention of Pauperism* (Boston: Office of the Christian World, 1843), 83. See also Lemuel Shattuck's endorsement of Rome's example, in Lemuel Shattuck et al., *Report of the Sanitary Commission of Massachusetts, 1850* (Cambridge, MA: Harvard University Press, 1948), 14–15.

66. *Daily Advertiser*, August 22, 1846. The spring on Mount Helicon was celebrated in Greek mythology as sacred to the Muses.

67. Johann Wolfgang von Goethe, *Letters from Italy*, trans. W. H. Auden and Elizabeth Mayer (New York: Penguin Books, 1995), 81. The letters were originally published in 1816 and 1817.

68. Channing, *A Plea for Pure Water*, 13.

69. *Daily Advertiser*, January 2, 1846.

70. *Daily Advertiser*, January 2, 1846; November 6, 1848.

71. *Daily Advertiser*, January 2, 1846.

72. *Daily Advertiser*, August 22, 1846.

73. City of Boston, *Statue of Josiah Quincy. Dedication Ceremonies, October 11, 1879, with Preliminary Proceedings*, City Document No. 115 (Boston: Boston City Council, 1879), 18, 42.

74. *Daily Advertiser*, August 22, 1846.

75. On the evolution of the time capsule in nineteenth-century America, see Nick Yablon, *Untimely Ruins: An Archaeology of American Urban Modernity, 1819–1919* (Chicago: University of Chicago Press, 2009).

76. *Daily Evening Transcript*, November 20, 1847.

77. *Chicago Tribune*, December 7, 1866.

78. Goethe, *Letters from Italy*, 81.

79. City of Philadelphia, *Report of the Watering Committee, upon the Present State of the Works for Supplying the City with Water, and the Several Other Plans Proposed for that Purpose, May 2, 1812* (Philadelphia: Jane Aitken, 1812), 12–13.

80. City of Boston, *Report [of the Standing Committee on the Introduction of Pure and Soft Water into the City]* (1838), 2.

81. Williams went on to present his version of the argument that watering the city would make it grow and would eventually pay for itself: "Beyond all question, for a long period hence, there will be a great and rapid increase of our population; thus creating an annual increased demand for water which cannot fail so to enlarge the income from water rents, as at no distant period to justify the reduction of the tax to such an extent that no one will feel it to be a burden." In Channing, *A Plea for Pure Water*, 39. At the Cochituate system groundbreaking two years later, Nathan Hale announced that the new works "was quite competent to furnish an ample amount for any probable future period and probable population which Boston might contain." *Daily Advertiser*, August 22, 1846.

82. *Daily Advertiser*, November 22, 1847. See Revelation 10:6.

83. City of Boston, *Celebration of the Introduction of the Water of Cochituate Lake into the City of Boston* (1848), 42.

84. *A Description of the Boston Water Works, Embracing All the Reservoirs, Bridges, Gates, Pipe Chambers, and Other Objects of Interest, from Lake Cochituate to the City of Boston* (Boston: Geo. R. Holbrook & Co., 1848), 27.

85. John B. Bigelow, cited in [Bradlee], *History of the Introduction of Pure Water into the City of Boston*, 83.

86. *Daily Evening Transcript*, November 17, 1849.

87. *Chicago Tribune*, January 30, 1854.

88. City of Chicago, *Eighth Annual Report of the Board of Public Works* (1869), 71.

89. Whether *The Course of Empire* can be read as a prediction is less clear. It depends on whether one believes that Cole was so convinced of the cyclical view of history and so skeptical of American exceptionalism that he was certain that the United States would meet the same fate as the civilization he depicts, or whether he thought it could yet escape the

destruction and desolation that are the fate of the city in the paintings. Howard S. Merritt claims that Cole "was a believer in the cyclical theory of history almost to obsession, and yet was also a Christian millenarian." Merritt, "A Wild Scene, Genesis of a Painting," *Baltimore Museum of Art Annual II: Studies on Thomas Cole, an American Romanticist* (1967): 24. See also Stow Persons, "The Cyclical Theory of History in Eighteenth Century America," *American Quarterly* 6 (1954): 147–63; and Joy S. Kasson, *Artistic Voyagers: Europe and the American Imagination in the Works of Irving, Allston, Cole, Cooper, and Hawthorne* (Westport, CT: Greenwood Press, 1982), 116.

90. "An Address to the Board of Aldermen, and Members of the Common Council, of Boston, on the Organization of the City Government, at Faneuil Hall, May 1, 1824," in City of Boston, *Inaugural Addresses of the Mayors*, 1:27–29.

91. On the relationship between massive water projects and totalitarian oppression, see Karl Wittfogel, *Oriental Despotism: A Comparative Study of Total Power* (New Haven, CT: Yale University Press, 1957), 49, 137.

92. Nehemiah Adams, *The Song of the Well: A Discourse on the Expected Supply of Water in Boston, Preached to the Congregation in Essex Street, on Thanksgiving Day, November 26, 1846* (Boston: William D. Ticknor & Co., 1847), 14.

93. *Daily Advertiser*, November 29, 1849. See also *Daily Evening Transcript*, November 30, 1849.

94. See Alain Corbin on sanitation as a way of concealing signs of mortality, in Corbin, *The Foul and the Fragrant: Odor and the French Social Imagination*, trans. Miriam L. Kochan, Roy Porter, and Christopher Prendergast (Cambridge, MA: Harvard University Press, 1986), 90.

95. This wish also underlay the language of public health. The Civic Cleanliness Committee of the 1859 National Quarantine and Safety Convention stated that the lesson taught by the aqueducts and sewers of Rome, "which, even in their ruins, excite our wonder and admiration," went beyond the fact that they preserved "the laws of health" and so made possible the gathering of a large population. The corollary message (and moral) was that "when those laws were neglected by the lawless democracy into whose hands the control of the city fell, her downfall and desolation began." It was "a difficult matter to persuade people to look forward to the comfort of generations to come after them, when they have to furnish the means for it," but this kind of prospective vigilance was required by those who had both a sense of history and a wish to escape it. Egbert L. Viele, *Report on Civic Cleanliness, and the Economical Disposition of the Refuse of Cities* (New York: Edmund Jones & Co., 1860), 15–16.

96. James Spear Loring, *The Hundred Boston Orators Appointed by the Municipal Authorities and Other Public Bodies, from 1770 to 1852* (Boston: John P. Jewett & Co., 1852), 498–99.

Chapter Seven

1. The urbanization of the United States, which began in earnest in the antebellum decades, became the dominant demographic trend of the century that followed. More and more Americans lived in larger and larger communities. In 1830 the nation could boast of only one city, New York, with a population of 100,000 or more. By 1860 there were nine, and by 1900 there were thirty-eight. Among the latter were six cities with over a half million people, including Philadelphia (1,293,697), Boston (560,892), and Chicago (1,698,575). The number of people living in Philadelphia, Boston, and Chicago peaked in 1950, and all

three were considerably smaller a half century later (Philadelphia went from 2,071,605 to 1,517,550, Boston from 801,444 to 589,141, and Chicago from 3,620,952 to 2,896,016). For these and related statistics, see Frank Hobbs and Nicole Stoops, *Demographic Trends in the 20th Century*, US Census Bureau (November 2002), 32–37, 44, http://www.census .gov/prod/2002pubs/censr-4.pdf (accessed April 26, 2012); Susan B. Carter et al., *Historical Statistics of the United States*, table Aa684–698, p. I-102, http://hsus.cambridge.org.turing .library.northwestern.edu/HSUSWeb/table/showtablepdf.do?id=Aa22–35 (accessed November 20, 2010); US Census Bureau, "Population, Housing Units, Area Measurements, and Density," table 2, http://www.census.gov/population/www/censusdata/hiscendata.html (accessed February 26, 2012); US Census Bureau, "Ranking Tables for Incorporated Places of 100,000 or More: Population in 2000 and Population Change from 1990 to 2000 (PHC-T-5)," table 2, http://www.census.gov/population/www/cen2000/briefs/phc-t5/index.html (accessed November 20, 2010); Campbell Gibson, "Population of the 100 Largest Cities and Other Urban Places in the United States: 1790–1990," United States Bureau of the Census Population Division Working Paper No. 27 (June 1998), http://www.census.gov/population/ www/documentation/twps0027/twps0027.html (accessed November 20, 2010).

2. Carter et al., *Historical Statistics of the United States*, table Aa684–698; Martin V. Melosi, *The Sanitary City: Urban Infrastructure in America from Colonial Times to the Present* (Baltimore: Johns Hopkins University Press, 2000), table 4.1, p. 74; p. 117. See also Melosi, *Precious Commodity: Providing Water for America's Cities* (Pittsburgh: University of Pittsburgh Press, 2011).

3. The source of the population figure for Philadelphia County is Elizabeth M. Geffen, "Industrial Development and Social Crisis, 1841–1854," in *Philadelphia: A 300-Year History*, ed. Russell F. Weigley (New York: Norton, 1982), 349. On the continuing politics of water in Philadelphia, see Michal McMahon, "Makeshift Technology: Water and Politics in 19th-Century Philadelphia," *Environmental Review* 12, no. 4 (Winter 1988): 20–37.

4. [Nathaniel J. Bradlee], *History of the Boston Water Works, from 1868 to 1876* (Boston: Rockwell & Churchill, 1896), 166–69.

5. Massachusetts Water Resources Authority, "Metropolitan Boston's Water System History," http://www.mwra.state.ma.us/04water/html/hist1.htm#summarytable (accessed November 22, 2010); Massachusetts Department of Conservation and Recreation, "Quabbin Reservoir," http://www.mass.gov/dcr/parks/central/quabbin.htm (accessed November 22, 2010). See also Wallace, Floyd Associates, "Task 18:20: A History of the Development of the Metropolitan District Commission Water Supply System," in *Metropolitan District Commission Water Supply Study and Environmental Impact Report—2020* (September 1984).

6. John Ericson, *Report on the Water Supply System of Chicago: Its Past, Present and Future* (Chicago: May 1905); City of Chicago, Bureau of Engineering, *A Century of Progress in Water Works: Chicago, 1833–1933* (Chicago: Department of Public Works, 1933); City of Chicago, *Chicago Water System* (Chicago: City of Chicago Department of Water and Sewers, 1974). Louis P. Cain, "Annexation," in *Encyclopedia of Chicago*, online edition, http://www .encyclopedia.chicagohistory.org/pages/53.html (accessed November 22, 2010).

7. City of Chicago, "Water Management," http://www.cityofchicago.org/city/en/depts/ water/provdrs/supply.html (accessed December 10, 2010).

8. The building is also the home of a city visitor center, as well as a public library branch and a theater company, while the Water Tower in the small park across Michigan Avenue has become an art exhibition space. http://www.explorechicago.org/city/en/travel_tools/ visitor_centers.html (accessed December 14, 2010).

9. American Water Works Association, "The AWWA's Story," http://www.awwa.org/About/Content.cfm?ItemNumber=3885&navItemNumber=1638 (accessed November 23, 2010). In 1981 the largest publicly owned water systems in the country formed the Association of Metropolitan Water Agencies, whose members serve more than a third of the residents of the United States. Association of Metropolitan Water Agencies, "About AMWA," http://www.amwa.net/cs/about_amwa/about (accessed November 23, 2010).

10. City of Chicago, *First Annual Report of the Department of Public Works, to the City Council of the City of Chicago, for the Fiscal Year Ending December 31, 1876* (Chicago: Clark & Edwards, 1877), 4.

11. Ericson, *Report on the Water Supply System of Chicago*, 16.

12. Melosi, *The Sanitary City*, tables 4.1–4.3, p. 74; table 7.1, p. 120. As of 2008, 51,888 public community water systems that served a stable population year-round delivered water to over 292 million Americans out of a total population of just fewer than 305 million. "Public Drinking Water Systems by Size of Community Served and Source of Water," in "Energy & Utilities: Water and Sewage Systems," *The 2010 Statistical Abstract: The National Data Book* (Washington, DC: US Census, 2010), http://www.census.gov/compendia/statab/cats/energy_utilities/water_and_sewage_systems.html (accessed November 23, 2010); table 2, "Population 1960 to 2008," in "Population," *The 2010 Statistical Abstract: The National Data Book* (Washington, DC: US Census, 2010), http://www.census.gov/compendia/statab/cats/population.html (accessed November 23, 2010).

13. Melosi, *The Sanitary City*, 120–23.

14. City of Chicago, *Thirteenth Annual Report of the Board of Public Works to the Common Council of the City of Chicago, for the Municipal Fiscal Year Ending March 31, 1874* (Chicago: J. S. Thompson & Co., 1874), 3.

15. Once cities like Philadelphia and Chicago began to sell their water to other non-annexed communities, the size of the collective grew even bigger, though individual customers in client communities did not have even a symbolic ownership stake in this public supply. Chicago's Department of Water Management sells water to 125 suburban communities as well as to individuals and businesses in the city itself, while Philadelphia's Department of Water and Water Revenue Bureau serve parts of Montgomery, Delaware, and Bucks Counties. City of Chicago, "Department of Water Management," http://www.cityofchicago.org/city/en/depts/water.html (accessed November 22, 2010); Philadelphia Water Department, Financial Statements for Fiscal Year Ended June 30, 2009 (Philadelphia, 2009), 3, http://www.phila.gov/water/ (accessed November 23, 2010).

16. As of 1863, Chicago's mayor was a member of the Board of Public Works, and starting in 1867 its members were appointed by the mayor and Common Council rather than being elected. Bessie Louise Pierce, *A History of Chicago*, vol. 2: *From Town to City 1848–1871* (New York: Knopf, 1937; reprint, Chicago: University of Chicago Press, 2007), 316. Harold L. Platt makes a strong case that "large-scale public works projects and continuous extensions of the networks of pipes and mains fit right in with [local ward bosses'] plans" to maximize their own control of resources and resist the efforts by the business elite to focus development on the downtown. In a time "of intense class and ethnic/religious conflict," Platt writes, "large public work projects figured prominently in the civic discourse that collectively gave definition to the political culture of the industrial city." While on one hand, "their technological scale and geographic scope seemed to express symbolically republican ideals of equality, community and uplift" that was celebrated at dedication ceremonies, "the politicians ran the waterworks and sewer department like a private business for their

own personal and partisan benefit." Contributing to this was that the waterworks was now operating well in the black, so that by the late 1870s "the city council quickly came to depend upon its water management utilities to produce an annual surplus of cash profits that bankrolled the council's slush fund for pet projects." Platt, *Shock Cities: The Environmental Transformation and Reform of Manchester and Chicago* (Chicago: University of Chicago Press, 2005), 171, 175, 178.

17. Fern L. Nesson sees the creation of the Cochituate Water Board as a depoliticization of water that was remarkable, given all the earlier controversy and debate that preceded the construction. "From that time forward," she writes, "it was the experts who controlled the system, monitoring all questions of demand and supply. What had been a popular, political issue became a technical issue initiated by an administrative request to the General Court for permission to add to the water supply system." Nesson, *Great Waters: A History of Boston's Water Supply* (Hanover, NH: University Press of New England, 1983), 7–8. See also Sarah S. Elkind, *Bay Cities and Water Politics: The Battle for Resources in Boston and Oakland* (Lawrence: University Press of Kansas, 1998).

18. In another organizational development that would become increasingly common, the Metropolitan Water Board merged in 1901 with the Metropolitan Sewerage Board, created in 1889, to form the Metropolitan Water and Sewerage Board, which took a more integrated approach to the provision of water and the removal of effluent. The Metropolitan Water and Sewerage Board was succeeded twenty years later by the Metropolitan District Commission and in 1984 by the Massachusetts Water Resources Authority. The Philadelphia Department of Water and the Chicago Department of Water Management are both responsible for their cities' rain and wastewater removal as well as their water supplies, and the Boston Water and Sewer Commission oversees both water delivery and sewerage within that city.

19. Melosi, *The Sanitary City*, table 7.2, p. 130; table 11.1, p. 215. As Nelson Manfred Blake explains, American cities have used much more water than their European counterparts. He points to the fact that before World War II per capita consumption in ten leading European cities was 39 gallons per day, while in ten major American cities it was 155 gallons. Blake, *Water for the Cities: A History of the Urban Water Supply Problem in the United States* (Syracuse, NY: Syracuse University Press, 1956), 271–72. Charles Fishman contends that the United States and the rest of the developed world have lived the last century "in a kind of aquatic paradise: our water has been abundant, safe, and cheap." He calls this period "a kind of golden age of water, when we could use as much as we wanted, whenever we wanted, for almost no cost." That age has now ended, he warns, and attitudes must change. Fishman, *The Big Thirst: The Secret Life and Turbulent Future of Water* (New York: Free Press, 2011), 3, 9.

While the national average per capita water use in the United States in 2006 was 575 liters, or about 152 gallons (compared to a little over 39 gallons in the United Kingdom, 51 in Germany, and about 99 in Japan), per capita water use in American cities has been decreasing in the last few decades. For national water use, see Data 360, "Average Water Use Per Person Per Day," http://www.data360.org/dsg.aspx?Data_Set_Group_Id=757 (accessed April 25, 2011).

20. In different times and places, either consumers or city officials (and sometimes both) have been wary of metering because they thought, whether rightly or wrongly, that they would do better with a flat rate. Platt argues that the absence of meters in Chicago was

at least partly driven by the opportunities this created for soliciting bribes in return for keeping particular customers' rates low. Platt, *Shock Cities*, 183–84, 378–80.

21. The city, however, established a program, called MeterSave, that encourages home owners to have a meter installed. "Because customers with metered homes pay only for the water they use," an online notice promoting the program reads, "they can save money while at the same time helping to protect Lake Michigan and save water." City of Chicago, "Meter-Save," https://www.metersave.org/ (accessed December 15, 2010).

22. "Chicago Mayor-Elect Rahm Emanuel: No More Free Water for Non-Profits," *Chicago Sun Times*, April 28, 2011, http://www.myfoxchicago.com/dpp/news/metro/chicago-mayor-rahm-emanuel-non-profits-community-groups-cutbacks-free-water-20110427 (accessed May 12, 2011).

23. City of Chicago, *Tenth Annual Report of the Board of Public Works to the Common Council of the City of Chicago, for the Municipal Fiscal Year Ending March 31, 1869* (Cambridge, MA: University Press, 1871), 77.

24. Philadelphia Fountain Society, *Annual Reports, Addresses, and Other Proceedings, during the Years 1873 & 1874* (Philadelphia: McFarland & Woodburg, 1875), 16–17.

25. Other historians of Chicago water have made note of this, but Platt offers the most sustained critique in *Shock Cities*. Platt is also the most outspoken in arguing that the much-praised Ellis S. Chesbrough (who did express concerns about the pollution of Lake Michigan by the Chicago River when he offered the lake tunnel system as an option in the early 1860s) should have known better and acceded too readily to political pressure. See Platt, *Shock Cities*, 137–38, 156, 188–89.

26. Chicago met with total failure the following decade when it dug what was called the Fullerton Conduit. Reversible propellers in the conduit were to direct water either back or forth, as needed, through this artificial channel between the North Branch and the lake in the hope of cleansing the river.

27. Lyman E. Cooley, quoted in G. P. Brown, *Drainage Channel and Waterway* (Chicago: R. R. Donnelley & Sons, 1894), 424, 429–32. In the late 1920s, the city altered the Chicago River in another way by straightening a downtown section to open the possibility of laying more north–south streets that would improve access to this area. The river exacted a measure of revenge for all the ways it had been altered when in April 1992 water began streaming through an opening in the bottom of the river at Kinzie Street that had been originally weakened months earlier by workers drilling a piling, apparently unaware that below the bottom was a turn-of-the-century freight tunnel network now being used to run utility lines. President George H. W. Bush declared Chicago a disaster area, the Loop was shut down for several days, and the estimate of damages was well over a billion dollars. "The Loop's Great Chicago Flood," *Chicago Tribune*, http://www.chicagotribune.com/news/politics/chi-chicagodays-flood-story,0,243150.story (accessed December 16, 2010).

28. State of Massachusetts Department of Conservation and Recreation, "Quabbin Reservoir," http://www.mass.gov/dcr/parks/central/quabbin.htm (accessed December 16, 2010); Lake Cochituate State Park provides a closer if smaller nature reserve, though its waters require the removal of sources of pollution by its shores, notably a former General Motors facility and a Department of Defense research complex known as the United States Army Soldier Systems Center, where a 1980s explosion sent PCBs into the lake. State of Massachusetts Department of Conservation and Recreation, "Lake Cochituate," http://www.mass.gov/dcr/parks/northeast/coch.htm (accessed December 17, 2010); Lake Cochituate, http://

www.lakecochituate.org/ (accessed December 17, 2010). United States Army, "Sediment Cleanup on Lake Cochituate," http://www.army.mil/-news/2010/06/24/41303-sediment-cleanup-on-lake-cochituate/ (accessed December 17, 2010).

29. "Boston Common Frog Pond," http://www.bostonfrogpond.com/ (accessed December 16, 2010). As was the case over 150 years ago, Boston fountains that do not recirculate their water may face restricted service. *Boston Globe*, "Boston's Flowing (and Not-So-Flowing) Fountains," http://www.boston.com/travel/boston/gallery/Bostons _flowing_and_sometimes_barren_fountains?pg=7 (accessed December 16, 2010).

30. Three years earlier, sculptor Emma Stebbins in a sense combined in one figure a woman and a bird in her winged *Angel of the Waters*, who alights on Central Park's Bethesda fountain, which was erected to acknowledge the many blessings brought by the Croton Aqueduct. The statue refers to the pool of Bethesda, from John 5:2–4, which the sick, the lame, and the blind visit in the belief that the first person in the pool after its surface is touched by the visit of an angel will be healed.

31. The fountain, which consists of three concentric basins atop one another, is 280 feet in diameter and 25 feet high. Its shifting displays can make use of up to 14,100 gallons of water per minute, while the overall capacity of the fountain is 1.5 million gallons. Chicago Park District, "Buckingham Fountain," http://www.cpdit01.com/resources/buckingham _fountain.cfm (accessed December 16, 2010).

32. Close by the high-tech fountain in Millennium Park, an artificial stream meanders through the park, providing a quiet place for hot and weary visitors to cool their feet. Millennium Park Chicago, "The Crown Fountain," http://www.millenniumpark.org/ artandarchitecture/crown_fountain.html (accessed December 16, 2010).

33. Melosi, *The Sanitary City*, 91.

34. Ibid., 16–18, 32–37; Charles Phillips Huse, *The Financial History of Boston from May 1, 1822, to January 31, 1909* (Cambridge, MA: Harvard University Press, 1916), 137; John Koren, *Boston, 1822–1922: The Story of Its Government and Principal Activities during One Hundred Years* (Boston: City of Boston Printing Department, 1923), 54, 69. In 1852, ten years after the opening of the Croton system, New York's Association for Improving the Condition of the Poor constructed a bathhouse for its target clientele. As in Liverpool, it included a laundry and swimming pool, as well as tub baths. Williams states that though it was successful at first, it closed by 1861, perhaps because the charge of three cents an hour for laundry and five or ten cents for a bath may have been too expensive. Marilyn Thornton Williams, *Washing "The Great Unwashed": Public Baths in Urban America, 1840–1920* (Columbus: Ohio State University Press), 16. As evidence for his argument that nineteenth-century city governments were more concerned with serving private interests rather than the public good, Sam Bass Warner Jr. cites Isaac Parrish's 1849 "Report on the Sanitary Condition of Philadelphia," which was part of the survey of several cities conducted by the American Medical Association. Parrish complained that there were only five public baths in Philadelphia, all with an entrance fee "which excludes a large proportion of the inhabitants who are without the facilities of bathing at their own houses." Parrish, quoted in Warner, *The Private City: Philadelphia in Three Periods of Its Growth* (Philadelphia: University of Pennsylvania Press, 1968), 110.

35. Williams, *Washing "The Great Unwashed*," 1. Williams states that by this time "personal cleanliness had become a necessity, not only for social acceptability and public health but also as a symbol of middle-class status, good character, self-respect, and membership in the civic community."

36. Jean-Pierre Goubert, *The Conquest of Water: The Advent of Health in the Industrial Age*, trans. Andrew Wilson (Princeton, NJ: Princeton University Press, 1989), 53–54; Blake, *Water for the Cities*, 258; Melosi, *The Sanitary City*, 85–87.

37. Harold Platt discusses the tension between engineers and public health physicians as dating back to as early as the 1860s. Platt, *Shock Cities*, 145.

38. Advertised prices ranged from $50 to $500. Boston Water Purifier Company, *Most Perfect Filter and Purifier of Water* [Boston, 1881], Rare Books and Manuscripts Department, Boston Public Library.

39. Between 1900 and 1922, the mortality rate fell from 35 to 2.8 per 100,000 in Philadelphia, 26 to 1.4 in Boston, and 39 to 3.2 in Chicago. Melosi, *The Sanitary City*, table 7.5, p. 138.

40. The 1872 report of the Chicago Board of Works (which devoted a great deal of attention to repairs necessitated by the fire) stated, "The propriety of establishing numerous public drinking fountains both for man and beast, throughout the thickly settled portions of the city is very apparent." It recommended placing them at half-mile intervals on major streets and at greater intervals on minor ones. City of Chicago, *Eleventh Annual Report of the Board of Public Works to the Common Council of the City of Chicago, for the Municipal Fiscal Year Ending March 31st, 1872* ([Chicago]: D. & C. H. Blakely, 1872), 15. The first nine of these were very well received. The following year the board stated, "The free drinking fountains that have been established have met with such favor that they are now considered almost a public necessity." It added, "There being but little wastage, the cost of the water supply, compared with the benefits derived, is but trifling." City of Chicago, *Twelfth Annual Report of the Board of Public Works to the Common Council of the City of Chicago, for the Municipal Fiscal Year Ending March 31st, 1873* (Chicago: National Printing Co., 1873), 9–10.

41. Women's Christian Temperance Union, "WCTU Drinking Fountains—Then and Now," http://www.wctu.org/fountains.html (accessed December 17, 2010); New York City Department of Parks and Recreation, "Tompkins Square Park," http://www.nycgovparks .org/parks/tompkinssquarepark/highlights/12753 (accessed December 17, 2010). On the removal of the Boston fountain, see "Weeding Out Bad Sculpture," *New York Times*, March 13, 1894.

42. On the bottled-water industry, see Elizabeth Royte, *Bottlemania: How Water Went on Sale and Why We Bought It* (New York: Bloomsbury, 2008); and Richard Wilk, "Bottled Water: The Pure Commodity in the Age of Branding," *Journal of Consumer Culture* 6 (2006): 303–25.

43. Philadelphia Fountain Society, *Annual Reports, Addresses, and Other Proceedings during the Years 1873 & 1874*, 19.

44. Cooley, quoted in Brown, *Drainage Channel and Waterway*, 432.

45. As Marshall Berman points out, Robert Moses, one of the twentieth century's great champions of infrastructural development, tried to identify this development as "the vehicle of impersonal world-historical forces, the moving spirit of modernity." Berman, *All That Is Solid Melts into Air: The Experience of Modernity* (New York: Penguin, 1982), 294.

46. *Los Angeles Times*, November 6, 1913.

47. *Los Angeles Times*, November 2, 1913. The story of the questionable methods that Los Angeles used to acquire the Owens Valley water and gain support for the aqueduct, and of Otis's complicity in the machinations, has been told many times. See Marc Reisner, "The Red Queen," in *Cadillac Desert: The American West and Its Disappearing Water*, rev. and updated ed. (New York: Penguin, 1987), 52–103.

48. *Los Angeles Times*, November 6, 1913; Mulholland, quoted in Los Angeles Department of Water and Power, "World Records," http://wsoweb.ladwp.com/Aqueduct/history oflaa/worldrecords.htm (accessed December 20, 2010).

49. *Los Angeles Times*, November 6 and 8, 1913.

50. *Bill of Fare, Anniversary Dinner of the Committee of Fairmount Water Works, at the United States Hotel, Philadelphia, April 28, 1851* [Philadelphia, 1851].

51. See, for example, City of Philadelphia, Department for Supplying the City with Water, *History of the Works, and Annual Report of the Chief Engineer of the Water Department of the City of Philadelphia* (Philadelphia: C. E. Chichester, 1860), 5–44; and *Annual Report of the Chief Engineer of the Water Department of the City of Philadelphia for the Year 1875* (Philadelphia: E. C. Markley & Son, 1875), 9–48; [Nathaniel Bradlee], *History of the Introduction of Pure Water into the City of Boston, with a Description of Its Cochituate Works* (Boston: Alfred Mudge & Son, 1868); [Desmond Fitzgerald], *History of the Boston Water Works, from 1868 to 1876* (Boston: Rockwell & Churchill, 1876); and City of Chicago, Bureau of Engineering, *A Century of Progress in Water Works. Chicago 1833–1933* (Chicago: Department of Public Works, 1933).

52. The Fairmount Works was similarly honored in 1977. See American Water Works Association, "AWWA Honors and Awards," http://www.awwa.org/Membership/awards.cfm ?ItemNumber=498&navItemNumber=55182 (accessed April 16, 2012).

53. *Los Angeles Times*, March 31, 1940.

54. Fairmount Water Works Interpretive Center, http://www.fairmountwaterworks .com/index.php (accessed December 19, 2010).

INDEX